AF559846

Sustainable Approaches for Environmental Conservation

Sustainable Approaches for Environmental Conservation

Editors
Prof. D.R. Khanna
Prof. A.K. Chopra
Dr. R. Bhutiani
Dr. Gagan Matta
Dr. Vikas Singh

Associate Editor
Mr. Chakresh Pathak

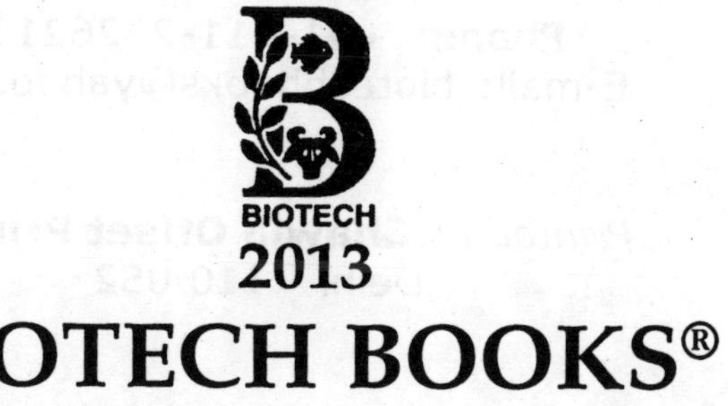

2013
BIOTECH BOOKS®

ISBN 978-81-7622-280-8

Published by: **BIOTECH BOOKS®**
4762-63/23, Ansari Road, Darya Ganj,
New Delhi - 110 002
Phone: +91-011-23262132
E-mail: biotechbooks@yahoo.co.in

Printed at: **Chawla Offset Printers**
Delhi - 110 052

PRINTED IN INDIA

Preface

The concept of sustainable development has in the past most often been broken out into three constituent parts: environmental sustainability, economic sustainability and sociopolitical sustainability. More recently, it has been suggested that a more consistent analytical breakdown is to distinguish four domains of economic, ecological, political and cultural sustainability. It helps us understand ourselves and our world. The problems we face are complex and serious-and we can't address them in the same way we created them. But we can address them.

At the 1992 UN Conference on Environment and Development (the Earth Summit), the Convention on Biological Diversity (CBD) was born. 192 countries, plus the EU, are now Parties to that convention. In April 2002, the Parties to the Convention committed to significantly reduce the loss of biodiversity loss by 2010.

Perhaps predictably, that did not happen. Despite numerous successful conservations measures supporting biodiversity, the 2010 biodiversity target has not been met at the global level.

The present book provides an introduction for sustainable approaches for environment conservation. The book is designed to incorporate both new and traditional methods to provide a foundation in the study of the conservation and development.

The present work is the compilation of various papers contributed by learned community from different part of the World working in the field of sustainable development this book describes the unpublished research of the author.

This volume brings together contributions from specialists in a wide range of fields, who examine issues of sustainability as they relate to sustainable conservation. The topics range in scale from individual buildings and sites to cities, landscapes, and other historic environments. The volume offers a global perspective and demonstrates that conservation must be a dynamic process, involving public participation, dialogue, consensus, and, ultimately, better stewardship. Through its dual focus on theory and case studies, the book also makes an important contribution to the larger debate on quality of life and the environment.

At last but not least we thank all our contributors who have devoted their precious time in contributing their research in the form of papers for this volume. The thanks are also due to M/s Biotech books publishing house, New Delhi in bringing out this volume in time with nice presentation.

Editors

Contents

2013, Sustainable Approaches for Environmental Conservation *Pages* **1–10**
Editors: **D.R. Khanna, A.K. Chopra, R. Bhutiani, Gagan Matta & Vikas Singh**
Published by: **BIOTECH BOOKS, NEW DELHI**

Chapter 1

Health Status of Women in Relation to Certain Socio-Biological Factors

D. Kumar, Anita Mishra and B.N. Pandey
Eco-Genetical Research Laboratory, P.G. Department of Zoology,
Purnia College, Purnia, Bihar

Terai belt of Bihar is a very remote area and majority of the populations are agriculturists. The economy of the area is based on agriculture. In this area, women work as daily labourers in agricultural fields building construction. In the present work health status of women was studied in relation to occupation, nutrition, sanitation, reproductive fitness and work load. The study clearly indicates that these women are victim of a large number of diseases (such as anaemia, T.B, malaria, kala-azar, menstrual disorders and nutritional disorders *etc.*) due to work load, poor nutrition, sanitation and high rate pregnancies. The study also showed high degree of foetal loss and infant mortality. Lower age of marriage and motherhood, malnutrition, work load, environmental sanitation and poor facilities of primary health centres are the possible reasons behind the poor health of the women. Besides this lack of safe drinking water and low rate literacy are the reasons of the poor health. There is a need of occupational health programme for these women.

***Keywords**: Terai belt, Bihar, Health status, Nutritional status, Foetal loss and Infant mortality.*

Introduction

Women occupy a nuclear position in the society. They are the backbone of the family. The status of women is an important factor affecting health status of the community. Women play an important role in their household economy. The status of women in a society is a significant reflection of the level of social justice in that society (Basu, 1993). Women's status is often described in terms of their level of income, employment, education, health and fertility as well as the roles they play within the family, the community and society (Ghosh, 1987).

Poor self-assessed health tends to be concentrated among women who are poor and belong to deprived ethnic or racial groups (Macran *et al.*, 1993; Cooper, 2002). Social inequalities in health have been explored in numerous contexts, although less attention has been paid to health disparities in India and other low-income countries (Braveman and Tarimo, 2002). Terai belt of Bihar is a very remote area and majority of the populations are agriculturists. The economy of the area is based on agriculture. In this area, women work as daily labourer in agricultural field and building construction. The status of any social group is determined by its level of health-nutrition, literacy-education and standard of living. So, the present work was designed, which deals with the health status of women in relation to certain socio-biological variables.

Materials and Methods

The present work was carried out in the villages of Terai belt of Bihar.

1. Data on food, food habits and food consumption pattern were collected with the help of questionnaires by personal interview with the subject by using recall method. The mean daily food intake was analysed with respect to different food group *viz.*, cereals, pulses, green leafy vegetables and other vegetables, fruits, milk and milk products, fats and oils and sweeteners. The mean nutrient intake was computed using table suggested by Gopalan *et al.* (1981), with respect to carbohydrate, protein and fat. The calculated values of food and nutrient intake were compared with the recommended dietary allowances in order to observe the extent of deviation from optimum nutrients intake as well as imbalance in group of food consumption.
2. Data on nutritional disorders and water borne diseases with the help of doctors and interviews.
3. A total of 155 families were selected for gathering data on foetal wastage and infant mortality. Women aged between 20 to 55 years were interviewed. A total of 750 pregnancies were noted. Information about the present age of the mother, the wetting period (gap in time between their marriage and first conception and their age at various succeeding conception), fertility, mortality and sex were calculated.
4. Literacy was measured as the number of individuals that can both read and write as a percentage of the total population.

Results and Discussion

Occupation

Occupation is the back-bone of socio-economic strata. It has great bearing on living standard. As the present work was carried out in rural areas, so majority of women are agricultural labourer or daily labourer in building construction. These women also prepare mats and jhadu to supplement their needs. Stress at work is a growing problem for all workers, including women. Many job conditions contribute to stress among women. Such job conditions include heavy workload, job insecurity, poor relationship with the supervisors, work that is repetitive and monotonous. Other factors such as work and family balance issues may also be stressors for women in the workplace (NIOSH, 2001). These women were reported with pain in neck, chest and back. However, majority of them reported with musculoskeletal pain. Various population based surveys have also shown positive associations between musculoskeletal disorders and work factors like awkward postures, high physical exertion and vibration (Liira *et al.*, 1996; Roy and Dasgupta, 2008).

Nutritional Status

Food is the prerequisite for survival and growth of any population. Population growth, food supply, health, mortality and nutrition are closely related with each other. A malnourished body become victim of a large number of diseases. Purnia is an agricultural zone. About 70 per cent of the population is of farmers. The chief foods of this area is rice, wheat and maize which vary according with the season. As majority of women are of weaker section and work as daily labourer in agricultural fields or building construction, so their food patterns also vary with the season. Due to poor economical condition use of pulses and green leafy vegetables are restricted. Average daily consumption of food by women in the present study is shown in the Table 1.1. From the Table 1.1 it is clear that use of cereals by these women varies with the season *i.e.* rice is widely used during winter and likewise wheat and maize are widely used during summer and rainy seasons respectively. It is quite clear from the table that diet of these women is inadequate by accepted Indian standards. The deficiencies in diet are both qualitative and quantitative. From the Table 1.2 it is quite clear that their diet is deficient in vitamins, carbohydrates, proteins, fats and the basic calorific requirement is not met. As a result they are victim of malnutrition and suffer from nutritional disorders such as anaemia, night blindness and bitot's spot *etc.*

Table 1.1: Daily Food Intake in the Studied Population of Purnia District

	Winter	*Summer*	*Rainy*
Rice	225	175	-
Wheat	-	225	102
Maize	-	-	300
Pulses	-	35	20
Green veg.	50	70	35
Other veg.	60	50	60
Fruits	Traces	Traces	Traces
Milk	Traces	Traces	Traces
Meat, Fish and Egg	40	15	30
Sugar	20	20	20
Roots and Tubers	40	60	50
Fats and Oils	20	12	15

Foetal Loss and Infant Mortality

The differential pace of abortions, still births and infant mortality have been related to differential pace of economic and social development as well as environmental deteriorations. So, it becomes imperative to identify different factors related with abortions, stillbirths and infant mortality.

According to Winikoff (1978) due to chronic malnutrition in expectant mothers the risk of death of foetus as well as mother may increase because lesser nutrients are available for the mothers. It increases the frequency of birth injuries and reduces overall efficiencies of reproductive processes. The higher percentage of loss of foetus during later stages in the studied women might be influenced by maternal nutritional status. As economical condition of these women is poor, they are unable to meet the expense of their balanced diet. From the study of nutritional status, it is quite clear that their diet is deficient in protein and vitamins; apparently these deficiencies in diet lead to abortions. Increased abortion rate has been found in women in jobs requiring heavy lifting, standing long hours (La Dou, 1993). In the present study women worked as daily labourers performing heavy lifting and remain standing long hours. Thus their occupation may have risk of spontaneous abortions. Such findings have been reported by Pandey *et al.* (2000 and 2002) in women of Purnia district and in tribal women of Jamshedpur (Jharkhand) who work as daily labourers in factories/agricultural fields. Deficiency of iodine is a common feature of this area which is manifested by the prevalence of the goitre leading to abortions and stillbirths (Hetzel, 1983). Smoking by women during pregnancy period may results in pregnancy complication *i.e.* higher prenatal mortality rates (Nasrat *et al.*, 1986; Butler and Goldstein. 1973). Majority of these women are addicted to biri or hukka and even local wine.

Table 1.2: Daily Nutrient Intake and Calorific Value of the Studied Population of Purnia District

	Carbohydrate	*Protein*	*Fat*	*K. Cal*
Rice	105.33	8.53	0.53	465.32
Wheat	75.64	13.18	1.85	371.69
Maize	66.20	11.0	3.66	342.0
Pulses	10.97	4.21	0.01	63.78
Green veg.	2.68	2.01	0.05	21.50
Other veg.	3.29	1.08	0.03	18.11
Fruits	-	-	-	-
Milk	-	-	-	-
Meat, Fish and Egg	15.58	6.23	1.13	34.46
Root and Tubers	10.50	0.70	0.05	42.62
Fats and Oils	-	-	15.66	140.94
Sugar	17.22	0.04	-	79.60
Total	307.41	46.98	22.97	1580.02

Table 1.3: Foetal Loss and Infant Mortality in Studied Population of Purnia District

	Foetal Loss		
Total Pregnancies	*Abortions*	*Stillbirths*	*Total*
750	58 (7.33 per cent)	50 (6.67 per cent)	98 (14 per cent)
Total birth	*Sex*	*Infant mortality*	*Total*
652	M – 327	40 (6.13 per cent)	87 (13.33 per cent)
F – 325	47 (7.20 per cent)		

Infant mortality in India is 70 per thousand (Nayar, 1998). Infant death within the first week are caused mostly by maternal factors related to delivery such as extreme of age, short birth intervals, malnutrition and multiparity (Puffer and Serrano, 1975). Such practises are common in the study area.

Poor quality of houses, lack of environmental sanitation and education are responsible for infant mortality (Dhanalakshmi and Murthy, 1993; Rao, 1985; Pandey *et al.*, 1998 and Wave, 1984). Of course, these women are illiterate and their standard of living is very poor. Malaria is prevalent in this area and is known to cause of placental barrier, parasite sequestration in placenta, suboptimal nutrition of the foetus, congenital malaria, intrauterine growth retardation, low birth weight, premature interruption of pregnancy, infant mortality and maternal death (Egwunyenga *et al.*, 1997; Kochar *et al.*, 1998 and Das, 2000).Women working outside the house, those who are poorly educated and those who are economically deprived have often been reported to have experienced greater infant and child loss (Ware, 1984; UN, 1985).When the mothers work away from the house the time for child care decreases leading to nutritional, physiological and pathological problems in child resulting in mortality. The workplace of these women is far away from their residing place. They have to cover a distance to 4 – 5 km to reach their workplace.

The high foetal wastage and infant mortality in the present study are due to:

1. Lower age of marriage and motherhood
2. Greater fertility rate of mother and short wetting period
3. Poor living condition and hard working during pregnancy period
4. Poor environmental sanitation and hygiene
5. Poor nutritional status
6. Poor literacy rate
7. Poor medical facilities and distance of hospitals or primary health centres from their residing areas
8. Getting services of untrained village doctors and untrained nurses
9. Addiction or smoking
10. Remote settlement
11. Consanguinity (short marriage distance)

Sanitation

The houses are generally insanitary. The same room is used for living as well as for keeping animals. In houses there is no provision for latrine and bathroom. Majority of houses are dark, ill ventilated and infested with cockroaches, leeches, mice and insects like mosquitoes and houseflies. There is no provision of sewerage in the villages. The household garbages are deposited in front of houses along with cow dung which becomes breeding ground of many insects that are carriers of many parasites. As there is no drainage system, logging of water around houses is very common feature. Both male and female go to field in the morning and evening for defection. Aged persons and children usually go for this purpose near the house. Children below the age of 6 – 7 years can urinate anywhere. Thus the entire areas around the house become dirty with the human waste. Retting of jute is done in water logged nearby houses. This water becomes breeding ground of mosquitoes.

Environmental factors as well as personal hygiene affect the community health. These women do not take bath regularly. 3 – 4 days gap in bathing is common in them. As a result these populations suffer from skin diseases. Generally cow dung and wood are used as fuel for cooking food. They never use soap to wash their hands. Plain water and soil is used for hand washing after defection. Cleaning of teeth is done by ash, mud or datun. Majority of women are addicted of hukka, biri, gutkha and even local wine.

The main source of drinking water in the study area is hand pumps. The water level is very high (10 – 15′). All water showed faecal contamination and at some places high incidence of iron and nitrate has been reported leading to health hazards (Pandey, 2000). Due to accumulation of sewage water near hand pumps, location of hand pumps near agricultural fields or cattle sheds and lack of concrete platforms around hand pumps the drinking water becomes contaminated (particularly in rainy season) leading to large number of water borne diseases (such as diarrhoea, cholera, amoebic dysentery, giardisis *etc.*).

Health Status

Health is an important aspect of development. The health of the people is determined by many factors. These factors largely reflect socio-economic, environmental conditions in which people live, nutritional status and political conditions. As these factors alter over a period of time, corresponding changes are seen in health status. It has been reported that whenever an environmental condition changes new diseases emerge out (Taylor, 1991).

The main diseases in these women are anaemia, malaria, filaria, kala-azar, T.B., gastrointestinal, leprosy, menstrual disorders and parasitic infestation (*Entamoeba, Giardia, Ascaris, Ancylostoma*). It has been reported that a large section (~90 per cent) of women in India suffer from anaemia and severe anaemia accounts for 20 per cent of maternal death in India, the maternal mortality being at least 450 per 100,000 live births in women of the age groups 15 – 49 (World Bank, 1996; National Human Development Report, 2001).

- ✰ Anaemia 18.8
- ✰ Malaria 7.2
- ✰ Kala-azar 7.6
- ✰ Filaria 5.0
- ✰ Water borne diseases 31.6
- ✰ T.B. 5.8
- ✰ Leprosy 2.4
- ✰ Joint pain 12.4
- ✰ Other 9.6

Education is an inclusive measure of socio-economic position because women are included regardless of their status in the labour market (Krieger *et al.*, 1997). Education can provide a woman with a broad set of cognitive resources, in addition to material gains and future financial security through better employment or marriage opportunities (Mohindra *et al.*, 2006). As far as literacy is concerned, it is almost negligible in the studied women. There is a relationship between socio-economic position and health status which is widely accepted among epidemological and public health researchers. People who are socio-economically better placed tend to be healthier; this relationship has been seen with measures of mortality, morbidity and self assessed health (Kosa *et al.*, 1969). As far as socio-economic position of these women is concerned, it is not satisfactory. Caste related economic and social differences suggest that women from lower castes will carry a higher burden of ill health (Sen *et al.*, 2002). Women of the present study belong to lower castes.

The health problems in the studied women can be categorised due to;

1. Pattern of settlement
2. Poor nutritional status

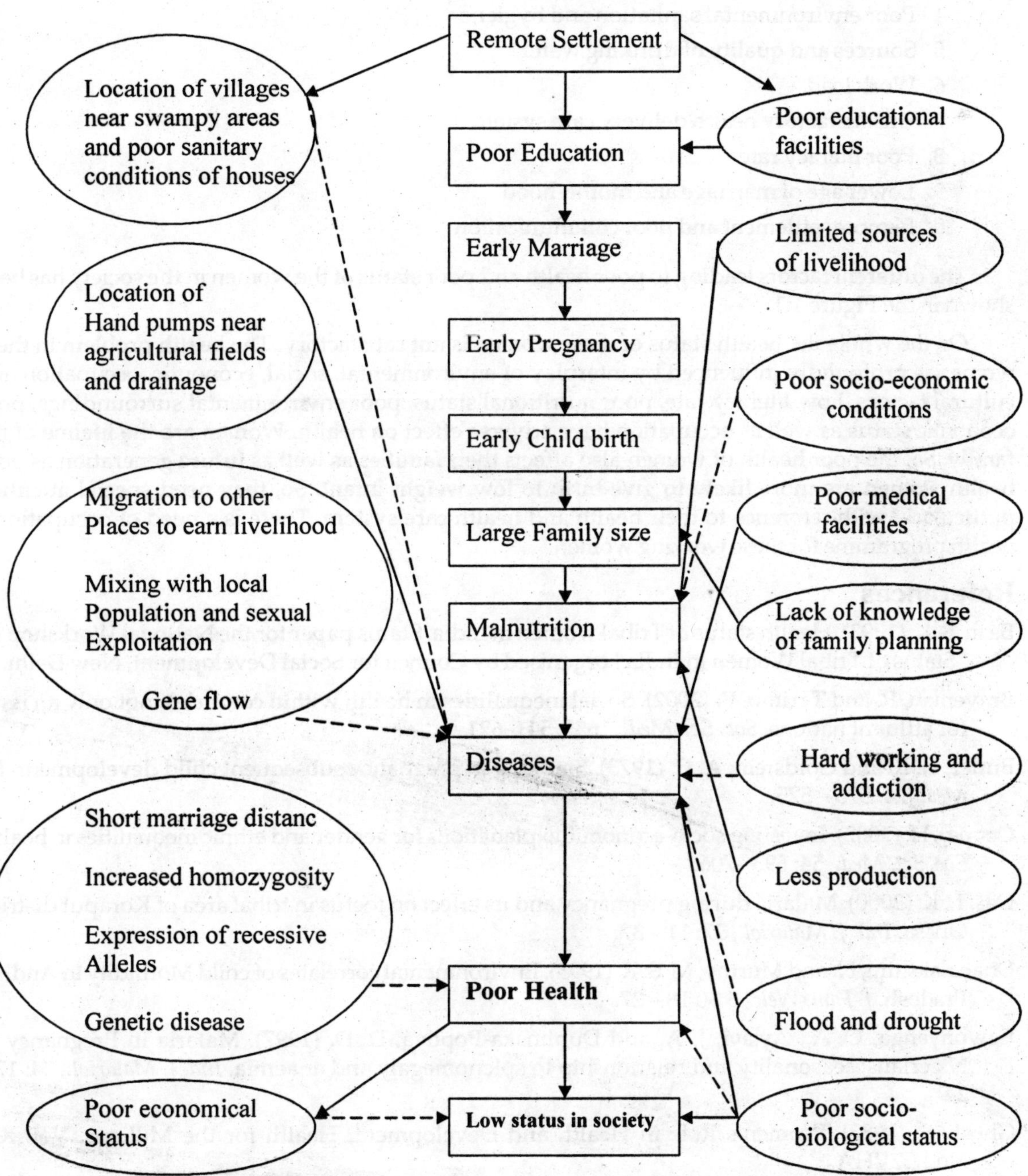

Figure 1.1: Poor Status of Women in Society

3. Poor living standard
4. Poor environmental sanitation and hygiene
5. Sources and quality of drinking water
6. Work load
7. Unsatisfactory health delivery care system
8. Poor literacy rate
9. Lower age of marriage and motherhood
10. Remote settlement and poor communication

The different factors leading to poor health and poor status of the women in the society has been shown in the Figure 1.1.

On the whole the health status of these women is not satisfactory. The health problem in these women is profoundly influenced by interplay of environmental, social, economic, occupation and cultural factors. Low literacy rate, poor nutritional status, poor environmental surroundings, poor economic status as well as occupation have adverse effect on health. Women are the lifeline of the family. So, the poor health of women also affects their families as well as future generation as poor health women are more likely to give birth to low weight infant. So, they need special attention particularly with reference to their health and health care system. There is a need of occupational health programme for these working women.

References

Basu, S. K. (1993). Health status of Tribal women in India. Status paper for the National Workshop on "Status of Tribal Women in India" organised by Council for Social Development, New Delhi.

Braveman, P. and Tarimo, E. (2002). Social inequalities in health within countries: not only an issue for affluent nations. *Soc. Sci. Med.*, 1635: 541–621.

Butler, N. R. and Goldstein, A. C. (1973). Smoking in pregnancy subsequent child development. *Br. Med. J.* 4: 573 – 575.

Cooper, H. (2002). Investing socio-economic explanations for gender and ethnic inequalities in health. *Soc. Sci. Med.*, 54: 693 –706.

Das, L. K. (2000). Malaria during pregnancy and its effect on foetus in tribal area of Koraput district, Orissa. *Ind. J. Malariol.*, 37: 11 – 37.

Dhanalakshmi, N. and Murthy, M. S. R. (1993). Environmental correlates of child Morbidity in Andhra Pradesh. *J. Fam. Welf.*, 3 30 18 – 27.

Egwunyenga, O. A., Ayjayi, J. A. and Duhlinska-Popova, D. D. (1997). Malaria in Pregnancy in Nigerians: seasonality and relationship to splenomegaly and anaemia. *Ind. J. Malariol.*, 34: 17 – 24.

Ghosh, S. (1987). Women's Role in Health and Development. Health for the Millions. Vol. XIII No.1&2VHA.

Gopalan, C.Ramsastri, B. V. and Subramanian, S. C. (1981). Nutritive values of Indian Foods. National Institute of Nutrition. ICMR, Hyderabad.

Hetzel, B. S. (1983). Iodine Deficiency Disorders (IDD) and their eradication. Lancet, 1: 1126 – 1129.

Kochar, D. K., Thanvi, I., Joshi, A., Subhakaran, S., Aseri, B. and Kumawat, I. (1998). Falciparum malaria and Pregnancy. Ind. J. Malariol., 35: 123 – 130.

Kosa, J., Zola,I. K. and Antoonovsky,A. (1969). A poverty and health: A sociological analysis. Cambridge: Harvard University Press.

Krieger, N., Williams, D. R. and Moss, N. E. (1997). Measuring social class in US public health research: concepts, methodologies, and guidelines. *Annu. Rev. Public Health*, 18: 341 – 378.

La Dou, J. (1993). International issues. In: Women workers, D. Headapohl (Ed). Occupational Medicine: State- of –the Art-Reviews. Hanley and Belfus, Inc. Philadelphia.

Liira, J. P., Shannon, H. S., Chambers, L. W. (1996). Long term back problems and Physical work exposure in the 1990. Ontario health survey. *Am.J. Public Health*, 86: 382

Macran, S. Clrke, L. and Sloggett, A. (1993). Women's socio-economic status and Self-assessed health; identifying some disadvantage groups. *Sociol. Health Illn.*, 16: 182 – 208.

Mohindra, K. S., Slim, Haddad and Narayana,D. (2006). Women's health in a rural community in Kerala, India: do caste and socioeconomic position matter?. *J. Epidemiol. Community Health*, 60 (12): 1020 – 1026.

National Human Development Report (2001). *Planning Commission*, Government of India. Oxford University Press, New Delhi, 2002.

Nasrat, H. A., Al-achim, G. M. and Mahmoud, F. A. (1986). Perinatal effects of nicotine. *Biol. Neonal.*, 49: 8 – 14.

Nayar, K. R. (1998). Old priorities and new agenda of public health in India: Is there a mismatch? *CMJ online*, 89: 1 – 7.

NIOSH, 2001. Looks for women's safety and health at work. Editorial. Nov.12 (cited on 2007 Mar). Available from: http://www.occupationalhazards.com.

Pandey, B, N., Jha, A. K., Das, P. K. L. and Ojha, A. K. (1998). Socio-ecology of tribes of Koshi zone in relation to its impact on health. In: Contemporary Studies in Human Ecology; Human Factor, Resource Management and Development. Bhasin, M. K. and Malik, S. L. (Eds). Kamala- Raj Enterprises, Delhi.

Pandey, B. N. (2000). Final Report on the Project – Biological Status of Tribal Groups Of Bihar submitted to I. C. M. R., New Delhi.

Pandey, B. N., Mishra, S. K. and Yadav, S. (2000). Foetal wastage and infant mortality In some endogamous populations of Purnia district, *Bihar. J. Hum. Ecol.*, 11(6): 477 – 481.

Pandey, B. N., Das, P. K. L., Jha, A. K. and Ojha, A. K. (2002). Foetal wastage, infant mortality and ecological and social factors in six tribal groups of Jharkhand. In: *Eco-degradation, Biodiversity and Health*. Pandey, B. N. (Ed.). Daya Publishing House, Delhi.

Puffer, R. R. and Serrano, C. V. (1975). *Birth weight, Maternal age and Birth order: Three important Departments of Infant Mortality*. Scientific Publ. No. 294. Pan American Health Organization, Washington, DC.

Rao, S. L. N. (1985). Factor associated with mortality decline in high mortality countries. Paper presented at the International Population Conference. Florence.

Roy, S. and Dasgupta, A. (2008). A study on health status of women engaged in Home-based "Papad-making industry in a slum area of Kokata. *Ind. J. Occup. Environ. Med.*, 12 (1): 33-36.

Sen, G., Iyer, A. and George, A. (2002). Class, gender and health equity: lessons from liberalizing India. In: Sen, G., George, A. and Ostlin, P. Eds. Engendering international health. The Challenge of equity. Cambridge: MIT Press.

Taylor, G. C. N. M. (1991). Human adaptation and the impact of disease. *J. Ind. Anthrop. Soc.*, 26 (122): 37 – 49.

United Nations. (1985). Socioeconomic Differentials in Child Mortality in Developing Countries. United Nations, New York, 149 – 158.

Ware, H. (1984). Effects of maternal education, women's role and child care on child mortality. *Population and Development Review, Suppl.*, 10: 191 – 214.

Wave, H. (1984). Effects of maternal education women's roles and child care on child mortality. In *Child Survival: Strategies for Research*. W. H. Mosley and L. C. Chen. Supplement to Population and Development Review, 10: 25 – 45.

Winikoff, B. (1978). Nutrition, population and health: Some implication for policy: The Population Council Working paper, 3: 1 – 32.

World Bank, 1996. *Improving Women's Health in India*. The World Bank, Washington, DC.

2013, Sustainable Approaches for Environmental Conservation *Pages 11–18*
Editors: **D.R. Khanna, A.K. Chopra, R. Bhutiani, Gagan Matta & Vikas Singh**
Published by: **BIOTECH BOOKS, NEW DELHI**

Chapter 2
Impact of Awareness Campaign on Energy Conservation: A Case Study Chandigarh

Meenu Wats[1], *Alka Grover*[1] *and Rakesh K. Wats*[2]
[1]*Department of Chemistry, DAV College, Sector-10, Chandigarh*
[2]*National Institute of Technical Teachers Training and Research, Chandigarh*

Whole world, today is facing two main alarming threats, energy shortage and environmental degradation. India too is facing these problems. Being a developing country, it is showing an increasing trend in energy consumption for its economic development. It will shortly become the third country in terms of high volumes of GHGs emission. Hence, environmental conservation along with energy is the prime concerns of the country. Similar is the case of the newest planned city of the country- Chandigarh, the beautiful city. Chandigarh was established in an area of 114 sq. km for a limited population of about 1.5 lacs, which has by now swelled to more than five times, thereby facing a great pressure in terms of its exponential increase in the demands of energy consumption for matching the life style of its populace. Although, Chandigarh administration is trying hard for its sustainable development but no such effort is fully successful without the involvement of its all citizens. Keeping the above objective in mind the Department of Science and Technology, Chandigarh Administration sponsored an energy conservation campaign, amongst its youngest populace *i.e.* school children, to create awareness amongst them about the importance of energy for the growth and development, its ever increasing demand leading to its scarcity and negative impact on environment, alternative sources of energy and their utilisation and finding ways and means for the conservation of energy resources. The broad aim of the project is to make energy conservation, a mass campaign starting with the young minds and through them reaching every home, street, office and nook and corner of the city.

The present chapter deals with the impact of this awareness campaign on school children, teachers, staff and other stake holders of the society. In the paper, a comparison has been drawn about the impact of this campaign between government and private schools, responses of girls and boys, male and female teachers and staff and the status of awareness in the pre and post campaign workshops.

Keywords: Environmental degradation, Energy conservation, Sustainable development.

Introduction

Global warming is a giant problem; the whole world is facing today, as it has the potential to change our lives and the environment of this planet and would in turn affect the whole mankind. The problem of climate change was identified way back in 1979 as an urgent world problem in the first 'World Climate Conference'. The threat of climate change is becoming more evident than ever before, for its impact is increasingly being felt in various parts of the world as environmental threats.

The rapid population growth and the increasing energy demand are the primary forces causing a huge release of harmful pollutants and GHGs in the atmosphere, resulting in environmental imbalances. As on July, 2011 CO_2 in earth's atmosphere is at a concentration level of 392 ppm, against its safe limit of 350 ppm (Wikipedia, 2011). This indicates that our earth is losing its natural capacity to soak this gas in forests and oceans - the natural sinks of CO_2. Human activities such as combustion of fossil fuel and deforestation have caused the atmospheric concentration of CO_2 to increase by about 35 per cent since the beginning of age of industrialization.

India, too, is facing a close tie between economic development and climate sensitivity. Climate change matters much to India as it is a home of world's $1/3^{rd}$ poor population. So, the economic, social and ecological price due to this change will be massive. In the recent years, India's energy consumption has been increasing at one of the fastest rates in the world. India rank's 5th in the world in terms of primary energy consumption. Despite of the overall increase in energy demand, per capita energy consumption in India is still very low (720 KWH) as compared to average global per capita (2340 KWH) consumption. Although, India is the 4th largest emitter of green house gases (GHGs) but the saving point is that it has the lowest per capita emission of GHGs. As per the predictions by the International Energy Agency, if India sustains its present high annual economic growth, then by the year 2015, it will become the 3rd largest emitter of GHGs.

In India, in its post independence era, the first planned city of the country, Chandigarh, the city beautiful city, was designed by the famous French architect- Le Corbusier. In just five decades of the time, since its establishment, this city has under gone a metamorphosis development in terms of flaring education sector, industrial growth, IT developments, tourism, medical facilities *etc.* The city feels proud to be the cleanest city in the country with a rating of 73.48 points of sanitation. In the recent past (year 2005), Chandigarh has also shown the highest per capita income in the country (Rs. 67,370/-) (Indian Express, 2006). The city administration boosts of taking a lead in making the city smoke free and also banning use of polythene bags to keep it clean. Even the green cover of this city is more (35.7 per cent) than the recommended (33 per cent) cover by the National Forest Policy of India (Times of India, 2009).

All the above features are more than enough for any person to get attracted towards this city and settle down. The swelling population of the city is posing a great threat to the administration to meet their demands, especially that of energy in terms of electricity. At present the assured power supply by

Chandigarh administration around the year is 184 MW but the summer demands go up to 300 MW. With no power generation system of its own, city administration is able to procure around 265 MW from neighboring states like Punjab, J&K and Himachal on short term purchase and power sharing basis. Chandigarh receives 67 per cent electricity from PSEB Mohali, Punjab; 23 per cent from Nalagarh, Himachal Pradesh and 10 per cent through BBMB Dhulkote (Indian Express, 2002).

Increasing energy demands of Chandigarh residents is also posing threat to its beautiful environment. At present Chandigarh emits 12,443,95 tonnes of CO_2/year due to electricity, petrol and LPG consumption and this is expected to increase to an alarming amount of 29,201,05 tonnes by the year 2018.

Realising the need for sustainable development for meeting the ever increasing demands of energy and the impact on environment, Chandigarh administration has initiated many projects, programmes, activities and campaigns for energy development and conservation. Setting up of solar city, which will be able to meet about 10 per cent of conventional energy load by 2012 and probably 20 per cent by 2018; Buchat Lamp Yojana (BLY) in which replacement of incandescent bulbs by CFLs in homes (@Rs. 15/- only); setting up of energy park in botanical garden by Tata BP Solar Ltd.; replacement of street lights in villages by solar street lamps; mandatory use of CFL in all government buildings *etc.* are few examples in this direction. The above initiatives will not only reduce the load on non renewable fossil fuels but also be a clean source of energy. Department of Science and Technology (DST), Chandigarh is helping the administration in generating awareness regarding energy conservation by sponsoring research projects. Zero Wastage and Non Conventional Energy Sources is one such project. The broad aim of this pilot project is to create awareness amongst young minds in schools regarding the need for energy conservation for better life and through them reach every section of society. The present paper deals with the findings of this pilot project.

Methodology

Awareness campaign was conducted in 10 specified schools, assigned by the DST. List of schools included both government and private schools. All schools were coeducational, so both girls and boys could be sensitized on the issue. Besides students, schools' faculty and staff members were also involved as energy and environment is a concern of each and every one. A sample of 1000 students and 300 teachers was taken for the purpose. The campaign included:

1. Interactive sessions for sensitization
2. Energy Audits (Electricity)
3. Power point presentations on the status of electricity in Chandigarh
4. Distribution of hand outs with energy saving tips
5. Conduct of intra and inter school competitions
6. Documentary shows
7. Exhibitions
8. Filling up of questionnaire
9. Feed back

Results and Discussion

Response of the campaign was drastically different in the beginning and at the end. Response of school administration, students and staff was pretty hesitant and non-accepting kind of in the beginning, but later on there was remarkable improvement in the participations in all sorts of competitions, interactions, workshops, shows *etc.* Various parameters were evaluated to study the impact on the students and staff of government and private schools, girls and boys, male and female staff members before and after the conduct of awareness campaign. The results of the study are as follow:

Level of Awareness

Students of government schools had a concept of energy conservation, being a topic in their curriculum, but ways and means were not clear. The students of private school still had better comprehension of the term. Initially, a large section of students thought energy conservation meant – saving of electricity only. A few respondents, mainly faculty and some students felt it as a means of saving fossil fuel in addition to electricity. A very small group added it being protection of environment. The response of concept and need of energy conservation was conceived properly after some interactive sessions. This step was very crucial for the success of the mission, as any effort by the government or any agency is futile until its stake holders understand the need of energy conservation and environment protection. Figure 2.1 presents the pre-interaction response in terms of the concept of energy conservation.

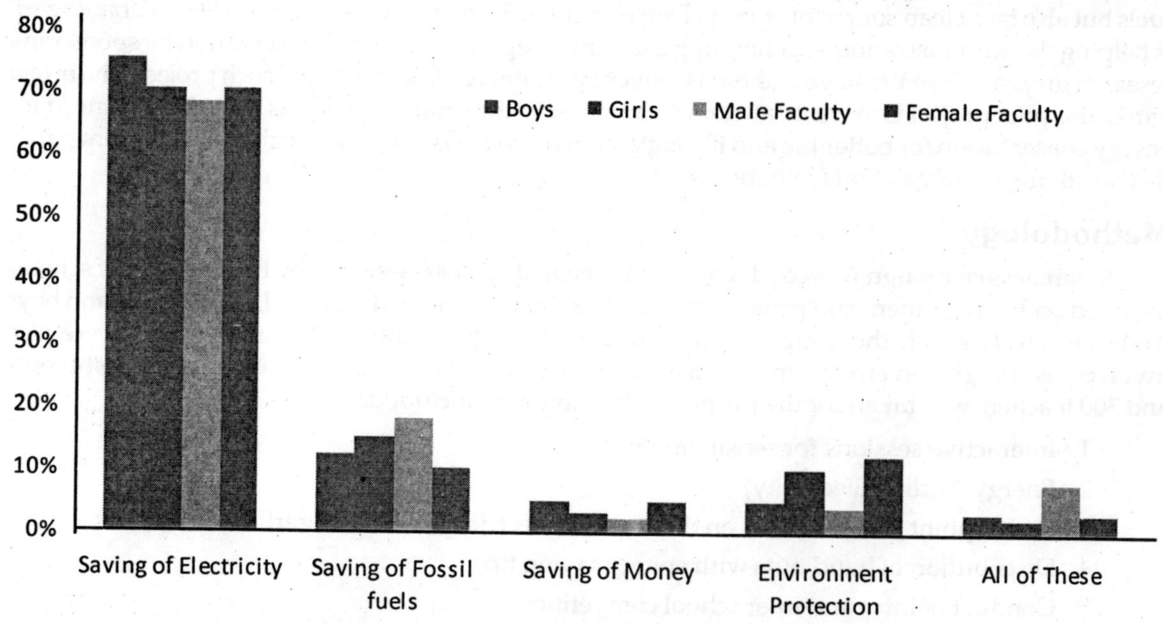

Figure 2.1: Concept of Energy Conservation

Response Towards Wastage

After some workshops and interactive sessions, almost everyone was convinced about their role in energy wastage both in homes as well as work place *i.e.*, schools. They were made fully aware of the sources of wastage of electricity in terms of misuse and overuse, inefficient electrical appliances and devices, leakage in supply and distribution system, electricity thefts *etc.* The campaign could sensitize

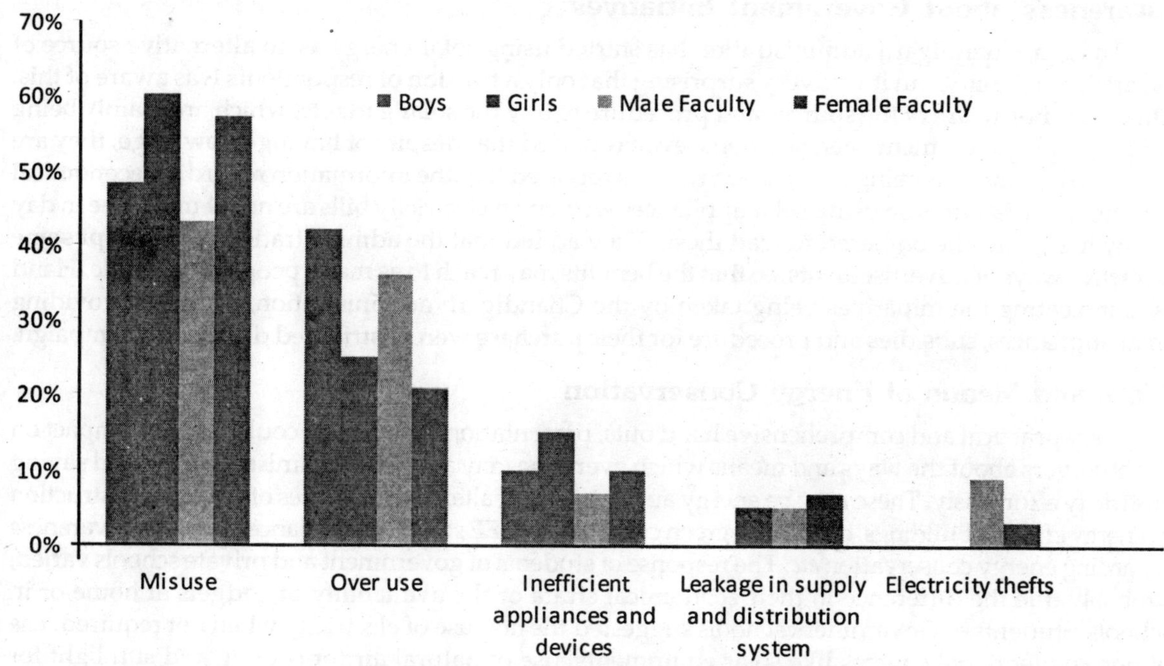

Figure 2.2: Response Towards Wastage

the target group to the need of the hour – energy conservation along with sustainable development. More emphasis was given to the students of private schools as most of them belonged to an economical class which can afford buying of more electrical gadgets, hence more wastage and more scope of saving. Schools energy audit report could make school authorities to realize their short comings too. Government schools were wasting electricity due to faulty circuits and letting fans and lights switched on even during off times like recess or any other outdoor class. Private schools showed over use of electricity in the race of making smart class rooms and also making most parts of the buildings air conditioned without taking care of the wastage points. Figure 2.2 presents the pre interaction response in terms of wastage in energy.

Awareness of Non-conventional Sources of Energy

Students as well as teachers had conceptual information of various types of non conventional sources of energy such as solar energy, wind energy, nuclear energy, biomass *etc.* but their use was almost nil. Very less number of students reported using any other type of energy source than the conventional electricity. To a great surprise, some government and private schools were already using solar lights and water heaters but neither the students nor the faculty was aware of these installations. This directly reflects the lack of awareness, even of the existing alternative sources of energy. Only a few students and female staff reported the use of solar equipment, which were mainly solar cookers and very rarely solar water heaters. The awareness of the use and benefits of various solar equipments was made through exhibitions and its impact was realised with the response of students and teachers in terms of their promises for the use of these appliances in near future.

Awareness about Government Initiatives

Though, Chandigarh administration has started using solar energy as an alternative source of electricity in the city, but it was very surprising that only a fraction of respondents was aware of this. They were not aware of the sources and procedure to buy the solar gadgets, which are mainly being sold through government agencies. Some even reported that despite of having knowledge, they are unable to buy these as being expensive. Students reported that the information regarding economical benefits and discounts on using solar appliances written on electricity bills are not of much use in day to day life, as no one bothered to read these. They added that the administration must adopt some attractive ways of advertisements, so that the benefits may reach to as many people as possible. Hand outs indicating the initiatives being taken by the Chandigarh administration, agencies providing solar appliances, subsidies and procedure for their purchase were distributed during this campaign.

Ways and Means of Energy Conservation

Very practical and comprehensive hand outs, presentations and shows could leave an impact on the observers about the ways and means which everyone may adopt to minimise wastage and saving electricity effortlessly. These may be energy auditing, use of alternative sources of energy, construction of energy efficient buildings, enhancing green cover, use of BEE standard appliances, creating awareness regarding energy conservation *etc.* The response of students of government and private schools varied, probably due the difference in their economical strata or the availability of gadgets at home or in schools. Students of Government schools suggested the non use of electricity when not required, use of non conventional sources like solar equipments, use of natural air for cooling and sun light for heating homes and schools, use of half covered windows to avoid heating in summers and use of CFLs *etc.* Students of private schools had inclination towards gadgets, so they opted for outdoor games than mobile or computer games, maintaining optimum temperature of ACs in summer to save

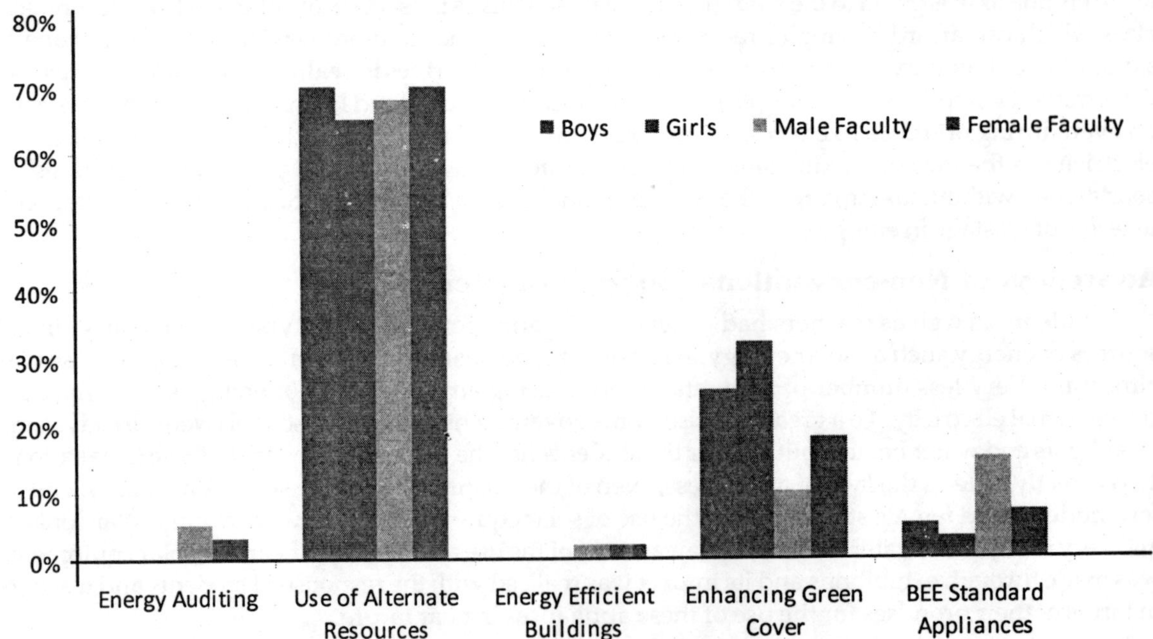

Figure 2.3: Ways and Means of Energy Conservation

electricity, more use of solar appliances *etc.* Male faculty thought that energy saving can be done by rotational cuts of electricity, use of BEE branded electrical appliances, least use of electrical gadgets *etc.* The surprising fact was that a meager fraction of respondents was aware of the concepts of green buildings or eco-friendly buildings or importance of landscaping or energy saving interiors. Figure 2.3 presents the pre interaction response in terms of ways and means of energy conservation.

Role of Agencies in Energy Conservation

Initially most of the respondents were of the view that conserving electricity and reducing its wastage is the sole responsibility of the government. Government should adopt varied strategies for the production of more energy and its optimum utilisation. A few faculty members were of the opinion that NGOs and other similar agencies can play a significant role in creating awareness in this direction. Most of the respondents were shirking in terms of their responsibility in this important mission of zero wastage and effective use of non conventional sources of energy. However, after varied interactive sessions, presentations, video films and demonstrations all were clear that the success cannot be achieved without the whole hearted participation of each and every member of society. Students of government schools could be easily convinced on this sensitive issue of energy conservation in comparison to that of private schools. Female teachers quickly agreed to their role in the same cause in comparison to males. Female faculty agreed to minimise the use of electric operated machines like washing machines, food processors, vacuum cleaners *etc.* while male faculty could be convinced on the less use of remotes for electronic items, more use of landlines in place of mobiles while at home or in offices, not letting electronic appliances on standby mode *etc.* Figure 2.4 presents the pre interaction response in terms of the role of agencies in energy conservation.

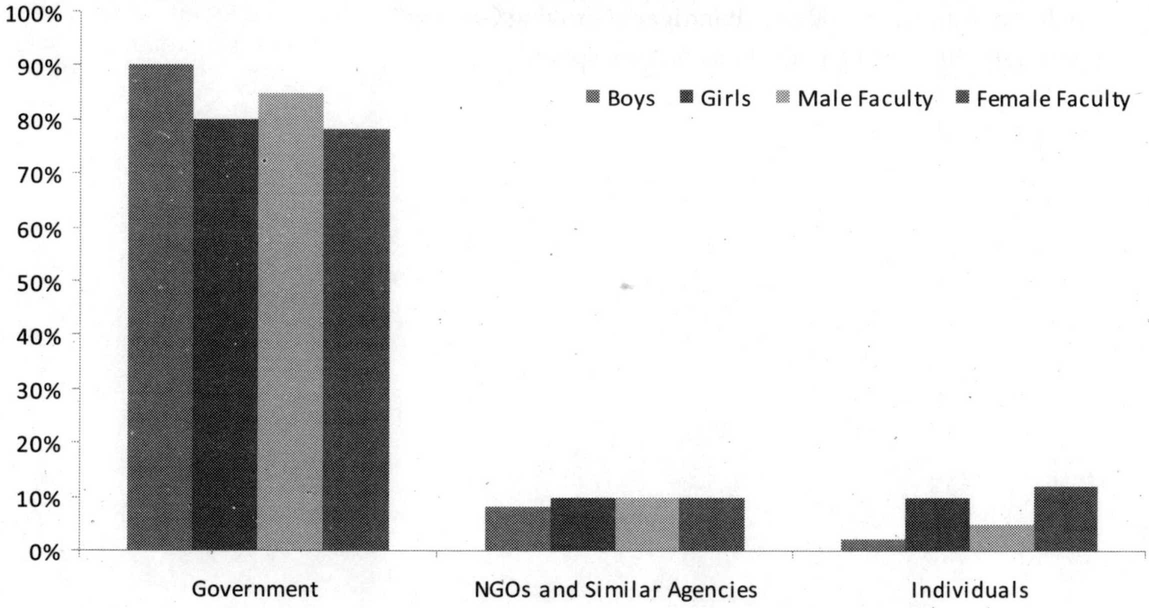

Figure 2.4: Role of Different Agencies in Energy Conservation

Conclusion

Looking at the global level climate changes and energy resources depletion, India is also realizing the need to pay attention to these aspects. On the similar lines the Chandigarh, which is growing at a very fast pace in every modern and technical sector, is bound to feel the same. Increasing demands of energy coupled with enhancing emission of carbon dioxide along with no energy generation system of its own has made Chandigarh administration to take some preventive measures for its sustainable development. The awareness campaigns for the young minds who will spread the message to each and every part of society can leave a remarkable impact in the times to come. Such projects and campaigns must be carried out regularly, as energy conservation requires a change in attitude, which is a long term process. Mass awareness is required for the sustainable growth of not only Chandigarh, but of the whole nation.

Acknowledgements

The authors are whole heartedly grateful to:

- ☆ Department of Science and Technology, Chandigarh Administration for sponsoring this Project on Zero Wastage and Non Conventional Sources of Energy.
- ☆ Management of DAV College, Chandigarh for its support and guidance in undertaking the project.
- ☆ Management, faculty, staff and students of the respondent schools for their cooperation and participation in the project.

References

Indian Express, April 2, 2002. " Chandigarh Looks to Become Solar City in a Decade".

Indian Express, June 9, 2006. " Chandigarh in Figures 2005".

Times of India, August 28, 2009. " Chandigarh Growing Greener".

Wikipedia, July, 2011. " CO_2 in the Earth's Atmosphere".

2013, Sustainable Approaches for Environmental Conservation *Pages* **19–23**
Editors: **D.R. Khanna, A.K. Chopra, R. Bhutiani, Gagan Matta & Vikas Singh**
Published by: **BIOTECH BOOKS, NEW DELHI**

Chapter 3

Eco-Degradation, Population and Health

A.K. Thakur[1], Ranjana Kumari[2], Anita Mishra[2] and B.N. Pandey[2]
[1]*K.K.M. College, Jamaui, Bihar*
[2]*Eco-Genetical Research Laboratory,*
P.G. Department of Zoology, Purnia College, Purnia, Bihar

Industrialization, urbanization, mining, deforestation, population explosion have resulted in certain hazardous effects on our ecosystem. Population explosion has resulted in the pollution of air, water and soil along with poverty. Some global issues of great concern have also originated from the same root, such as global warming, ozone layer depletion and acid rains. Heavy use of insecticides, pesticides, fertilizers and poisonous gases released from auto exhausts and industries have contributed some deleterious effect on health of human being and their live stocks. The bio diversity in the forests has already been alarmed and struggling for its survival due to large scale of deforestation, industrialization, urbanization and expanded agriculture. Besides this draught, flood, earthquakes, tsunamis and tornadoes regularly damages the humans as well as their live stocks. All these environmental issues, which are alarming to mankind, are issues of great concern and proper effort should be taken for their management before it becomes too late

Keywords: *Industrialisation, Urbanization, Mining, Deforestation, Population explosion, Health.*

Survival of any organism depends upon its surrounding environment. The relationship between man and his environment has been never static. Men's extensive interaction with the environment has resulted in Eco-degradation. Eco-degradation has posed to come up as one of the major concerns for the mankind. Increasing rate of urbanization, industrialization, mining, deforestation and population explosion have resulted in highly polluted milieu and formidable health hazards not only for mankind but for the entire living beings. No doubt, environmental degradation and destruction, because of ecological relationships, inevitably leads to negative social, economical and political changes. The present communication gives a broad spectrum on Eco-degradation and its impact human health with a brief account of its management and conservation.

Human being always strives for a better life. People do not hesitate to leave their mother land for a foreign land that assures them a better standard of living. Mineral ores extracting sites and industrial plants attract people from all directions. All established industrial towns in the map today were either jungles or barren lands for which people were not aware of in the past. But today these are densely populated area. This may in part be due to migration of populations to work in industries and mines as well as different trades. All industrial areas once upon a time were full of jungles, where the industrialization has introduced large scale migration of people in search of livelihood from different corners of the state as well as of the country. This has resulted in increased population growth. With the growth of the industrialization and urbanization, the forest area has been reduced to a large extent and an ever increasing biotic pressure can be felt. This has eventually resulted in multifarious problems like housing, sanitation and health problems. Industrial activities are known to disturb the ecological system by creating pollution of different kinds. Production of materials and consumption have always led to environmental degradation, but these become a problem when industrialization and population enormously increased the scale operation. Many production processes such as mining, extraction, refining or smelting pollute air and land. Wastes or residuals discarded from production result in pollution of various kinds. It is to be noted that, human activities have and will continue to modify the patterns and biological impact of diseases. Environmental changes are invariably associated with the emergence of new diseases and other problems.

Industrial enterprises have been, in general, accused of upsetting the ecological balance comprising air, water, soil, flora and fauna. Every day auto emissions, industrial smoke stack and utility plants hurl tonnes of pollutants into the air. Simultaneously, factories dump enormous quantity of waste materials into the rivers and streams. It is not the ecological balance which is at stoke, the very existence of mankind faces a threat from the menace of industrial pollution. This is because industries have introduced harmful substances in our environment. As a result, the highly the highly delicate balance between the components nature such as air, water and soil, is on the verge of collapse. The reckless falling of trees has resulted in climatic changes, reduced rainfall, enhanced soil erosion and reduced availability of wood. Similarly, the release of harmful chemicals into the environment has threatened the thin ozone layer which surrounds the earth. This layer of ozone absorbs the ultraviolet radiations from the sun. Reduction of this layer may result in extinction of a large number of biological species and can cause harm to human beings. A coal based thermal plant pollutes the atmosphere by gaseous emissions of sulfur dioxide, caused acid rain which is known to damage soil, vegetation and aquatic lives of the region and also produces a tremendous amount of solid wastes, fly ash. A super thermal plant using, even normal or low sulfur coal emits about 100 tonnes of sulfur dioxide a day. The gaseous pollutants have an impact on biotic as well as abiotic components of the ecosystem.

Increasing human population, coupled with ever rising standard of living and consumerism would necessarily cause increasing demand on all natural resources – land, water, forests, minerals *etc.* In India, in the last 50 years; per capita availability of total as well as agricultural land has been reduced to about one – third to one – fourth, threatening food security, causing hunger and malnutrition, shortage of irrigation and drinking water leading to non sanitation and diseases. Naturally, there will be more human and animal waste discharge which along with the destruction of forests, would lead to severe atmospheric changes; increase in numbers and ranges of vectors, pests and pathogens as well as other problems which would form a vicious cycle, threatening human survival (Figure 3.1). Deforestation, growing human activities, including habitation in flood prone zones and river beds

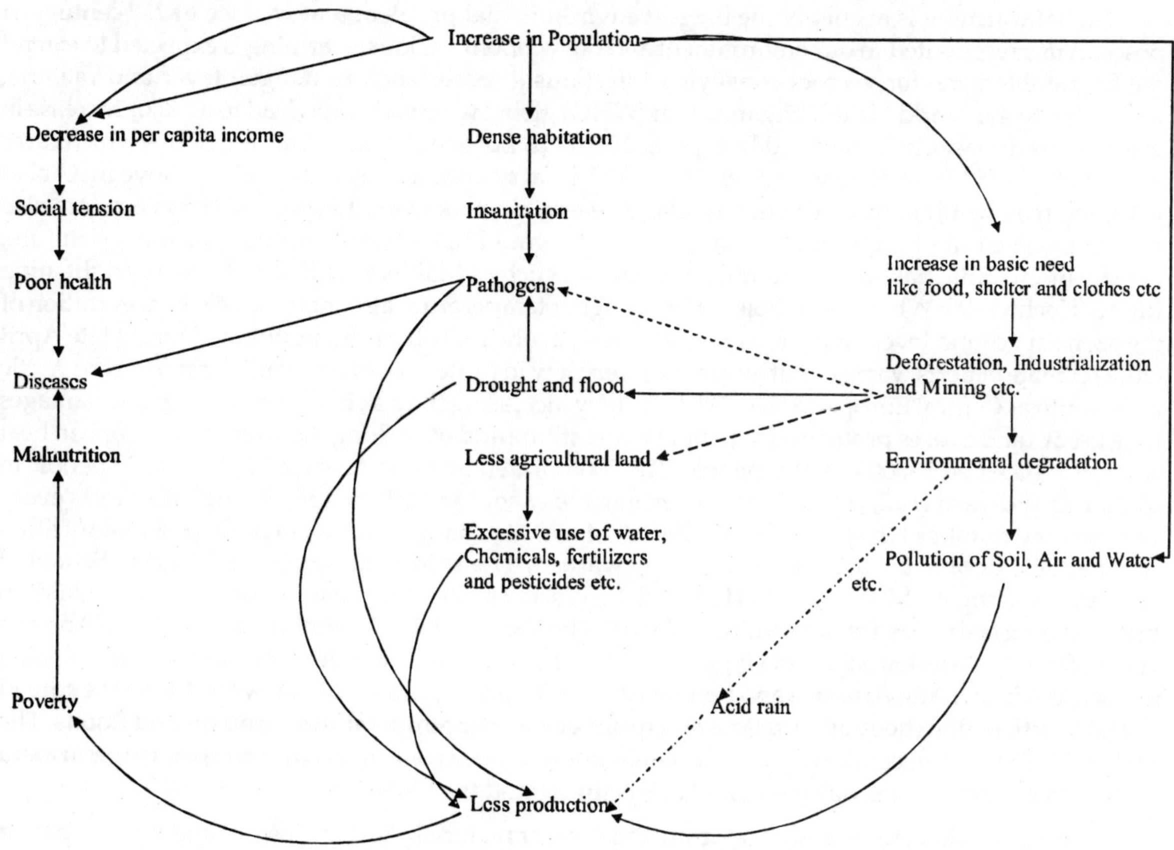

Figure 3.1: Effect of Eco-Degradation on Health

have led to higher flood damage in terms of loss of human and cattle life. In 1990, about 50,000 hectares of paddy crop was damaged by serious paddy pest, *Sodoptera mauritia* in the Purnia district of Bihar (India). It has been suggested that pest migrated to this area from Nepal along with flood water (Pandey and Mishra, 1991). Likewise, many epidemic diseases spread out causing damage to livestock. The flood disturbed the whole ecology of the area and adversely affected the production of fishes and other aquatic animals.

Climate change is certain to add the diseases caused by pathogens. According to WHO report (1996) climate changes may have a major impact on water resources and sanitation by reducing water supply. This would in turn reduce water for drinking and washing as well as reduce efficiency of local sewerage system leading to increased concentration of pathogens in raw water supply. This did happen in Peru in 1991 when a cholera epidemic swept across Latin America, killing over 5000 persons in 18 months. The Inter-governmental Panel on Climate Change (IPCC) in its April 2007 report on the impact of climate changes warned that rising temperatures may result in the altered spatial distribution of some infectious vectors as well as have mixed effects, such as the decrease or increase of the range and transmission of malaria in Africa. Disease carried by insects and ticks are likely to be affected by environmental changes because these creatures are themselves sensitive to vegetation type, temperature, humidity *etc.*

Global warming is arguably the biggest environmental problem that we face in 21st century. It poses an unprecedented to our environment and our economy. Global warming is expected to extend the favourable zone for vectors conveying infectious diseases such as dengue fever and malaria. According to the World Health Organization (WHO) global warming could lead to a major increase in insect borne diseases in Britain and Europe, as Northern Europe becomes warmer, ticks – which carry encephalitis and Lyme disease and sand flies which carry leshmaniasis are likely to move in. Global warming may result in more cardiovascular diseases, doctors warn. Respiratory problems are also known to arise out of high temperature. Further, it would have major implications on agriculture, fishing and wildlife habitats. A number of countries such as Maldives will have to face a frightening future (Hecht, 1990; White, 1991; Foley, 1991). Higher temperature also increases the concentration of the ozone at ground level. In the lower atmosphere, ozone is the harmful pollutant. During late April (2011) climate experts warned that an air mass very low in protective ozone could drift from the Arctic as far south as Central Europe or New York, thereby increasing the risk of skin-cancer. Ozone damages lung tissues and causes problems for people with asthma and other lung diseases. The European heat wave of 2003 killed 25000 – 70,000 people. The 2006 United States heat wave has killed 139 people in California (Edward *et al.*, 2006). Global warming kills about 160,000 people through its effects every year and the number lying from" side effects" of climate change such as malaria and malnutrition could almost double by 2020, scientists have warned. The study of scientists at London School of Hygiene and Tropical Medicine (LSHTM) and the World Health Organization concluded that children in developing countries are most vulnerable to the impact of Global Warming. According to Andrew Hains of the UKs London School of Hygiene and Tropical Medicine most deaths would be in developing nations in Africa, Latin America and Southeast Asia. These regions would be worst hit by the spread of malnutrition, diarrhoea and malaria as a result of warmer temperatures, drought and floods. The spring of 2011 was exceptional, unfortunately not in a positive sense. Earthquakes, tsunamis and tornadoes inflicted enormous losses and hardship around the globe.

Release of radioactive substances from nuclear war materials, test explosions and nuclear power plants drastic effect the exposed biota (Ramana, 1999). Radiation exposure may cause serious disease problems like leukemia and bone tumours, genetic damage and infant mortality. Diseases such as leukemia, bone cancer, thyroid cancer and lung cancer manifest somatic effects of radiations (ReVelle and ReVelle, 1974). Pandey (2000) has reported large number of genetical abnormalities, musculo-skeletal abnormalities, high rate of abortions and still births, disturbance in menstrual cycles as well as sterility in males in the tribal populations residing near the Uranium Corporation of India, Jadugoda (Jharkhand), India. Wastewater of UCIL is discharged into the nearby Gudra river which is, incidentally, is the only source of irrigation water for farmers. Villagers say that use of this water for irrigation resulted in much less production and due to this, the area once rich in cultivation, is now a dry patch of land. Besides these large number of seedless fruits were also found in the area around the UCIL.

In conclusion man's action with environment has resulted in land degradation, soil erosion, pollution of air, water and soil as well as loss of biodiversity which has resulted in a large number of diseases. Besides, increasing rate deforestation has adversely affected the availability of products like fruits, wood, fodder, medicinal herbs and plants. As a result a number of diseases have now becomes threat to life status of human beings. Increasing rate of urbanization, industrialization, mining, deforestation and population explosion have now reached to point of no return. No doubt, this situation appears grim but man never loses hope and always thinks for better tomorrow. With vigorous efforts to implement education for all, especially the girl child would surely check the rate of population growth. Govt. of India has initiated programmes on population check by celebrating the World

Population Day on 11th July every year. Besides, Biotechnology and Bioinformatics are opening new doors to solve the problems of food shortage, nutrition and environment as well as medical treatment, Techniques are on the way to meet the challenges of water scarcity, water salinity, pest and pathogens resistance *etc.* For millions of years trees kept Earth protected from overheating by absorbing carbon dioxide. There is a Chinese proverb – If you want to plan for a year, plant crops and if you want to plan for 10 years plant trees but if you want to plan for 100 years, educate people. So educating the people in the real sense about the use of trees should be taken under priority, if we want to save ourselves from the losses of the deforestation. Fossil fuels have overwhelmed this natural carbon cycle. Geophysicist Klaus Lackner from Columbia University wants artificial trees to restore the balance. The global issues like ozone layer depletion and global warming are becoming horrid nightmares. The UNEP had appropriately chose the slogan: "Global warming: Global Warning" to alert and caution the people on World Environment Day *i.e.* June 5th every year.

References

Edwards, L.M., Kozlowski, D., Bair, A., Juski, J., Blier, W. and O' Hara, B. (2006). A review of the July heat wave in California. *American Geophysical Union, Fall meeting* (2006).

Foley, G. (1991). *Global warming, who is taking heat*? Panos Publications Ltd. London.

Hecht, J. (1990). Global warming; back to the future. *New Scientist*, December,38 - 41.

IPCC (2007). Climate Change 2007: Impacts, Adaptation and Vulnerability. .

Pandey, B. N. (2000). Final report submitted to the Ministry of Environment and Forests, Govt. of India, on the project entitled – "Pattern of Settlement in and around the North Gangetic belt of Bihar with particular references to certain Scheduled Tribes.

Pandey, B. N. And Mishra, S. K. (1991). A note on heavy infestation of *Sodoptera mauritia* in Purnia District of Bihar. *J. Curr. Biol. Sci.*, 9(2): 47 – 48.

Ramana, M. V. (1999). Underground test, Ravaging nature. *The Hindu Survey of Environment* (Ed. N. Ravi, Chennai), 127 – 133.

Re Velle, G. and Re Velle, P. (1974). *Sourcebook on the Environment*, Houghton Mifelin, Co, Boston.

Robine J. M., Cheung, S. L., Le Roy S., Van Oyen H., Griffiths, C., Michel, J.P. and Herrmann, F.R. 2008. Death toll exceeded 70,000 in Europe during the summer of 2003. C R Biol. 2008; 331(2):171–8. dol: 10.1016/j.crvi.2007.12.001. Epub 2007 Dec 31.

Sardon, J.P. (2007). The 2003 heat wave. *Euro Survelli*, 12(3):226 (Pub Med).

White, R.M. (1991). The great climatic debate. *Scientific American*, July, 263 (1):18 - 25.

World Health Organization (2003). Climate change and Human health – risk and responses.

2013, Sustainable Approaches for Environmental Conservation *Pages* **25–29**
Editors: **D.R. Khanna, A.K. Chopra, R. Bhutiani, Gagan Matta & Vikas Singh**
Published by: **BIOTECH BOOKS, NEW DELHI**

Chapter 4

Deterioration of Chickpea Seeds Due to Bruchids Infestation Stored Under Ambient Environment of Nagpur, M.S.

Rajesh Gadewar[1], Ashish Lambat[1], Sanjeev Charjan[2] and Pravin Charde[1]
[1]Sevadal Mahila Mahavidyalaya, Nagpur, M.S.
[2]Dr. P.D.K.V's College of Agriculture, Nagpur, M.S.

Chickpea (*Cicer arietinum* L.) is an important source of protein. The seed infested by bruchid during storage is a major problem. This paper gives a brief account of certain physiological, biochemical and mycological changes in qualities of Chickpea seeds due to infestation of bruchid during storage. In the present study, it was found that the percentage of moisture content, total ash, crude fiber, crude protein significantly increased and crude fat, total carbohydrate, total sugar, reducing sugar and non-reducing sugar significantly decreased in bruchid infested seeds of Chickpea during storage under ambient environment of Nagpur, Maharashtra State. Increase in protein content is attributed to bruchid metabolites like uric acid, which is nitrogenous in nature. The incidence percentage of fungi such as species of *Alternaria, Aspergillus, Curvularia, Fusarium, Penicillium* and *Rhizopus* were predominant over all other fungi on infested Chickpea seeds and it is increased with increase in bruchid infestation during storage. The physical and physiological qualities of Chickpea *i.e.* 100-seed weight, germination, seedling vigour and field emergence percentage decreased with increase in infestation of bruchid during storage.

Keywords: Infestation, Storage, Metabolites, 100-Seed weight.

Introduction

Pulses are the most important source of protein in Indian diet. Storage of pulse seeds is a major problem and it is estimated that about 10 per cent of stored pulse seeds are lost due to biological factors of which insects and rodents alone account of 5 per cent. In severe cases the infestation was observed to be about 90 per cent. Pulse beetles of various species belong to the family Bruchidae are important insect pest attacking variety of pulses in store. Adult female stick their eggs on the pulse seeds and the emerging grubs and bore into the seeds. The grubs remain inside the seed and appearance of a capped exit hole on the seed indicates the pupil stage. After a few days the adult emerges from the seed. About one month is required to complete one generation.

The stored grain insect's pest's infestation also encourages fungus growth by increasing the moisture content of the seeds which decreased the quality and viability of the seeds. Christensen and Kaufmann (1969) reported that the fungal pathogen associated with stored seed are chiefly responsible for seed deterioration and reduction in germination potential. Apart from this the seedling vigour is also adversely affected. Among the storage fungi species, many were well known toxin producers. The present work was carried out to investigate the post harvest losses in qualities of Chickpea (*Cicer arietinum* L.) seeds due to pulse beetle infestation.

Materials and Method

Chickpea (*Cicer arietinum* cv chaffa) freshly threshed seeds by multi crop thresher which were then cleaned and sieved with 10/64 inch (3.96 mm) diameter sieve to remove small fraction of seeds or insects produced in 2008-2009 (Kharif season). The seeds were dried upto the safe moisture level (10 ± 1 per cent wb). The experiment conducted in glass bottle of two litre capacity. The glass bottle was then filled with 1,000 grams of Chickpea (*Cicer arietinum* L.) cv. C-11 seeds. There were four replications. Ten pairs of 2-3 days old pulse beetles (*Callosobruchus analis*) were released in glass bottles covered with muslin cloth. The set of experiment was kept in well ventilated wire mesh almirah in mesonary building having cemented walls, roof and floor under ambient temperature (10.2 to 46.0°C) and relative humidity (27 to 80 per cent) from January 2009 to June 2009, for determination of physical, physiological, biochemical and mycological changes in stored seeds of Chickpea (*Cicer arietinum L.*) at interval of 3 months. The initial observations also taken at the start of experiment. The physical qualities of Chickpea (*Cicer arietinum* L.) seeds *i.e.* seed infestation percentage, moisture content and 100 seeds weight were studied. 100 seed weight was tested in quadruplicated with 100 seeds in each replication. The infested seeds were counted and total damaged seeds were reported in percentage. Moisture percentage was estimated according to International rules for seed testing (ISTA, 1985). The physiological qualities of Chickpea (*Cicer arietinum* L.) seeds *i.e.* seed germination, seedling vigour and field emergence were studied. The germination percentage was evaluated on the value for percent normal seedlings (ISTA, 1985). The seedling vigour index was worked out following the method of Abdul-Baki and Anderson (1973). For field emergence test, sowing of Chickpea (*Cicer arietinum* L.) seeds was done in randomized block design with four replications with inter and intra-row spacing of 1 feet and 6 inches, respectively. Observations for field emergence were recorded daily and finally the established seedlings were counted after one month of sowing.

To assess the biochemical qualities of the seeds of Chickpea (*Cicer arietinum* L.) *i.e.* protein, fat, total ash, crude-fibre, reducing and non-reducing sugars according to the standard procedures of A.A.C.C. (ISTA, 1962). Values for carbohydrate and total sugar were calculated (Joslyn, 1970). The fungal flora of the seeds were detected by the standard moist blotter and agar medium techniques as prescribed by (ISTA, 1976). The different types of fungal growth on the seeds were expressed in percentage. The experimental data was statistically scrutinized as per Panse and Sukhatme (1967).

Results and Discussion

It was observed from the Table 4.1 that the moisture content of the seeds increased with increasing storage periods *i.e.* 3 months (10.91 per cent) and 6 months (12.26 per cent). A significant increase in moisture content was observed; this might be due to the activities of pulse beetles on seeds during storage. Similar observations also reported by Shrivastava *et al.* (1989). Seed damage increased with increasing the storage periods of 3 months (25.10 per cent) and 6 months (59.28 per cent), respectively. Charjan *et al.* (2006) reported that the infestation of pulse beetles increased with increasing the storage periods. The 100-seed weight of seed decreases with increasing the storage periods. Similar observation also reported by Charjan (1995). Similarly the germination, seedling vigour and field emergence per cent decreases with increasing the storage periods. In costal region of Andhra Pradesh percent germinability of Chickpea (*Cicer arietinum* L.) was found to decrease from 81 per cent to 65 per cent

Table 4.1: Effect of Pulse Beetle Infestation on Physical, Physiological, Biochemical and Mycological Qualities of Chickpea (*Cicer arietinum* L.) during Storage

Sl.No.	Seed Quality Parameters	Initial	After 3 Months	After 6 Months	SE (m)	CD at 5 per cent
A.	**Physical seed quality**					
	1. Seed moisture (per cent wb)	10 ± 1	10.91	12.26	0.5	1.5
	2. Seed damage (per cent)	0	25.1	59.28	2.1	6.5
	3. 100-seed weight (gm)	12.84	10.05	7.92	0.4	1.2
B.	**Physiological seed quality**					
	1. Germination (per cent)	96.12	69.5	50.25	2	6.2
	2. Seedling Vigour Index (SVI)	4619	2992	2004	n/a	n/a
	3. Field emergence (per cent)	86.4	55.75	33	2.5	7.8
C.	**Biochemical seed quality**					
	1. Total ash (per cent)	4.92	5.46	6.02	0.03	0.1
	2. Crude fibre (per cent)	7.03	7.59	8.27	0.04	0.12
	3. Crude protein (per cent)	25.42	25.71	26.01	0.9	2.8
	4. Crude fat	2.9	2.6	2.14	0.02	0.07
	5. Total carbohydrate	70.09	66.33	63.04	2.01	6.1
	6. Total sugar (per cent)	7.41	7.04	6.12	0.04	0.14
	7. Reducing sugar (per cent)	11.41	7.19	1.04	0.05	0.17
	8. Non-reducing sugar	6.57	6	5.18	0.04	0.15
D.	**Mycological observation**					
	1. *Alternaria* sp. (per cent)	7.25	6.75	4	n/a	n/a
	2. *Aspergillus* sp. (per cent)	1.25	7	14.25	n/a	n/a
	3. *Curvularia* sp. (per cent)	1.75	2.25	10.25	n/a	n/a
	4. *Fusarium* sp. (per cent)	0.75	3	8	n/a	n/a
	5. *Penicillium* sp. (per cent)	0	2.75	8.75	n/a	n/a
	6. *Rhizopus* sp. (per cent)	0	1.5	7	n/a	n/a
	7. Total incidence (per cent)	11	23.25	52.25	n/a	n/a

within 4 months of storage (Vimla and Pushpamma, 1993). Charjan and Tarar (1994) reported that the germination percentage, seedling vigour and field emergence percentage decreases with increasing storage periods in moth bean infested by pulse beetles during storage.

Pulse beetles feed on the cotyledonous portion of the Chickpea (*Cicer arietinum* L.) seed leaving the seed coat intact and that is one reason that higher values for crude fibre and total ash have been obtained in infested seed, as seed coat is rich in crude fibre and minerals (Singh *et al.*, 1968 and Shrivastava *et al.*, 1989). Increase in protein content is attributed to insect metabolites like uric acid, which is nitrogenous in nature (Shrivastava *et al.*, 1989). Increase in reducing sugars and decreasing in non-reducing sugars has been shown in stored Chickpea (*Cicer arietinum* L.) seeds. Similar results have been reported by Khare (1972), Shrivastava *et al.* (1989), Charjan and Tarar (1994). The following fungi were found to be associated with stored seeds of pigeon pea. The present pulse beetles damaged seeds yielding a particular fungus *viz., Alternaria* sp., *Aspergillus* sp., *Curvularia* sp., *Fusarium* sp., *Penicillium* sp. and *Rhizopus* sp. irrespective of storage periods. In the present study, the incidence percentage of storage fungi increases with increasing seed damage by pulse beetles and storage periods. The results are in conformity with the results of Charjan *et al.* (2006).

Conclusion

Thus from the present study, it can be concluded that infestation of pulse beetle in Chickpea (*Cicer arietinum* L.) increases the moisture content which is favorable for multiplication of fungal flora and decreases the 100-seed weight, germinability, seedling vigour and field emergence percentage of seeds during storage. The decrease of the crude fat, total carbohydrates and total sugars and increase of total ash, crude fibre and crude protein in infested pigeon pea was also observed. Increase in protein content is attributed to insect metabolites like uric acid, which is nitrogenous in nature. The percentage of storage fungi increases with increasing pulse beetles damage and storage period. Among the identified fungi species, many were well known toxin producers. The pulse beetle infested Chickpea (*Cicer arietinum* L.) seeds should be avoided for sowing or consumption purposes.

References

Abdul Baki, A. A. and Anderson, A. A. 1973. Vigour determination in soybean seed by multiple criteria. *Crop Science*. 13: 630-633

Charjan, S.K.U. 1995. Efficacy of *Acorus calamus* L. rhizome powder against the pulse beetle, *Callosobruchus chinensis* L. infested stored Chickpea (*Cicer arietinum L.*)(*Cajanus cajan*). In: Herbal, medicine, biodiversity and conservation strategies (Ed. Rajak and M. K. Rai) International Book Distributor, Deharadun: 242-246.

Charjan, S.K.U. and Tarar, J. L. 1994. Biochemical changes in jowar grains due to infestation of rice weevil and khapra beetle during storage. *Proceeding National Academy of Science*. 64(B) IV: 381-384.

Charjan, S. K. U., Wankhede, S. R. and Jayade, K. G. 2006. Viability, vigour and mycoflora changes in pulse beetle (*Callosobruchus* sp.) damaged seeds of cowpea produced in dryland condition. In: Economics of sustainability of dryland agriculture, Nagpur: 5-6.

Christensen, C. M. and Kaufmann, H. H. 1969. Grain Storage – The role of fungi in quality loss. 153 pp. Univ. of Minnesola Press, Minneapolis.

ISTA. 1962. Approved methods of American Association of Cereal Chemist, 8th rev. ed. St. Paul (Minnesota)

ISTA. 1976. International rules for seed testing. *Seed Sci. and Technol.* 4: 108

ISTA. 1985. International rules for seed testing. *Seed Sci. and Technol.* 13: 299-513

Joslyn, M. A. 1970. *Methods of food analysis: Physical, chemical and instrumental methods of analysis.* Academic Press, New York: 845.

Khare, B. P. 1972. Insect pests of stored grain and their control in Uttar Pradesh. Research Bulletin. No.5 G.B.U.A.T., Pantnagar: 132-139.

Panse, V. G. and Sukhatme, P. V. 1967. *Statistical Methods for Agril. Workers*, I.C.A.R. Publication, New Delhi.

Shrivastava, S., Gupta, K. C. and Agrawal, A. 1989. Japanese mint oil as fumigant and its effect on insect infestation, nutritive value and germinability of Chickpea (*Cicer arietinum* L.) seed during storage. *Seed Res.* 17(1): 96-98.

Singh, S., Singh, H. D. and Sikka, K. C. 1968. Distribution of nutrients in anatomical parts of common Indian pulses. *Cereal Chem.* 45: 13-18.

Vimla, V. and Pushpamma, P. 1983. Storage quality of pulses stored in three agroclimatic regions of Andhra Pradesh II. Viability Changes. *Bull. Grain* 21(3): 217-22.

2013, Sustainable Approaches for Environmental Conservation *Pages* **31–35**
Editors: **D.R. Khanna, A.K. Chopra, R. Bhutiani, Gagan Matta & Vikas Singh**
Published by: **BIOTECH BOOKS, NEW DELHI**

Chapter 5
Efficacy of Indigenous Organic Preparation on Viability, Vigour, Field Emergence and Seed Mycoflora of Mungbean

***Rajesh Gadewar*[1], *Ashish Lambat*[1] *and Sanjeev Charjan*[2]**
[1]*Sevadal Mahila Mahavidyalaya, Nagpur, M.S.*
[2]*Dr. P.D.K.V's College of Agriculture, Nagpur, M.S.*

Quality of seeds is very important for any crop to grow better in the future and to give higher yield and also to grow vigorously. These could also enhanced by treating with many poisonous chemicals and growth factors. But these are always hazardous to human beings, animals, microorganisms, environment and natural habitat. There are some facts like, if the crops are grown under the indigenous organic preparations, cultivation are less prone to diseases attack and safe to health. In the view of the above facts, the present study was conducted to evaluate the effect of different organic preparation which were indigenously prepared on seed quality parameters *viz.*, germination percentage, length of plumule, length of radicles, seedling dry weight, seedling vigour, field emergence percentage and incidence percentage of mycoflora. The results revealed that the seeds soaked in Brahmastra showed significantly highest germination percentage, length of plumule, length of radical, seedling dry weight, seedling vigour, field emergence percentage and lesser invasion of mycoflora followed by Beejamrutha, Panchgavya and Jeevamrutha. The control which was soaking in water recorded significantly lowest seed quality parameters and highest invasion of mycoflora.

Keywords: *Germination percentage, Length of plumule, Length of radicles, Seedling dry weight, Seedling vigour, Field emergence percentage.*

Introduction

Pulse crops are major source of protein, minerals and vitamins and forms important part of predominantly vegetarian diet of Indian people. Among all grain legume, mungbean (*Vigna radiata* L.) is the most important kharif and summer crop. The modern agriculture depends more on chemical, which cause several environmental concerns, as well as hazardous to human beings, livestock's, microorganisms, increased resurgence of pathogens. Such concerns and problems posed by modern agriculture paved the way for rebirth to organic farming. Moreover, science behind organic farming is to cultivate crops with environmental harmony, enhance soil fertility, promote soil biota, manage pests, diseases and weeds in cm eco-friendly manner and promote livestock and human health. In view f the above facts, the present investigation was therefore initiate to study the efficacy of indigenous organic preparation on viability, vigour, field emergence and seed mycoflora of mungbean.

Materials and Method

The laboratory and field experiments were carried out during 2010 in Department of Agril. Botany, College of Agriculture, Nagpur. Seeds of mungbean (*Vigna radiata* cv. Kopergaon) were treated with four organic preparation *viz.* Beejamrutha, Jeevamrutha, Panchagavya and Brahmastra. Their efficacy was compared with control (soaked in water) by studying viability, vigour, field emergence and seed mycoflora.

Methodology of Beejamrutha Preparation

Cow dung (50 g) was taken in a cotton cloth and tied firmly and clipped in a glass beaker containing 1 litre of water overnight. Meanwhile lime solution was prepared by dissolving 1.2 g in 25 ml of water. Next day morning, lime solution and 125 ml of cow urine were poured into the glass beaker. Mixture was stirred well for proper mixing and solution is used for mungbean seeds soaking for five minutes (Swamy, 2006a).

Methodology of Jeevamrutha Preparation

This organic preparation was made by mixing 250 g cow dung, 250 ml cow urine, 50 g jaggery, 50 g gram flour, 2 g farm soil mixed well with 5 litre water in a plastic barrel. Barrel was kept in shade for one week undisturbed. It was stirred in anticlockwise direction daily twice and solution is used for mungbean seeds soaking for five minutes (Swamy, 2006b).

Methodology of Panchagavya Preparation

This organic preparation was made by mixing 50 ml cow milk, 50 ml curd, 25 ml ghee, 70 ml cow urine and 125 ml fresh dung slurry in wide mouthed vessel and was kept in a shady place. The solution was stirred everyday during morning and evening. The Panchagavya solution was ready on ninth day and was used for soaking mungbean seeds for five minutes (Swamy, 2006c).

Methodology of Brahamastra Preparation

This organic preparation was made by mixing 100 ml cow urine, 30 g cow dung, 30 g neem leaves, 20 g castor leaves, 20 g lakki (*Calatropsis* sp.) leaves, 20 g custard apple leaves, 20 g parthenium leaves, 10 g pongamia and 10 g bitter gourd in 2 litre of water in a container. After this, contents in the container were boiled and were kept for 30 days for fermentation. After filtration, the solution is used for soaking mungbean seeds for 5 minutes.

The mungbean seeds soaked in above four solution (treatment) for five minutes were followed by two minutes shade drying and about 100 seeds were used to inculcate in blotting paper for all the solution treatments separately in four replications for determining viability percentage (germination per cent). The germination percentage was evaluated on the value for percent normal seedling (Anonymous, 1985). Seedling were taken randomly for radical and plumule length measurement (cm) from each treatment. The seedling vigour index was workout following the method of Abdul-Baki and Anderson (1973). The fungal flora of the mungbean seeds were detected by the standard moist blotter and agar medium techniques as prescribed by International rules for seed testing (Anonymous, 1985). The different types of fungal growth on the seed were expressed in percentage.

Results and Discussion

Data presented in the Table 5.1 showed that brahmastra was significantly superior over other treatment, where 98 percent germination was observed and recorded highest seedling vigour index (3432) followed by Jeevamrutha with 96 percent germination and seedling vigour index 3169, Panchgavya recorded germination of 94 percent with seedling vigour index 2915 followed by Beejamrutha which recorded 91 percent germination with 2549. The control seed which was soaked in water recorded lowest germination of 87 percent with seedling vigour index of 1918. It is clear from the present investigation that seed soaked in brahmastra was found to be significantly superior with 98 percent and seedling vigour index 3432 than Jeevamrutha, Panchgavya, Beejamrutha and control. This must be due to Brahmastra solution which contain plant and animal products having anti-pathogenic principles. It protects the mungbean seeds from seed borne pathogens which could affect them during the germination process. Cow dung and cow urine would provide the nutrition for seeds which could give good germination and seedling length. The similar results were also reported by Jahagirdhar *et al.* (2001), Sugha (2005), Sumangala (2007) and Sridhar *et al.* (2011).

Table 5.1: Impact of Organic Oreparations (Treatment on seed germination per cent, length of seedling (cm), seedling vigour index (SVI) and field emergence (per cent)

Sl.No.	*Treatments (Organic preparation)*	*Mean Germination (per cent)*	*Mean Length of Seedling (cm)*	*Mean Seedling Vigour Index*	*Mean Field Emergence per cent*
1.	Beejamrutha	91	28.2	2549	81
2.	Jeevamrutha	96	33.1	3169	87
3.	Panchgavya	94	31.5	2915	84
4.	Brahmastra	98	35.8	3432	91
5.	Control	87	22.5	18	75
	SE ±	0.78	0.14	19	0.77
	CD at 1 per cent	3.12	0.59	—	3.09

Field emergence percent confirmed the superiority of Brahmastra (91 per cent) over the Jeevamrutha (87 per cent), Panchgavya (84 per cent), Beejamrutha (81 per cent) and control (75 per cent). This might be due to the high seedling vigour index. Bharadwaj (1995) and Bansal (2011) reported that animal waste product improve the soil physical conditions and environmental quality as well as provide nutrients for plants.

It might be seen from the Table 5.2 that species of *Alternaria*, *Aspergillus*, *Cladosporium*, *Curvularia*, *Fusarium*, *Penicillium* and *Rhizopus* were isolated from the seeds of Mungbean. The incidence percentage of isolated fungal flora was higher in seeds soaked in water (control). The seeds soaked in brahmastra provided much protection to mungbean seeds in preventing the development of fungal colonies both quantitative and species wise as compared to Panchgavya, Jeevamrutha and Beejamrutha treatments. This might be due to anti-pathogenic properties of plant and animal products of Brahmastra. The results obtained were in conformity with the findings of Bhaskara (1994), Jahagirdhar *et al.* (2001) Sumangala (2007) and Mane *et al.* (2001).

Table 5.2: Impact of Organic Preparations on Incidence Percentage of Mycoflora on Mungbean Seeds

Sl.No.	*Treatments (Organic preparation)*	*Percentage of Fungi Encountered on Mungbean Seeds*													
		A		*B*		*C*		*D*		*E*		*F*		*G*	
		1	*2*	*1*	*2*	*1*	*2*	*1*	*2*	*1*	*2*	*1*	*2*	*1*	*2*
1.	Beejamrutha	3	27	3	29	2	20	2	28	2	19	2	20	1	6
2.	Jeevamrutha	2	18	3	20	1	4	1	6	2	10	1	6	1	4
3.	Panchgavya	2	20	3	29	2	10	2	10	2	16	2	18	1	6
4.	Brahmastra	1	4	2	10	-	4	-	6	-	4	-	4	1	2
5.	Control	7	32	15	61	3	15	4	35	7	39	6	12	4	29

A: *Alternaria* sp; B: *Aspergillus* sp; C: *Cladosporium* sp; D: *Curvularia* sp; E: *Fusarium* sp; F: *Penicillium* sp; G: *Rhizopus* sp; 1: Standard blotter paper method; 2: Agar plate method.

Conclusion

Thus from the present study, it can be concluded that seeds of mungbean should be soaked for five minutes in Brahmastra solution before sowing. Because this treatment showed significantly maximum germination per cent, length seedling, seedling vigour index, field emergence per cent and minimum invasion of fungal flora on the mungbean seeds.

References

Abdul-Baki, A. A. and Anderson, J. D. 1973. Vigour determination in soybean seed by multiple criteria. *Crop Sci.* 13: 630-633.

Anonymous. 1985. International rules for seed testing. *Seed Sci. and Technol.* 4: 299-513.

Bansal, Amrit Kaur 2011. Effect of different (Cattle poultry-horse and goat dung) on nutritional status of vermicompost. *Green Farming* 2: 114-115.

Bhaskara, Padmodaya. 1994. Biological control of seedling disease and wilt in tomato caused by *Fusarium oxysporum*. Unpublished. *Ph.D. Thesis*, University Agric. Sci. Bangalore, India.

Bharadwaj, K. K. R. 1995. Recycling of crop residues oilcakes and other plant products in agriculture. In: Tandon, HLS, Editor. 1995. Recycling of crop animal, human and industrial waste in agriculture. Fertilizer Development and Consultation Organization, New Delhi, PP 9-30.

Jahagirdhar Shamarao, Siddaramaiah, A. L. and Ramaswamy, G. R. 2001. Influence of biocontrol agent and MPG-3 on *Fusarium oxysporum*.

Mane, V. A., Belwadkar, V. N. Gawade, B. V., Medhe, N. K. and Randive, S. N. (2001). *In vitro* evaluation of bio-agents and botanicals against *Alternaria* leaf blight of chilli. *Green Farming*. Z: 97-98.

Sugha, S. K. 2005. Antifungal potential of Panchgavya. *Pl. Dis. Res.* 20: 156-158.

Swamy, Anand. 2006 a. *Preparation of Beejamrutha,* In: Subhash Palekarara Shoonya Bandavalada Naisargika Krushi (Local language), 8th Ed. Published by Agni Prakashana, Bangalore, India, P: 8-9.

Swamy, Anand. 2006b. *Preparation of Jeevamrutha,* In: Subhash Palekarara Shoonya Bandavalada Naisargika Krushi (Local language), 8th Ed. Published by Agni Prakashana, Bangalore, India, P.13-20.

Swamy, Anand. 2006 c. *Natural pesticides,* In: Subhash Palekarara Shoonya Bandavalada Naisargika Krushi (Local language), 8th Ed. Published by Agni Prakashana, Bangalore, India, P.91-92.

Sumangala Koulagi. 2007. Studies on grain discolouration in rice. Unpublished *M.Sc. (Agri.) Thesis,* University Agric. Sci. Dharwad.

Sridhar, D. Mahanthesh, B. Patil, Sumangla Koulagi and Ravi Kumar H. S. 2011. Influence of indigenous organic preparation on germination and vigour index of paddy. *Green Farming.* 2: 302-304.

2013, Sustainable Approaches for Environmental Conservation *Pages 37–45*
Editors: **D.R. Khanna, A.K. Chopra, R. Bhutiani, Gagan Matta & Vikas Singh**
Published by: **BIOTECH BOOKS, NEW DELHI**

Chapter 6

Role of Banks in Low Carbon Economy

S.K. Sharma

School of Commerce,
HNB Garhwal (Central University) Campus-Badshahithaul
District Tehri Garhwal, Uttarakhand

India, the world's fourth-largest greenhouse gas emitter, though still low on per-capita emissions, is under pressure to cut emissions to battle climate change. Sustainable development denotes development that does not reduce the possibilities and choices for the future generations, at the same time ensuring continuity of economic progress for the present generation. India ranked 10th among G-20 nations in 2009 with clean- energy investments of $2.3 billion. Over the next decade it is projected to rise to number four with a projected investment of $118 billion to $169 billion. Banking sector is one of the major sources of financing investment for commercial projects such as steel, paper, cement, chemicals, fertilizers, power and textiles *etc.*, which cause maximum carbon emission. The banking sector can play an intermediary role between economic development and environmental protection for promoting environmentally sustainable and socially responsible investment. Indian banks must immediately adopt green strategies *i.e.* carbon credit business, green financial products and carbon foot print reduction *etc.* Managing carbon emission is one of the fastest growing segments in financial services in the world wide and carbon will be world's biggest market overall.

Under Kyoto Protocol and Clean Development Mechanism (CDM), industries are now in favour of adopting cleaner and energy efficient technology. Projects related to renewable energy such as solar energy, wind energy, hydropower, biomass and alternative fuels are getting high priority. As India has committed to cut its carbon (CO_2) intensity by 20-25 percent from 2005 levels by 2020, banks have an important role towards developing a strong and successful low carbon economy. In a low carbon economy, there will be many challenges and opportunities for Indian banks. Banks should be proactive and prompt in identification of new green business opportunities to be a dominant entity in conserving green and clean environment.

Keywords: *Green banking, Sustainable development, Environmental risks, Clean Development Mechanisms (CDM), Corporate Social Responsibilities (CSR).*

Introduction

Climate change is already evident from observations of increases in global average temperature, sea level rise, precipitation changes and extreme weather ever. Since the early 1990s, there have been international efforts to gain consensus on how to tackle climate change. Central to this is the United Nations Framework Convention on Climate Change (UNFCCC), which produced the Kyoto Protocol* in 1997. India has set voluntary targets to reduce carbon emissions intensity by 20-25 per cent by 2020. The targets are supported by legislation that will require mandatory fuel efficiency norms for all vehicles, introduces green building codes and provides an amendment to the Energy Conservation Act to make it necessary for an initial group of 714 energy intensive businesses to take part in a scheme to cap energy usage and subsequently trade energy efficiency certificates. India is keen to be part of the solution and is actively involved in discussions that will lead to a 'post Kyoto' regime that tackles the problem.

What is Low Carbon Development?

"Low carbon development" and "climate change mitigation" are generic terms to indicate the transition towards development options that reduce human interference with the climate system. This means reducing the release of greenhouse gases into the atmosphere from chemicals and fossil fuels. It also includes emission reductions from changes in land use, such as deforestation and more energy intensive methods of agriculture.

Environmental protection and sustainable ecological balance have emerged as significant theme in twenty first century, with an increasing number of 'green technologies' finding their way into various functional areas including banking. Sustainable finance promotion is the call of the hour. Banks are starting to invest in low carbon technologies and developing new products and services that will address risks and opportunities of climate change. Carbon market have been effective in reducing global green house gases (GHGs) in the atmosphere and have promulgated an 'MRV' culture promoting transport monitoring and reporting of emission reductions.

Economists are clear that substantial funding from the private sector is needed to achieve the level of investment required to control the effects of climate change. The World Bank estimates that the cost of mitigation in developing countries alone ranges from US$140 billion to US$175 billion annually until 2030. Although this is a great challenge, Indian banks are now in a strong position for the leveraging and channelling of this investment and to use the opportunity this provides as the low carbon economy.

---

* *Kyoto Protocol*: An agreement negotiated in 1997 that commits developed countries to binding greenhouse gas emissions targets but provides flexible market based mechanisms such as Clean Development Mechanism (CDM) to minimise the cost to participants to meet their targets. 39 countries within European Union, Economies in Transition and OECD (known as Annexe B countries) agreed to binding commitments on the emissions from 1990 to 2008-2012.

The National Action Plan on climate change has led to opportunities for the financial sector and there have been proactive initiatives across the spectrum of public, private and foreign banks in India.

Environmental Liabilities of Banks

Banks also have a direct liability to act in accordance with the related environmental regulations while conducting businesses. The direct impact including carbon emission of all the activities if not taken care of may lead to environmental degradation. Every single decision, if not as per the criteria may lead to various risks

Credit Risk

It has been observed that due to global warming there have been severe weather conditions which severely damage both the financial as well as physical assets of banks. Extreme weather conditions also affect economic assets financed by banks, which may lead to a high incidence of credit default. It can arise indirectly where banks are lending to customers whose businesses are adversely affected by the cost of cleaning up pollution or due to changes in environmental regulations. The cost of meeting new requirements on emission levels may force the polluting companies to close down.

Legal Risk

Banks like other business entities face legal risks if they do not comply with relevant environmental legislation. They may also face a risk of direct lender liability for cleanup costs or claims for damages, if they have actually taken possession of contaminated or pollution causing assets. An environmental management system helps a bank to reduce risks and costs, enhance its image and take advantage of revenue opportunities.

Reputation Risk

Due to increasing awareness about environment, banks are more prone to loose their reputations if they are involved in projects, which are viewed as socially and environmentally damaging. Reputation risks involved in the financing of environmentally and ethically objectionable projects. Besides these risks other important related issues need to be discussed.

The small and medium enterprises (SMEs) on account of their financial constraints may not be easily able to install the necessary equipment to meet the emission standards prescribed by the competent authorities. Also because of their small scale of operation, the SMEs escape from the eyes of the concerned authorities. SMEs constitute India's manufacturing base and do not have the capital to quickly shift to clean technology. Similarly the industries like cement steel, which are more carbon intensive, could come under stress.

National Action Plan on Climate Change (NAPCC)

The Government of India has planned reductions in carbon intensity of its economy by launching the NAPCC. Under the NAPCC, several initiatives have been planned which will propel low carbon investments in India as well as better management of water, natural habitats and human settlements. There are provisions within the NAPCC that oblige Indian businesses as well as incentivise them to be more energy-efficient and purchase or invest in renewable energy. The NAPCC will affect a range of Indian businesses; equally, a variety of businesses can contribute towards one or more missions within the NAPCC. The priority missions within the NAPCC show a number of characteristics that make them attractive to investors:

- They can meet an essential demand for India's future prosperity (*i.e.* energy).
- They encourage tried and tested technologies, thus investments are relatively low risk.
- The government has provided a number of incentives to investors and businesses.
- They have the potential to attract carbon finance and generate CERs (Certified Emissions Reductions) tradable in the international market, at least until 2012.
- They can have a high impact on improving resource management within strategic industries such as power, cement, fertilisers, metals (aluminium, iron and steel), railways, pulp and paper and textiles. This will result in cost savings and a better reputation for these industries. The national government has shown consistent progress towards providing a stable platform for business growth in this area.

Indian Business: Low Carbon Investment in India

Indian businesses have an ever increasing choice of sources and mechanisms for investment into low carbon development activity and projects. They can be broadly categorised as follows:

- Conventional finance mechanisms- debt and equity.
- Carbon finance – mainly Clean Development Mechanism (CDM) funding.
- Emerging low carbon mechanisms – Renewable Energy Certificates (RECs), Energy Saving Certificates (ESCerts), Energy Services Companies (ESCos).
- UN, multi-lateral and bi-lateral sources.
- Other relevant mechanisms – government policies and incentives, guarantees, insurance.

To promote for investing in renewable energy technologies in wind, solar, bio-mass, small-hydro – government incentives includes

- Financial incentives from Ministry of New and Renewable Energy (MNRE), such as interest and capital subsidy.
- Soft loans through:
  1. IREDA
  2. Nationalised banks and other financial institutions for identified technologies/systems.
- Fiscal Incentives:
  1. Direct taxes – 100 per cent depreciation in the first year of the installation of the project.
     a. Accelerated 80 per cent depreciation on specified RE-based devices/projects.
     b. Section 80IA – Industrial undertakings set up in any part of India for the generation or generation and distribution of power – 100 per cent deduction is allowable from profits and gains for the first five years. undertakings on profits derived from the business of generation and distribution of electricity.
     c. Section 80JJA – 100 per cent deduction in respect of profit and gains from business of collecting and processing bio-degradable wastes.
  2. Exemption/reduction in excise duty.
  3. Exemption from Central Sales Tax and customs duty concessions on the import of material, components and equipment used in RE projects.
  4. Generation-based incentives for solar and wind power projects.

5. Government policies covering wheeling, banking, buy-back, and third-party sale of power.
6. Other incentives for preparation of feasibility reports and detailed project reports (DPR).
7. Research and development (R&D) subsidy to the tune of 100 per cent of project cost in government R&D institutions and 50 per cent in the case of private institutions.

India is a signatory to both the UNFCCC and the Kyoto Protocol. The latter's Clean Development Mechanism (CDM) has provided India with a significant opportunity for reducing carbon emissions at a relatively low price through renewable energy projects and energy efficiency projects, among others. This provides India with the ability to make money through creating and trading carbon credits through regulated carbon emission trading schemes. The CDM allows developed countries to invest in emission reductions where it is cheapest globally. From its inception in 2001 and up to 2012, the CDM is expected to produce 1.5 billion tonnes of carbon dioxide equivalent emission reductions.

The Clean Development Mechanism (CDM) and Carbon Finance

The Clean Development Mechanism (CDM) is one important instrument of the Kyoto Protocol designed for GHG emission reduction. It refers to project based cooperation's between two countries, where GHG emission reduction take place in the country with lower marginal abatement costs. In other words a country that has adopted a quantified GHG emission reduction or limitation commitment under the Kyoto protocol can fulfill part of the commitment on the territory of another country where the costs are lower. The industrialized countries (sponsor countries) pay for the projects that cut or avoid emission in developing nations for which they are allowed credits that can be applied to meetings its own target. The recipient countries (host nations) benefit from free infusions of advance technology that allow their factories or electrical generating plants to operate more efficiently at lower costs and at reduced emissions. If a company emits 20000 tonnes of carbon dioxide, it could negate its emissions planting sufficient trees which can absorb 20000 tonnes of carbon dioxide annually (Butzengeiger, 2005). Alternatively the company can sponsor and invest in various eco-friendly non conventional sources of power generation. This investment is popularly known as 'Carbon finance'. Clean development mechanism can be viewed not only as a market but also as a subsidy and a political mechanism (Wara, 2007). In terms of numbers of projects, India's CDM projects are second only to China, facilitating an investment of Rs 151,397 crores. The type of projects in India tend to be smaller in scale, with a low intensity of CERs generated and have been in the sectors of energy efficiency, renewable energy, forestry, fuel switching, industrial processes and municipal solid waste. The potential for CER generated inflow in India is estimated by National Clean Development Mechanism Authority (NCDMA) at 25,500 crores [5.73 billion USD] by 2012. However, while this could provide a significant funding stream, the future of the CDM is not clear beyond 2012, though international efforts are being made to improve the CDM process, its impact and secure its future beyond 2012. Indian companies are increasingly measuring, reporting and managing their GHG emissions. In a survey carried out as part of the Carbon Disclosure Project (CDP), a significant percentage of the responding companies acknowledged the various risks and opportunities climate change presents. They reported investment into research and innovation to reduce their risks and enable them to comply with future regulations.

Firstly, the 'project design document' is submitted to National Clean Development Mechanism Authority for getting its validation. After getting validation the project is registered in the host country. After this the project design document is submitted to United Nations Frame Work Convention on Climate Change (UNFCCC) for review and validation. After favorable validation the project is all set for operationalisation. After periodic verification and certification the project is awarded with Certificates Emission Reductions (CERs) which is known as Carbon Credits. In India, carbon finance is interpreted to mean a financial mechanism that is derived from the carbon credit revenue stream generated from a clean project registered as a Clean Development Mechanism - CDM project under the Kyoto Protocol. The Kyoto Protocol has resulted in market-based mechanisms whereby a country with an emission reduction or emission-limitation commitment can implement emission reduction projects in developing countries including India. Such projects can earn suitable Certified Emission Reduction (CER) credits, each equivalent to one tonne of carbon dioxide which can be counted towards meeting Kyoto targets. The CDM's primary goal of supporting development whilst creating cost-effective greenhouse gas emissions reduction is achieved through the buying and selling of CER credits.

Carbon credit is the key component for carbon emission trading scheme. Companies that are interested in reducing their carbon emissions purchase carbon credits from a climate investment fund. CERs are either long term (lCER) or temporary (tCER), depending on like duration of their benefit. CERs can be purchased from the original party that makes the reduction or secondary market. The mechanism enables to ensure environmental outcomes with low cost. The reduction in the cost compliance would make tighter environmental targets possible and certainly more politically feasible. (Hepburn, 2007).

Adoption of Green IT and its Relevance to Banking Sector

The biggest environmental issue for banks is waste of papers. Banks, especially public sector banks (PSBs), annually huge quantity of paper. With the adoption of information technology on a large scale, Indian banks can now switch over to paper less banking, which saves cost, time and storage space. Some studies show that demand for information technology has been more several times but has contributing to carbon emission and e-waste. Paper waste caused by ATM receipts, paper checks, bank statements, tax returns and much more pollute the environment every day. Due to climatic changes, there have been various issues like droughts, cyclones, raising the sea levels *etc.* Generally, such issues have negative effect on the economic assets financed by the banks, which may lead to the risk of default.

The term "Environment Friendly Banking or Green banking" is becoming popular as more citizens look for ways in which they can help the environment. While, green banking encompasses a wide variety of banking services, many banks are promoting their online banking services as a form of green banking. The environment and the banking industry can both benefit if more bank customers start to use the online banking services that are available. Benefits of online banking include less paperwork and less driving to branch offices by bank customers, which will have a positive impact on the environment. Interestingly, online banking can also increase the efficiency and profitability of a bank. A bank can lower their own costs that result from paper overload and bulk mailing fees if more of their customers use online banking. Green banking also can reduce the expenditure of branch banking.

Green computing or green IT refers to environmentally sustainable computing or IT which is more environments friendly. The green computing is also defined as the study and practice of designing, manufacturing, using, and disposing of computers, servers and associated subsystems - monitors, printers, storage devices and networking and communications systems - efficiently and effectively with minimal or no impact on the environment. Modern IT system rely upon a complicated mix of people, networks and hardware, as such, a green computing initiative must cover all of these areas as well. The other aspect of Green IT comprises of using the IT services to reduce environmental impact of other industries.

Indian Banks Towards Low Carbon Economy

This initiative of green banking is mutually beneficial to the banks, industries and the economy. Moreover Green banking will ensure that the asset qualities of banks are improved in future. Contrary to the belief, environmental friendly technologies make economic sense for the industries and actually lessen their financial burden as well. The polluting industries face more resistance and often forced to close down or face massive resistance from the public. This adds to their cost enormously. So adopting environmentally sustainable technologies or modes of production is no more considered as a financial burden, rather it brings new business opportunities and higher profit. Green banking optimises costs, reduces the risk, enhance banks reputations and contribute to the common good of environmental sustainability. So it serves both the commercial objective of the bank as well as its social responsibility.

This is one among the most novel and innovative green banking initiatives introduced by banks - novel because the bank balance is getting increased and innovative because it is very simple to understand. Every time there is a paperless transaction in the bank, the bank is able to save some money and as a part of green banking initiative a portion of this amount is given to its customer. This will encourage more paperless transactions by the customers and in the process help to reduce the carbon footprint as well. It is a win-win-win situation for the customer, the bank and the environment. Many Indian banks have started to realize the importance and they are taking up various GREEN initiatives:

IndusInd Bank introduced "Human aur Hariyali" campaign which initiated for solar powered ATM`s. The bank expects to save 1980 Kwh of energy annually and will be accompanied by a simultaneous reduction in CO_2 emissions of 1942 kgs. They are also supporting environment friendly finance program and providing incentives to go green. *IDIBI Bank* has an exclusive group for working on climate change and more specifically carbon credits advisory to the clients to deal with Clean Development Mechanism (CDM), Carbon Credits of Kyoto protocol and Voluntary Emission Reduction (VERs) authority. *State Bank of India* the largest lender of the country, is planning to spread its green banking operation across the country in as fast pace as possible. It aims to bring 700 of its branches under its Green Channel Counter (GCC) in the upcoming few months. GCC's working mechanism is based on SBI debit card. When the card is swiped on a point of sale terminal (POS), the transaction takes place and acknowledgement received on a small paper note. The bank has more than 318 GCC branches all over India. SBI is also promoting 'Green Housing' projects, which reduce carbon emission and promote renewable energy by offering them financial benefits such as concession in margin, interest rate and waiver of processing fees on new loans. There are exceptions, though. The new Green Home Loan Scheme from SBI, for instance, will support environmentally-friendly residential projects and offer various concessions. These loans will be sanctioned for projects rated by the Indian Green Building Council (IGBC) and offer several financial benefits - 5 per cent concession in margin, 0.25 per cent concession in interest rate and processing fee waiver. Green home loan scheme which supports environment friendly housing projects and offer subsidy and interest rates reduction. *Union Bank of India* under takes electrical energy audit annually. Energy audit helps in optimizing energy costs and controlling pollution. The bank has also installed solar water heaters at their premises. It also finances SMEs projects which are energy efficient.

ICICI Bank has been assisting various organizations to undertake clean energy and environmentally sustainable projects/initiatives. ICICI Bank has assisted projects that would specifically promote energy efficiency, renewable energy, biomass co-generation, biomass gasification, demand side management (by utilities), waste heat recovery, energy service companies (ESCOs) that demonstrate substantial savings in energy on a shared savings basis as well as projects that lead to pollution prevention and waste minimization at source. 50 per cent waiver in processing fee of cars that use alternate mode of energy like electricity and CNG Cars like Reva, LPG or CNG versions of Tata and Maruti variants. For reducing the carbon footprints, bank brings together all the alternate channels under one umbrella and gives customer the convenience of banking. It offers 50 per cent waver on processing fee on car models which make use of alternate mode of energy *i.e.* of LPG, CNG and Solar batteries versions. It also provides incentive to customers purchase homes in 'Leadership in Energy and Environmental Design' (LEED) certified buildings. It conducts green themed events during festivals and celebrates the World Environment Day on June 5 on every year. Bank is encouraging energy conversation practices within the organization and providing car pool transport facility for up and down to its employees. Bank has initiated a program to sensitize corporate bodies, institutions, banks and government agencies involved in project planning on issues like biodiversity, wildlife habitats and environmental laws. *Axis Bank* provides facility of e-statement and for each e-statement registration by a customer; Axis bank will donate a note book to the needy and poor. *Yes Bank* is incorporating community development initiatives such as clean and green drives, energy efficiency practices, workplace health and safety and the development of local disaster management plans through its "Yes Community" initiatives.

The Opportunities

Being a major source of fund provider, banks can play a crucial role in ensuring environmentally sustainable and socially responsible investments in the economy. Banks should try and reduce the increase in carbon footprint caused by them either directly and indirectly and should play a vital role in ensuring sustainable and environment friendly development. The Green Agenda, on one hand it provides challenges to the banks and on the other hand it provide many opportunities as well. It is strongly believed that within the foreseeable future carbon will have an effective price tag attached to it. And it gives banks a role to play in transition to a low carbon economy. Banks may formulate innovative financial solutions and redesign the existing ones so as to incorporate environmental perspectives. Some areas of development could be:

Green Financial Products

Loans with financial concessions to companies which undertake environmental friendly projects such as manufacturers of fuel efficient automobiles, solar and wind equipment *etc.* Banks can also introduce a 'Green Fund' to provide climate conscious customers the option of investing in environmental friendly projects. Besides introducing specific green banking products, banks can incorporate an Environmental Impact Assessment (EIA) in their project appraisal while financing any project to measure the nature and magnitude of environmental impact as well as suggest environmental risk mitigation measures.

Carbon Credit Business

Indian banks can involve themselves in carbon credit business, wherein they can provide all the services in the area of Clean development Mechanisms (CDMs) and carbon credits including services of identification and funding of CDM projects, advisory services for registration of CDM projects and commercialization of CERs under different structures to meet the requirements of its customers, acting as an intermediary for buying Certified Emission Reductions (CERs) on behalf of end-users or carbon funds, financing against CERs and CERs receivables and other related banking services. As India has huge potential for carbon credit business, Indian banks can set up dedicated carbon credit cells to capture a major share of this carbon credit business.

Conclusion and Suggestions

Banks play an indispensable role in mobilizing financial resources across the economy in particular, providing capital for large-scale infrastructure and low carbon technology deployment. Banks can act as catalysts by supporting, coaxing, coercing and motivating entrepreneurs to invest in environment friendly projects. India's leading banks are recognizing and seizing opportunities in an enlarging low carbon economy. Banks indicate that integrating sustainable development into the organization's policies and management approach improves morale of employees and provides a strong and confident long-term relationship with stakeholders. Being the key driver of the climate change agenda within their respective organizations and that the cost of introducing new processes and products banks are increasingly aware of the opportunities that are available to stimulate investment–such as through low carbon funds. However, the correct financial incentives are essential to make this a reality and the banks need to proactively engage with the Government India to ensure that the right incentives are in place. Pricing carbon is an effective means of closing the gap between the cost of energy produced from fossil fuels and the cost of energy produced from a renewable and low carbon sources.

Banks should expand the use of environmental information in the credit extension and investment decisions. The endeavor will help them to proactively improve their environmental performance and creating long term value for their business.

Environmental issues are on the top of agenda of government around the world as rightly said 'No Planet, No People, No Profit.' Banks and financial institutions in particular, have an important role in this context by contributing to the creation of a strong and successful low carbon economy.

References

Bihari, S.C. (2011). Green Banking, Socially Responsible in India, The Indian Banker Journal of IBA Vol.VI, No.1 pp32-37.

Butzengeiger, Sonja. (2005). Voluntary compensation of GHG-emissions: will it master? An analysis of interaction with major international climate policy systems." Hamburg Institute of International Economics, June.

The Climate Group Report, www.theclimategroup.org

Economic Times. Various issues.

Hepburn, Michael (2007). Carbon trading: a review of the Kyoto Mechanisms, Vol. 32, November. pp 375-393.

http://india-reports.in/earth-solutions/government-policies-for-renewable-energy-in-india/The Carbon Disclosure Project is an independent not-for-profit organisation holding the largest database of primary corporate climate change information in the world. www.cdproject.net

Sharma, S.K. (2011). Green Banking: An Indian Perspective, Journal of Commerce and Information Technology vol.11 No.1, Jan. - June, 2011, Pp.54-60.

Wara, Michael. (2007). Is global carbon market working? Nature, pp.495-496.

World Development Report, 2010. www.worldbank.org

www.rbi.org.in

www.unepfi.org

2013, Sustainable Approaches for Environmental Conservation *Pages* **47–50**
Editors: **D.R. Khanna, A.K. Chopra, R. Bhutiani, Gagan Matta & Vikas Singh**
Published by: **BIOTECH BOOKS, NEW DELHI**

Chapter 7

Environmental Imperatives of Sustainable Development in India

Tushar Kumar Gandhi
Department of Environmental Science, Singhania University, Rajasthan

The concept of sustainable development is still being developed and the definition of the term is constantly being revised, extended and refunded; its main components are the economic, social, and environmental factors. Sustainable development is the concept of environmental protection for the sake of future generation. Sustainability is the process of change in which exploitation of resources, direction of investments orientation of technological changes *etc.* Sustainable development is the preservation of the production possibilities of an economy to provide the same goods and services obtained from nature. Sustainable development involves disciplines such as ecology, biology, ethics, economics, chemistry, physics, statistics and engineering. Sustainability is not an opinion but imperative. For a better India to live in; we need good air, pure water, nutritious food, healthy environment and greenery around us. Without sustainability environmental deterioration and economic decline will be feeding on each other leading to poverty, pollution, poor health, political upheaval and unrest. The environment is not to be seen as a stand-alone concern. We need to tackle the environmental degradation in a holistic manner in order to ensure both Economic and Environmental sustainability. Forests play an important role in environmental and economic sustainability. They provide numerous goods and services and maintain life support systems for life on earth. Some of these life support systems of major economic and environmental importance.

Keywords: *Sustainable development, Environment and development, Social foresting, Green economics of India.*

Introduction

Sustainable development is playing important role to all societies, but critical for poor ones, which depends more heavily on natural resources such as soils, rivers, fisheries and forests than do the richer nations. It is argued that environmental problems in developing countries are predominantly driven by poverty, while those of richer nations are driven by affluence and over consumption. This is the subject of the future, especially when considered in a long-term generation perspective. Sustainability is about the possibility that the things we value in the present will continue to exist in the future. Sustainable development is the concept of relationship between economic growth and the environment. This paper addresses some pertinent issues concerning economics of sustainable development in the present context through literature.

Objectives of Sustainable Development

Economic Objectives

- Growth
- Efficiency
- Stability

Social Objectives

- Full employment
- Equity
- Security
- Education
- Health
- Participation
- Cultural identity

Environmental Objectives

- Healthy environment for humans
- Rational use of renewable natural resources
- Conservation of non-renewable natural resources

Review of Literature

As suggested by Barbier (1987) sustainable economic development is likely to require simultaneous account to be taken of at least three systems, namely the biological system, the economic system and the social system. He contends that sustainable development is indistinguishable from the total development of society and cannot effectively be analyzed separately as sustainability depends upon the interaction of economic changes with social, cultural and economic transformations. Human ascribed goals apply in relation to each of these systems and must be taken into account simultaneously in determining a path or path of sustainable developments. He suggested that goals in relation to each of these three systems must be:

- *For the biological system*: maintenance of genetic diversity, resilience and biological productivity
- *For the economic system*: the satisfaction of basic needs, equity enhancement, increasing useful goods and services
- *For the social system*: ensuring cultural diversity, institutional sustainability, social justice and partition.

Georgescu Roegen (1972) suggested that actual human population need to be reduced rather than stabilized to ensure the maximum period of existence of human species. He based his view on the Entropy low of physics. Economic growth relies on high entropy which it uses and disperses. Eventually the stock of such resources will be completely dispersed and no longer available for production. In his view unlimited economic growth is in principle impossible and there is no possibility of human intervention making it so because of operation of the basic principles of entropy.

United Nations Conference on Environment and Development held in Rio de Janeiro, Brazil in 1992, United Nations Conference on Sustainable Development 1993; and World Summit for Sustainable Development in Johannesburg 2002.

Brundtland Report 1987: Sustainable development is development that meets the needs of the present without compromising the ability of future generations to meet their own needs.

Global Warming–Climate Change

We have to pay the ecological and human costs of globalizing agriculture as well as industry. It is a major signal that we need a shift from production oriented conventional modern agriculture to ecologically resilient agriculture.

Besides these phenomena, the emission of green house gases through thermal plants, chemical industries, nuclear, arsenal and petrochemical units, leads to climate change. With countries habitual to periodic epidemics facing the menaces of bird flue fraught with disastrous consequences for human life.

Its spread in winter appears a trickle unusual. The recent surfacing to chikungunya in Italy, a disease so far confined to warmer tropical climate change which emerges critical to horizon. These are cases of global warming. Experts had been warming that global warming is fuelling the spread of epidemics in areas higher to unaffected. Large scale emission of carbon dioxide from industries is responsible for these climate devastations and the adversely affected 262 million people between 2000 and 2004 by these climatic disasters have been mostly the poor.

Thus, globalization enthused conventional productivity oriented growth models with excessive consumerism and comfortable life style has led to climatic change through emission of green house gases, mainly carbon-dioxide which trap heat in atmosphere leading to large scale climatic disasters causing miserable life conditions for poor in developing world.

Environment and Development

The United Nations General Assembly through its Agenda 21 has provided a compressive picture of inter links related to environment and sustainable development. Agenda 21 suggests an action plan is to link national and international policies for revitalizing economic agro with sustainability. Combating poverty iimprovement in demographic structure; change in consumption patterns health, human settlements, pollution control, energy management, treatment of industrial wastes, control of hazardous materials and after the input sustainability are vital requirements for overall sustainable development of nations.

- ☆ Changing consumption patterns: Less wasteful life styles sustainable consumption levels; informed consumer choices.
- ☆ Health: Pollution health risks, urban health, Basic needs; communicable diseases; vulnerable groups.

- ☆ Human settlements: shelter, land and settlement management, environmental infrastructure, energy and transport, human resources and capacity building, disaster prone areas.
- ☆ Urban water supplies: drinking water, sanitation, intercultural planning, monitoring
- ☆ Fresh water resources: integrated assessment, development and management, production of quality and resources, drinking water, sanitation, water for agriculture.
- ☆ Energy: Sustainable energy development and consumption, house hold, transport, industry.

India´s Green Economy

Economics is going green. Its latest concept is sustainable development which can only be achieved if economic theory can be utilized to determine sustainable natural resources. Development can be sustainable if it has roots in its own people, culture, soil and heritage rather than the glamour of the others. Sustainability is not an option but imperative. For a better world to live in; we need good air, pure water, nutritious food, healthy environment and greenery around us. Without sustainability environmental deterioration and economic decline will be feeding on each other leading poverty, pollution, poor health, political upheaval and unrest.

India´s Environmental Ethics

Ecological economics accounts not only for the financial constraints on consumption, as in conventional economics, but also for the natural constraints implied by the limited ability of the environment to provide natural resources and to absorb the wastes of production and to absorb the wastes of production and consumption. Sustainable management of the economic and ecological system is one of the major focuses of ecological economics.

Conclusion

Social and environmental stresses are the failures of institutions to manage and provide public goods to correct the spillovers. Getting socially preferred outcomes require institutions that can identify who bears the burden of social preferred outcomes require institutions that can identify who bears the burden of social and environmental neglect and who benefits and who balance these diverse interests within society. This perspective helps in understanding why technically sound policy advice is so seldom taken up. Thus, sustainable development is about enhancing human well being thorough time.

So, is enjoying physical security and basic civil and political liberties. And so is it appreciating the natural environment, breathing fresh air, drinking clean water, living among an abundance of plant and animal varieties, and not irrevocably undermining the natural processes that produce and renew these features.

2013, Sustainable Approaches for Environmental Conservation *Pages* **51–55**
Editors: **D.R. Khanna, A.K. Chopra, R. Bhutiani, Gagan Matta & Vikas Singh**
Published by: **BIOTECH BOOKS, NEW DELHI**

Chapter 8

Effect of Distillery Effluents on Germinability Seedling Vigour and Field Emergence in Soybean

***Ashish Lambat*[1], *Sanjeev Charjan*[2], *Rajesh Gadewar*[1] *and Pravin Charde*[1]**
[1]*Sevadal Mahila Mahavidyalaya, Nagpur, M.S.*
[2]*Assistant Professor, Dr. P.D.K.V's College of Agriculture, Nagpur, M.S.*

Laboratory and field work has been carried out to study the effect of distillery effluent (coming out of biomethanation plant of distillery) on seed germinability, seedling vigour and field emergence of soybean. In the present investigation the germination paper were soaked in effluent at concentrations ranging 1 to 100 percent and for control in distilled water. The soybean seeds germination percentage in control was 90 per cent. In 100 per cent effluent treatment, the germination percentage was zero. From this experiment it was observed that 1 per cent and 5 per cent effluent treatment showed increase in germination percentage, seedling vigour and field emergence over control. The germination percentage, seedling vigour and field emergence in soybean significantly decreased with increase in concentration of effluent above 5 per cent.

Keywords: *Germination percentage, Seedling vigour, Field emergence percentage, Effluent, Soyabean.*

Introduction

Effluent is resulting liquid flow from the wastewater treatment system of sugar factory or distilleries. For the production of 1 liter of alcohol, nearly 12 to 14 liters of effluent are discharged as wastewater. Every distillery unit is generating 5 to 10 lakh liters of raw effluent every day. The effluent is disposed off into water bodies and soil as well as water well which has been causing major pollution problem.

Recycling effluent through irrigating plant is a must for future generations to avoid ecological disasters on earth, because the plant roots and associated bacteria remove polluting chemicals from wastewater and the photosynthesizing plant leaves help to remove atmosphere gases which are increasing in the atmosphere causing the earth temperature to increase (Wolverton, 1989).

According to World Bank untreated distillery contains suspended soil (10 to 60 mg/l), BOD (1000 to 1500 mg/l) and COD (800 to 3000 mg/l). Sugar factory effluent and distillery effluent constitute number of Physico-chemical elements such as suspended and dissolved solids with high amount of biological oxygen demand, chemical oxygen demand (COD), chloride, sulphide, calcium, magnesium, nitrate as well as some trace amount of heavy metals such as zinc, copper and lead. The presence of these chemicals in large quantities in the effluent, which is used for irrigation not only affect seed germination and plant growth but also deteriorate the physico-chemical characters of soil. The present work deals with the study of effect of distillery effluents on germinability, seedling vigour and field emergence in soybean.

Materials and Method

The experiment was conducted with the distillery effluent coming out of the distillery after treatment. The physico-chemical properties of effluents were analyzed following the standard procedures (Anonymous, 1965). The aerobic and anaerobic population of various groups of microorganisms in the effluents was enumerated using the standard serial dilutions plating technique (Waksman and Fred, 1992). Seeds of soybean were treated with 0.05 percent of Mercuric Chloride to disinfect it and then washed several times with sterilized distilled water. The evaluation of germination on the basis of normal seedlings by blotter germination paper method (Anonymous, 1985). The effluents were studied for their effect on germination, length of radical, length of plumule and seedling vigour of soybean seeds. In the present investigation, the blotter germination paper were soaked in following concentrations ranging 1 to 100 percent for control in distilled water. In each treatment, three replication each of 100 seeds were maintained.

For soaking seeds in effluent, six treatment with control as a seventh treatment.

T_1 – 1 per cent effluent/1 ml. made upto 100 ml. with distilled water.

T_2 – 5 per cent effluent/5 ml. made upto 100 ml. with distilled water.

T_3 – 10 per cent effluent/1 ml. made upto 100 ml. with distilled water.

T_4 – 15 per cent effluent/1 ml. made upto 100 ml. with distilled water.

T_5 – 20 per cent effluent/1 ml. made upto 100 ml. with distilled water.

T_6 – 100 per cent effluent/1 ml. made upto 100 ml. with distilled water.

T_7 –Control-blotter germination paper soaked in distilled water.

For field emergence test, sterilized neutral soil was taken in tea cups and one seed sown in each tea cup. For irrigation with this effluent the above seven treatment were fixed and irrigated daily. The data was statistically analyzed with the help of Panse and Sukhatme (1967). The observation on germination, seedling length, seeding vigour index and field emergence were taken on 7th day from the start of experiment. The seedling vigour index were calculated according to Abdul-Baki and Anderson (1973). The data was statistically analyzed with the help of Panse and Sukhatme (1967).

Results and Discussion

The values of different physical characteristics in distillery effluent are given in Table 8.1.

Table 8.1: The Physical Characteristics of the Effluent

Name of Effluent	*Colour*	*Odour*	*Temp (ºC)*	*Turbidity*	*Total Solids (CPPM)*	*Suspended Solids (PPM)*	*Dissolved Solids (PPM)*
Distillery effluent	Dark brown	Decaying Molasses	29 to 58ºC	Highly Turbid	1612 to 1906 PPM	531 to 719 PPM	1201 to 1340 PPM

The values of different chemical characteristics estimated in distillery effluent are given in Table 8.2.

Table 8.2: Chemical Characteristics of the Effluent

Sl.No.	*Chemicals Characters*	*Estimation*
1.	pH	7.0 to 6.0
2.	Electrical conductivity (mhos/cm)	30 to 40.0
3.	Total nitrogen (ppm)	1512 to 1850
4.	Amoniacal nitrogen (ppm)	1179 to 1189
5.	Total phosphate (ppm)	407 to 451
6.	Total potassium (ppm)	9600 to 10612
7.	Calcium (ppm)	1091 to114
8.	Magnesium (ppm)	2199 to 2301
9.	Sodium (ppm)	530 to 549
10.	Chlorides (ppm)	9680 to 10700
11.	Sulphates (ppm)	2200 to 3600
12.	BOD (ppm)	5332 to 5914
13.	COD (ppm)	33000 to 41100
14.	Dissolved oxygen (ppm)	0.006 to 0.009

Table 8.3: Biological Characteristics of Effluent

Sl.No.	*Biological Characters*	*Estimation*
1.	Bacteria ($X10^5$/ml)	13 to 25
2.	Fungi ($X10^3$/ml)	37 to 59
3.	Actinomyctes ($X10^4$/ml)	8 to 11
4.	Yeast ($X10^2$/ml)	23 to 31

The values of different physical, chemical and biological characters estimated in distillery effluents are given in Tables 8.1–8.3 respectively.

Data from Table 8.4 revealed that the control (T_7) showed a seed germination of 90.1 per cent while in other treatments the range was 50.1 to 92.4 per cent; the minimum being in T5, T4 and T3 (20 per cent, 15 per cent and 10 per cent effluent). T_6 (100 per cent effluent) recorded zero percent germination. There was maximum germination percentage in T1 and T2 (1 per cent and 5 per cent effluent) treatment over the T7 (control). From the experiment it was observed that 1 per cent and 5 per cent effluent treatment showed increase in germination percentage over control. Sahai *et al.* (1983) studied the effort of the various concentrations of the distillery effluent on the germinability of seeds, seedling growth, pigment content and biomass of *Phaseolus radiatus* L. They reported that the distillery effluent was found beneficial upto 15 per cent for the overall growth of the plant. The germination percentage in soybean significantly decreased with increase in concentration of effluent above 5 per cent. This may be due to phytoxic effect of effluent on soybean seeds. Sahai *et al.* (1983), Kannan (2001), Pande (2008) and Gaikwad *et al.* (2010) reported that the germination percentage decreased as the effluent concentration increased.

Table 8.4: Effect of Distillery Effluent on Germination, Seedling Length, Seeding Vigous and Field Emergence

Sl.No.	Treatment	Germination (per cent)	Plumule Length (cm)	Radical Length (cm)	Seedling Vigour Index (SVI)	Field Emergence (per cent)
1.	T_1 – 1 per cent effluent	92.4	54.91		4405	85.2
2.	T_2 – 5 per cent effluent	91.9	52.94		4391	83.4
3.	T_3 – 10 per cent effluent	76.3	37.19		3084	65.3
4.	T_4 – 15 per cent effluent	63.3	35.94		2101	51.4
5.	T_5 – 20 per cent effluent	50.1	30.98		2002	35.6
6.	T_6–100 per cent effluent	0.0	-		-	-
7.	T_7 – Control	90.1	50.76		4182	82.4
	S.E.D.	4.81	0.60		-	4.00
	C.D.	14.45	2.00		-	12.00

The length of seedling and seedling vigour index also found maximum in T1 and T2 (1 per cent and 5 per cent effluent) treatment as compared to other treatments. The seedling length and seedling vigour index was reduced significantly with increase in concentration of effluent above 5 per cent. Similar observations also reported by Sahai *et al.* (1983), Sahai *et al.* (1987), Kannan (2001) and Gaikwad (2010). In the seedling length, radical growth was more adversely affected than the plumule growth. This is conformity with the observations of Sahai (1983) and Kannan (2001) in rice and *Phaseolus aureus* respectively.

Similarly the superiority of T1 and T2 (1 per cent and 5 per cent effluent) treatment in case of field emergence percentage were observed maximum as compared to other treatments. The field emergence percentage in soybean significantly decreased with increase in concentration of effluent above 5 per cent. This might be due to inhibitory effect of effluent on soybean seeds in an outcome of its high salt and toxic metals content. Similar results also observed by Pendias and Pendias (1992) and Lauchli and Luttge (2000).

The present investigation suggests that 1 per cent and 5 per cent effluent treatment showed increase in germination percentage, seedling length, seedling vigour index and field emergence over control. The above parameters significantly decreased with increase in concentration of effluent above 5 per cent as it showed phytotoxic effect.

References

Anonymous (1965). Standard methods for the examination of water and wastewater. American Public Health Association, A.W.P.C.F., Broadway, New York.

Abdul-Baki A and Anderson, J. D. (1973). Vigour determination in soybean seed by multiple criteria.

Anonymous (1985). International rules for seed testing. *Seed Sci. and Technol.* 13: 299-513.

Gaikwad, S. S., Ahad Najam, K. A., Wanule, D.D. and Bhowate, C. S. (2010). Phytotoxic effect on sugar factory effluent and distillery effluent on tukishbean. *Bionano Frontier*. 3(2): 256-258.

Kannan, J. (2001). Effect of distillery effluents on crop plants. *Ad. Plant Sci.* 14(1): 127-132.

Lauchli, A. and Luttge, U. (2000). *Salinity-Environments-Plants Molecules*. Kluwer Academic Publisher, Dordrecht, the Netherlands.

Panse, V. G. and Sukhatme, P. V. (1967). *Statistical Methods for Agricultural Workers*. I.C.A.R. Publication, New Delhi.

Pandey, A. K., Dutta, S. and Sharma, K. C. (2002). Impact of marble slurry on sub surface water– A case study of Kishangarh. Dist-Ajmer. *Envirn. Pollut. Technol.* 1: 5 to 11.

Pendias, Kobata and Pendias, H. (1992). *Trace Element in Soils and Plants*, 2nd Edn., CRC Trace Boca Raton Fl.

Sahai, R. S., Jabeen and Saxena (1983). Effect of distillery effluent on seed germination, seedling growth, pigment content and biomass of *Phaseolus radiatus* L. *Indian J. Ecol.* 14: 21-25.

Waksman, S. A. and Fred, E. B. (1992). A tentative outline of the plate method for determining the number of micro-organisms in soil. *Soil* Sci. 14: 27-28.

Wolverton (1989). Treatment of industrial wastewater. *Biocycle*. 30: 48-50.

2013, Sustainable Approaches for Environmental Conservation *Pages* **57–64**
Editors: **D.R. Khanna, A.K. Chopra, R. Bhutiani, Gagan Matta & Vikas Singh**
Published by: **BIOTECH BOOKS, NEW DELHI**

Chapter 9
Role of Education in Environment Conservation

Jatesh Kathpalia, Rashmi Tyagi and Savita Vermani
Department of Sociology,
CCSHAU, Hisar, Haryana

Environment is an interrelationship of certain components such as man, nature and interaction of the two. This relationship is as old as old mankind. Man has drawn his subsistence from the environment. It's only when human needs increased and as humans multiplied that man's demands on environment became more than could be borne by it.

Urban settlements and developments caused environment degradation in the form of overuse of land and forest resources, pollution of water systems and by upsetting the delicate atmosphere balance so very essential for the survival of all living systems.

Education has always played a crucial role in the society because it disseminates knowledge, provides necessary skills and help in forming certain attitudes. So there is a need for an environmentally oriented education system which would resolve the environmental crises by preparing environmentally conscious citizens. Hence there is a need to educate man as how to protect the environment? How to utilize resources rationally? How to maintain ecological balance and human relationship. Education has emerged as an integrated discipline which helps the learner starting from an integrated discipline early education to tertiary level to understand nature and to live a balanced life. 'Environment education is not a separate branch of science or subject of study. It should be carried out according to the principle of lifelong integral education'.

Therefore, attempt has been made in this paper how to educate the students about the environment conservation/protection to improve the quality of environment to reduce the environment pollution.

Keywords. *Environment, Environmental degradation, Education, Conservation, Ecological balance.*

Friends let us stand together
To preserve our mother nature
By fighting environmental pollution
As it's a question of survival

Introduction

With increasing population, there is heavy pressure on our available resources. It backs to long time that East-west stretch of the foothills of the Himalayan ranges from Himachal to Arunachal Pradesh, once full of bush green forests has already disappeared. Trees have been felled indiscriminately for fuel wood, timber, plywood, paper pulp and packing bones. Reduction of forest cover on earth is causing unprecedented change in her climatic conditions. Drought prevails now for periods longer than before in different parts of the country causing diversification of land. The area of grazing ground is shrinking to meet the demand for more land for cultivation than ever before. So for these stated reasons for degraded environmental into view, environment protection, conservation and development should therefore, be our primary concern.

Education is a positive response to meet the challenges for the present environment we are facing. Education and awareness are tools which are to be employed to understand, evaluate and to mitigate the challenges and it also play a crucial role in moderating the human behaviour. Education at its first step involves the accumulation of information, gathered from vast oceans of knowledge in which man is on a voyage for unveiling the secrets of 'nature' in a little boat of discovery. The second step of education pertains to the transfer of this acquired knowledge from one individual or one generation to the other generation. Education thus has two distinct entitles *i.e.* provider and receiver. The environmental education must therefore, be capable of meeting all challenges and problems arising as a result of exploitation of 'nature' to use, overuse, unwise and misuse of resources, inflicting imbalance in ecosystem.

Students of any country are the future of that country. Education has always played a crucial role in the society because it disseminates knowledge, provides necessary skills and help in forming certain attitudes. Hence the Stockholm conference on Human Environment from 5 to 16 June 1972 emphasised the need for an environmentally oriented education system, which would resolve the environmental crises by preparing environmentally conscious citizens. Stockholm conference also initiated the idea of observing World Environment Day on 5th June every year. This was meant for creating an awareness among people specially students who are the backbone of any nation through various activities of different organizations and institution like schools and colleges concerning the Environment.

Environmental Education

Environmental education is the education through the environment about the environment and for the environment. Its an action process related to the work of almost all subject areas. It is concern with the dynamic relationship between man and nature. It aims at improving the environmental quality. It is the process of recognizing value and clarifying concepts in order to develop skills and attitudes necessary to understand and appreciate the inter relatedness of man, his culture and his biophysical surroundings. Also it entails practice in decision making and self formulation of a code of behaviour, about issues concerning environment quality.

Now most countries, including India, have realized the need of environmental education. The aim of environmental education is that the individual and social groups should acquire awareness and knowledge, develop skills and abilities and participate in solving real environmental problems. The perspective should be integrated, inter disciplinary and holistic in character. According to organization of American States Conference on Education (1971), "Environmental education involves techniques about the value judgements and the ability to think clearly about complete problems – about the environment which are political, economical and philosophical and technical".

Environmental Education: Importance and Scope

According to UNESCO the goals of environmental education are to develop a world population that is aware of and concerned about total environment and its associated problem and commitment to work individually and collectively towards solution of current problems and the prevention of new ones. Consequently, UNESCO launched the international programme in environmental education in January, 1975.

- ☆ To facilitate the coordination, joint planning, pre-programming activities of environmental education.
- ☆ To formulate, pertaining to environmental education.
- ☆ To coordinate research to understand better the various phenomena involved in teaching and learning about environment.
- ☆ To promote and evaluate new methods, materials and programmes (both in school and out of school for youth and adults) in environmental education.
- ☆ To train and retrain personal and staff for the environmental education programmes.
- ☆ To prepare advisory services to member-states relating to environmental education.

Objectives of Environmental Education

The objectives of EE as established by the first inter government conference held at Tbilesi, USSR in 1977 under the agies of UNESCO are as under:

1. *Awareness* – to help social groups and individuals acquire an awareness and sensitivity to the total environment and its allied problems.
2. *Knowledge* – to help social groups and the individuals gain a variety of experience in and acquire a basic understanding of the environment and its associated problems.
3. *Attitude* – to help social groups and individuals to acquire a set of values and feelings of concern for the environment and the motivation to act in protecting and improving it.
4. *Skills* – to help social groups and individuals, acquire the skills for identifying and solving environmental problems.
5. *Participation* – to provide social groups and individuals with an opportunity to be actively involved at all levels in working towards resolving of environmental problems.

This conference also stressed that environmental education should consider the environment in its totality natural and man made, sense audit should be a life-long process and be interdisciplinary in the approach.

Guiding Principles of Environment Education

The principles of environmental education are as follows:

1. To appreciate the gifts of the nature.
2. To consider the environment in its totality- nature and built economic, political, technological, cultural, historical, moral and aesthetic values.
3. To examine major environmental issues from local regional, national and international points of view.
4. To help in understanding the inter dependence of life at different tropic levels.
5. To help in developing interest in the flora and fauna of the environment.
6. To enable learners to have a role in planning their bearing experience and provide an opportunity for making decisions and accepting their consequences.
7. To relate environmental sensitivity, knowledge, problems solving skills and values to every age.
8. To help in developing interest in the community and the problems of the people and society.
9. To develop tolerance towards different casts, creeds races, religions and cultures.
10. To hep in understanding the effect of unchecked population growth or unplanned resource utilization of the world.
11. To value equality, liberty, fraternity, truth and justice.
12. To examine trends in the growth of population and interpret them for the socio economic development of the country.
13. To help learners discover the symptoms and real causes of environmental problems.
14. To evaluate the utilization of physical and human resources and to suggest remedial measures.
15. To help diagnose the different cause of environmental pollution and to suggest remedial measures.

16. To help diagnose the causes of social tensions and to suggest methods for avoiding them.
17. To promote the value and necessity of local, national and international corporation in the prevention and solution of environmental problems.
18. To utilize diverse learning environment and a broad array of educational approaches to teaching/learning.

In the formal school system, environmental education concepts may be carefully integrated with different subject areas in a creative and functional manner.

Approaches to Environmental Education (EE)

The focus in EE should be to create awareness of the role of environment in our real life. It can be developed in the following two ways:

1. To underline the environmental aspects of each element of the students bearing processes.
2. To bring in EE as a synthesis of various elements of the secondary school curriculum through an integrative study of interrelations.

In the first approach, elements need to be identified corresponding to stage of learning *e.g.* at primary stage of learning individual, family school, city *etc.*

At the later stage the same approach can be maintained in the field of socio-economic, urbanisation, transportation, agriculture, ethics, law *etc.*

Further treatment in teaching EE in this regard would vary in the geo-cultural background of the region. Since our country is a vast country having different topography and hence climatic conditions including temperature, rainfall *etc.* vary from region to region such as hilly region, plain region, coastal region and desert areas.

Local environment also affect the EE programmes and teaching learning process. Children in the rural areas depend more on nature than the children in urban areas. Hence their perception undergoes a change. In urban areas, children are subject to environmental pollution, industrial hazards, over population, crowd *etc.* It results in changing their perceptions, values, attitudes and behaviour.

While rural children are enjoying natural environment like plant, animals birds, vegetation, rocks, sand, hills *etc.* Even social and psychological environment may also vary.

Components of Environmental Studies

The area of EE is very comprehensive and deep. It is both a medium and process that link man's relationship with his natural, social and cultural environment. It also includes the relationship of population, industrialization, pollution, conservation, depletion, technology, transportation and urban-rural planning to total bioshere. So EE is multi disciplinary in character. It should not aim at imparting

knowledge, only but should aim at inculcation of skills, values, attitudes. Also EE is not an end in itself rather forms and integral part of lifelong education.

Following are the components of curriculum in EE

(a) Natural System

1. Environment, earth, biosphere.
2. Abiotic components
3. Biotic component (plant and animal)
4. Process (weather and climate, evolution and extinction)
5. Biological system (ecosystem, food, community, population, habitate and niche).

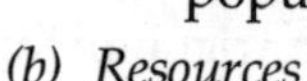

(b) Resources

1. Natural resource (distribution and exemption, management and conservation, sustainable development).
2. Abiotic resource (forests, wildlife, fisheris and biodiversity).
3. Degradation of resources of use (lime stone system and pollution)

(c) Human System

1. Human and environment (human as a part of environment, human adaptation to environment, human influence on environment, population problems)
2. Technical system (agriculture, settlement, manufacturing and technology).
3. Environmental awareness and protection (values, ethics, education and communication, participatory works, legislation and enforcement).

It also includes the learning for the students how to start a waste reduction program. It also shows how the program cash benefit the school, the community and the environment by reducing, reusing and recycling waste. Many materials that are generated in schools can be recycled.

Environmental Education at Different Levels

EE at Primary Level

EE is local surrounding oriented where studies are integrated with different school subjects like social studies, science in the form of nature study and methodology is through practical activities like:

1. Nature study/observation.
2. Games related to environment
3. Quiz on environment
4. Collection of specimens from nature
5. Visits to Zoo and animal sanctuaries

EE at Secondary Level

EE is environmental oriented. Its content is incorporated with subjects like General Science, Social Studies, Languages; Methodology is again practical oriented. Practical activities includes:

1. Observation of natural phenomena
2. Activities related to plants, birds, animals, rocks, sand and sun
3. Games related to environment
4. Quiz on environment
5. Small experiment related to nature
6. Field trips to special centers related to natural phenomenons.
7. Identification of local environment related problems and issues like air, water, soil, round pollution, deforestation, uncontrolled grazing desert, reclaimation, marine environmental pollution.

College/Tertiary Level

EE at College/Tertiary stage has wide scope with regard to teaching, research and extension of knowledge. The following practices can be worked out:

- Integrated EE in each course in all the faculties/departments
- Diversity in courses like environmental engineering, civil/architectural engineering, town planning human settlement, industrial designing landscape, social forestry *etc.*
- Related to conservation, courses like utility of land energy, waste management, wild life management national park, biosphere reserves, water resources, mining management, non-polluting energy development *etc.*
- Environmental health, public health and hygiene sanitation should be given a place in Medical and Para-medical courses
- Courses like bio-medical sciences, toxicology, nutrition, drug uses *etc.* should also find a place in higher education courses of studies.
- In the faculty of humanities and social sciences courses like human ecology, social planning, community organization and services, pregchology and conselling, environmental ethics *etc.* should be included.
- Organisation of clubs and societies related to different studies in environment should form an important component of higher education.
- Membership and defilations to national and international organization need to be encouraged for developing need based courses and research areas.
- Identification and discussion on environment related issues in the form of symposium, debate, seminars, conferences, workshops and research need to be organized.

Conclusion

Education and Environment are interrelated for man's quality living and happy life. Environment degradation is very much related to man's over use of natural resources. Over exploitation of natural resources to meet the needs results in degradation of environment and pollution of various types. Hence men have to face new problems. In this background, there is need to educate him as how to protect the environment ?; how to conserve it ?; how to utilise resources ? and how to maintain ecological balance and human relationship. Environmental education had emerged as integrated discipline which helps the students starting from early education to tertiary level to understand nature and to live a balanced life.

References

Dasgupta, Partha, 2001. Human Well-being and the Natural Environment, New York: Oxford University Press.

Gruenewald, D.A., 2004. A Foucauldian analysis of environmental education: toward the sociological challenge of the Earth charter, Curriculum Inquiry 34 (1): 71-107.

http://www.ukessays.com/essays/economics/environmental pollution.php

http://web.mit.edu/polisci/mpepp/Reports/eier.pdf

http://www.epa.gov/epawaste/education/pdfs/toolkit/tols.pdf

http://www.epa.gov/enviroed/pdf/goalsobjectives./pdf

http://www.treepeople.org/education-and-environment-initiative

http://www.ecoliteracy.org/teach

Sharma, K.K. and Sachdeva M.S., 2003. Education in the Emerging Indian Society, Parkash Book Depot Ludhiana.

Smyth, J.C., 2006. Environment and education; a view of a changing scene, Environmental Education Research 12(3,4): 247-264.

Yadav and Yadav. Education in the Emerging Indian Society, Tandon Publications, Ludhiana.

2013, Sustainable Approaches for Environmental Conservation *Pages* ***65–68***
Editors: **D.R. Khanna, A.K. Chopra, R. Bhutiani, Gagan Matta & Vikas Singh**
Published by: **BIOTECH BOOKS, NEW DELHI**

Chapter 10

Need to Innovate Low Cost Alternative Building Material for Under Privileged Population

G.C. Mishra, Dudhesh Bhola and Chaitanya Yadav
Department of Civil Engineering,
Lingaya's University, Faridabad, Haryana

As the civilization of human being has progressed from stone age towards modernity, the need for three basic essentialities such as roti, kapra and makaan (food, clothes and shelter) became inevitable. There after depending upon availability of resources and skill of individuals or groups, a kind hierarchy developed in the society in terms of development. Then the choices for better, bad or worst arose with the passage of time for housing to be utilized by human being. After Independence the population of India exploded and has also categorized it into various income groups in terms of affordability for houses. Post industrialization impact resulted into urbanization which was supplemented by mass migration of rural population towards urban centric areas in various parts of India, leading to create an acute shortage of houses. The government as well as the industrial installations or corporate sectors also started to contribute by providing houses to its workers in various sectors like services, industries *etc.* According to a survey in 2005, the demand for houses was 209.5 million and the supply was 189.5 million. In order to narrow the gap between demand and supply, lots of efforts has been made by government to device certain policies to provide houses to under privilege population which is re- categorized under BPL (Below Poverty Line).Yet due to resource crunch at the government level and bureaucratic bottlenecks, the target population has not been provided houses for their development. Apart from that, there are several other factors such as lack of effective regularity mechanism, high cost of land, stringent land acquisition process, double taxation system creates hurdles in facilitating benefits to target population. Therefore, it is need of ours to expedite the approval of housing projects, rationalization of transaction cost, value engineering in construction by providing alternative building materials for low cost housing, becomes an evolving methodology for Public Private Partnership (PPP Model), in order to fulfill the desire of millions of populations to have their own houses.The objective of this paper is to develop some innovative technology as well as utilization of existing resources in sustainable manner to provide low cost housing for underprivileged population for holistic growth of nation especially in developing countries like India.

Introduction

With the increase in population and affordability there is tremendous increase in solid waste generation, there are varieties of solid wastes in which the construction and demolition waste contributes major part in India. It is approximate 10-12 million tons of waste produced. Projection for building material requirement of the housing sector indicate a shortage of aggregates to the extent of about 55000 million c.u.m. and additional 750 million c.u.m. aggregates would be required for achieving the targets of the road sector. Recycling of aggregate material from construction and demolition waste may reduce the demand-supply gap in both the sectors.

While retrievable items such as bricks, plastics, wood, metal, tiles are recycled, the concrete and masonry waste, accounting for more than 50 per cent of the waste from construction and demolition activities, are not being currently recycled in systematic manner in India. Recycling of concrete and masonry waste is, however, being done abroad in countries like U.K., USA, France, Denmark, Germany and Japan. Concrete and masonry waste can be recycled by sorting, crushing and sieving into recycled aggregate. This recycled aggregate can be used to make concrete for road construction and building material. Work on recycling of aggregates has been done at Central Building Research Institute (CBRI), Roorkee and Central Road Research Institute (CRRI), New Delhi. The study report stresses the importance of recycling construction waste, creating awareness about the problem of waste management and the availability of technologies for recycling. According to a study commissioned by Technology Information, Forecasting and Assessment Council (TIFAC), 70 per cent of the construction industry is not aware of recycling techniques. The study recommends establishment of quality standards for recycled aggregate materials and recycled aggregate concrete. This would help in setting up a target while maintaining product quality for producers and assure the users of a minimum quality requirement, thus encouraging them to use it. These waste materials need a large area to dump and hence the disposal of land fill sites has become a severe social and environmental problem.

Reuse of materials is an efficient way to reduce the use of energy intensive materials. Instead of discarding tonnes and tonnes of wastes from the factories and homes, some part of it could be used for creative construction. This will help in achieving "Waste Reduction" and help us to move one step forward towards the sustainable environment.

Any construction activity requires several materials such as concrete, steel, brick, glass, stone, clay, mud and so on. However, the cement concrete remains the main construction material used in construction industries. For its suitability and adaptability with respect to the changing environment, the concrete must be such that it can conserve resources, protect the environment, economize and lead to proper utilization of energy. The utilization of recycled aggregate is particularly very promising as 75 per cent of concrete is made of aggregates. In that case, the aggregates considered are fly ash, power plant waste, blast furnace by product and recycled concrete. The enormous quantities of demolished concrete are available at various construction sites, which are now posing a serious problem of disposal in urban areas. This can easily be recycled as aggregate and used in concrete. Research and development activities have been taken up all over the world for proving its feasibility, economic viability and cost effectiveness.

In this research, the focus is on finding ways to produce cheaper building materials. There is an urgent need to accelerate the research process and start manufacturing cheaper eco-materials, especially for the construction of low cost housing.

Appropriate Construction Technology and Local Materials

While fixing technical specifications, effort should be made to utilize, to the maximum possible extent, local materials and cost effective technologies developed by various institutions. The implementing agency should contact various organizations/institutions for seeking expertise and information on innovative technologies, materials, designs and methods to help beneficiaries in the construction of durable and economical houses. State Governments may also arrange to make available information on cost effective environment friendly technologies, materials, designs *etc.* at block/district level. Technology using bricks, cement and steel on large scale should be discouraged. As far as possible, cement should be substituted by lime and lime *surkh*i manufactured locally.

Discussions

Economical Sustainability

The definition of low-cost (housing) depends on the economical capacity of the target group. If the house and services are too expensive, the poor cannot afford to live there and the programme will fail. One concept may be affordable in one part of the world, but too expensive for another group. Financial schemes could be linked to reduce the problem of affordability. However this subject may need definition in terms of political economy as well as also due to the tradition of non-payment of services and loans. Therefore a programme dealing with willingness to pay must be incorporated. Job creating activities such as labour-intensive construction methods and the creation of small workshops could present another way of obtaining local economical sustainability.

Environmental Sustainability

Modern housing development has a major impact on the ecological system. The master plan of a specific housing project has to be adjusted to the surroundings. The damage to sensitive landscapes, including scenic, cultural, historical and architectural glare must be minimized.

The use of sustainable sound building materials has to be incorporated in order to serve dual purpose of cost efficiency and sustainable solid waste management. Local traditional materials often have a minor impact on environment than modern materials such as bricks, concrete and corrugated iron sheets. A life-cycle-analysis must be utilized to determine the sustainability of the conventional building materials and modern building materials.

Water supply and sanitation must be designed and maintained appropriately to minimize the impact on the local environment. To minimize the environmental impact from the inhabitants, renewable energy such as solar and wind energy must be integrated to the highest extent in low income housing projects.

In general, environmental sustainability is a matter of minimizing the pollution from the consumption of energy, water, materials and land and maximizing the use of recycled materials and renewable resources.

Conclusion

The fact that low cost alternative building technologies have not taken root in provincial housing projects is indicative of the complexity of changing people's mindsets and multiplicities in dealing with these technologies. Efforts to up-scale alternative building technologies in provincial housing projects should therefore be done with due consideration of these intricacies and an awareness that it demands a comprehensive multi-pronged strategy to provide houses to under privileged population for holistic growth of society.

References

Ahn, W.Y. and Moslemi, A.A. (1980). SEM examination of wood-Portland cement bonds. *Wood Sci.*, 13(2), 77-82.

Demirbas, A. and Aslan, A. (1998). Effects of ground hazelnut shell, wood and tea waste on the mechanical properties of cement. *Cement Concrete Res.*, 28(8), 1101-1104.

Hachmi, M., Moslemi, A.A. and Campbell, A.G. (1990). A new technique to classify the compatibility of wood with cement. *Wood Sci. Technol.* 24(4), 345-354.

Hong, Z. and Lee, A.W.C. (1986). Compressive strength of cylindrical samples as an indicator of wood- cement compatibility. *For. Prod. J.* 36(11/12), 87-90.

Karade SR, Irle M, Maher, K. (2003). Assessment of wood-cement compatibility: A new approach. Holzforschung, 57: 672-680.

Sandermann, W. and Kohler, R. (1964). Studies on mineral-bonded wood materials. IV. A short test of the aptitudes of woods for cement-bonded materials. *Holzforschung* 18, 53:59.

Weatherwax, R.C. and Tarkow, H. (1964). Effect of wood on setting of Portland cement. *For. Prod. J.* 14(12), 567-570.

2013, Sustainable Approaches for Environmental Conservation *Pages* **69–78**
Editors: **D.R. Khanna, A.K. Chopra, R. Bhutiani, Gagan Matta & Vikas Singh**
Published by: **BIOTECH BOOKS, NEW DELHI**

Chapter 11

Water Conservation Projects in the Villages of Hoshangabad and Indore Districts under Corporate Social Responsibility by National Fertilizers Limited: A Case Report

S.K. Ghai, Sunita Vivek, Jyoti Pande and J.K. Narang
National Fertilizers Limited, Sector-24, Noida, U.P.

National Fertilizer Limited (NFL), a Government of India Undertaking, a Schedule A and Mini Ratna Company, is the second largest producer of Nitrogenous Fertilizers in the Country with 15.8 per cent share in domestic production of Urea achieved in the country. The Company also manufactures and markets Bio-fertilizers and a wide range of industrial products which include Methanol, Sodium Nitrate, Sodium Nitrite, Nitric Acid, Sulphur, Liquid Oxygen, Liquid CO_2, Liquid Nitrogen *etc.* National Fertilizers Limited is also engaged in farmer friendly social activities that helped in improving socio economic status of farming community who are engaged in agricultural production, which is key to the national food security. The major focus of these programmes is creating awareness about health and hygiene, women education and self-employment, water conservation, water harvesting, groundwater recharging, soil and water conservation *etc.*

Keywords: NFL, CSR, Water conservation, Stop dam.

Introduction [About National Fertilizers Limited (Company Profile)]

National Fertilizer Limited (NFL), a Government of India Undertaking, a Schedule A and Mini Ratna Company, is the second largest producer of Nitrogenous Fertilizers in the Country with 15.8 per cent share in domestic production of Urea achieved in the country. Kisan Urea NFL's popular brand is sold over a large marketing territory spanning the length and breadth of the country. The Company is pioneer in the manufacturing of Neem coated Urea which has been manufacturing and selling from its manufacturing plants since 2002-03. The Company also manufactures and markets Bio-fertilizers and a wide range of industrial products which include Methanol, Sodium Nitrate, Sodium Nitrite, Nitric Acid, Sulphur, Liquid Oxygen, Liquid CO_2, Liquid Nitrogen *etc.* The company has also taken initiative to make available other agro inputs under single window like quality seeds, Insecticides and Bio-pesticides by collaboration with other reputed organizations. R&D trials are under way for testing the efficacy of Bio-pesticides, Urea Ammonium Nitrate (Liq. Nitrogenous Fertilizer) and liquid biofertilizer in collaboration with Agriculture institutes.

National Fertilizers Limited is also engaged in farmer friendly social activities that helped in improving socio economic status of farming community who are engaged in agricultural production, which is key to the national food security. The major focus of these programmes is creating awareness about health and hygiene, women education and self-employment, water conservation, water harvesting, groundwater recharging, soil and water conservation *etc.* Accordingly 6 villages were selected and baseline survey has been conducted for each village by a specialized agency to identify developmental activities required in each village and accordingly action has been initiated to implement the identified activities. The company in the adopted villages under district Indore and Hoshangabad in Madhya Pradesh have successfully implemented water conservation projects.

Water Conservation and its Importance

Water is essential for all life and is used in many different ways - for food production, drinking and domestic uses and industrial use. It is also part of the larger ecosystem on which bio diversity depends. Precipitation, converted to soil and groundwater and thus accessible to vegetation and people, is the dominant pre-condition for biomass production and social development in dry lands. Lack of water is caused by low water storage capacity, low infiltration capacity, large inter-annual and annual fluctuations of precipitation and high evaporative demand. Groundwater source is of crucial importance in semiarid regions especially in agriculture country like India where large parts of population depend on groundwater. National Fertilizer Limited is consuming water for making fertilizer in their units located at Nangal, Bhatinda, Panipat and Vijaypur, therefore company understands its responsibility towards society for water conservation and groundwater recharging and has undertaken several water conservation projects especially construction of stop dams.

A stop dam is a masonry barriers built across the direction of water flow on shallow rivers and streams for the purpose of water harvesting for irrigation as well as for domestic and animal use. Stop dam also reduces erosion and gullying in the channel and allow sediments and pollutants to settle.

Status of Water Conservation in Madhya Pradesh

Madhya Pradesh rain fed state comprising an area of 308,000 sq.kms and population of 60 million. In total 78 percent of the total population is engaged directly in agriculture. About 73 per cent agriculture in the state is rain-fed and agricultural production gets severely affected in the event of untimely or erratic rains or a dry spell. The state receives an average annual rainfall of 1150 mm concentrated in the brief monsoon season; most water bodies remain dry from January to June. There is also the problem of water conservation and storage in most of the areas of the state due to undulated topography and lack of water storage structures.

It was, therefore, very necessary to plan and efficiently execute a community based movement to find an abiding solution to water problem in the surveyed areas through water conservation activities with proper coordination between local community and government.

Objectives

The present case study presents change in socioeconomic scenario covering irrigation benefits, food security, status of migration and poverty reduction of the adopted villages Lalpura under district Indore and Bhamuria under district Hoshangabad. The study covers post construction benefits of three stop dams. First stop dam was constructed during 2009-10 at Lalpura village district Indore and two stop dams were constructed in 2011 at village Bhamuria district Hoshangabad.

Project Area (Stop Dam area) and its Characteristics

Characteristics of Lalpura Stop Dam Area

The project area comprises of three villages Biagram, Gajinda and Lalpura, 40 km away from Indore district. These villages together comprise 830 families and a population of 5000. The average land holding per family is about 0.58 ha but considering the undulating topography, poor soil health (low soil depth, gravelly and low in nutrition) and lack of irrigation, this is not considered enough for providing stable income source for a family of 5- 6 members. Soybean is the main kharif season crop and Wheat and Gram are the main crops of Rabi season. Table 11.1 below provides base line information of the project area before the project implementation. Major crops grown in the project area and their average productivity are shown in Table 11.2.

Table 11.1: Land, Water and Agriculture Features of the Project Area

1.	Average land holding	0.58 hact.
2.	Per cent of Net sown area to total geographical area	25.59 per cent
3.	Per cent of Net irrigated area to net sown area	15.54 per cent
4.	Fertilizers consumption Kg/Ha	123.33
5.	Cropping intensity per cent	136.9
6.	Double cropped area to Net sown area (Ha)	50.7
7.	Seed Replacement Rate	5-10 per cent
8.	Average Rainfall	96 cms
9.	Area under irrigation	50.6 ha
10.	Sources of Irrigation	Tube well, Well, Taalabs

Table 11.2: Major Crops and their Productivity (Qtls/ha) of the Project Area

Sl.No.	*Crop*	*Per cent of the Crop Aown to the Net Sown Area*	*Av. ProductivityProject Area*
1.	Maize	0.2	10.36 Quintal/ha
2.	Black gram	0.1	6.92 Quintal/ha
3.	Soyabean	40.73	12-15 Quintal/ha
4.	Paddy	-	-
5.	Cotton	-	-
6.	Wheat	35.23	24.89 Quintal/ha
7.	Gram	12	14.28 Quintal/ha

As per the baseline information about 55 per cent of the families of the project area did not have the food security for the whole year from own agriculture sources and out of which 32 per cent reported severe shortage of food grain for 6-7 months in a year. The rest 23 per cent had just enough to meet the food requirements for the whole year with marginal marketable surplus, but in a drought year they too had to face the shortage of food grain from own sources. Lal pura Dam site was known as Kutta Nalla, during rainy season massive water flows and meet into minor river, only 10 percent water stores in Kutta Nala. This area is totally rainfed area and crops mainly depend on rainwater. Villagers face acute problems of drinking water during the months of March to June as dug wells and hand pumps became dry at the end of February. Drought like conditions prevailing during *Rabi* season for the crops force farmers to take only *Kharif* crops. Sufficient water was not available even for cattle and animals for their normal survival during summer season.

Basic Details of the Lalpura Stop Dam

The lalpura stop dam can be considered small as far as size is concerned. The length of the surveyed stop dam is 6 meter and average height was within 5.97 meter (Table 11.3). This stop dam is built on the small rivulets to harness the post-monsoonal flow, primarily for the reason of irrigation during Rabi season and groundwater recharge with supplementary use like drinking water for cattle for their survival during summers. The main objective of constructing stop dam in Lalpura village was to ensure rabi crop as due to failure of rabi crop farmers in the nearby villages use to took only *Kharif* crops. The average water storage capacity of the stop dam is to 36000 cubic meters.

Table 11.3: Basic Details of the Lalpura Stop Dam

Sl.No.	*Parameters*	*Details*
1.	Name of District	Indore
2.	Name of Stop Dam	Lalpura Stop Dam
3.	Name of Department	National Fertilizers Limited
4.	Year of Construction	2009-2010
5.	Water Storage Capacity	36000 cubic meter
6.	Length	6 m
7.	Height	5.97 m
8.	Back water	
9.	Openings	
10.	Total Cost (Rs. In Lakhs)	3.54

Characteristics of Shiv-Shankar Stop Dam and Umashankar Stop Dam Area

Total population of the village Bhamuria is 550. Majority of househlds are doing agriculture. Total area under agriculture is 185 ha that is more than 80 per cent of total village geographical area. Beauty of agriculture in this village is both *Rabi* and *Kharif* area is equal that is rarely found in any other village (Table 11.4). Soybean is the main *Kharif* season crop and Wheat and Gram are the main *Rabi* season crops in the area. The area under different crops and average productivity of the crop is shown in Table 11.5. Few farmers also take third crop but area is very less only 12 ha. Majority of land, 150 ha is irrigated through canal irrigation but tale end farmers are not getting timely water for their crops thus their production is not as good compared to head end farmers. Only 10 ha of land is irrigated through open well. Some farmers also lift water from river flowing from the village.

Figure 11.1: Lalpura Stop Dam District Indore, MP

Table 11.4: Land, Water and Agriculture Features of the Project Area

1.	Average land holding	4-5 hact.
2.	Per cent of Net sown area to total geographical area	89
3.	Total area under crops	185 Ha
4.	Kharif crops area	185 ha
5.	Rabi crops area	185 Ha
6.	Jayad crops area	12 ha
7.	Per cent of Net irrigated area to net sown area	84
8.	Fertilizers consumption Kg/Ha	50
9.	Cropping intensity per cent	206
10.	Double cropped area to Net sown area	90
11.	Seed Replacement Rate	8-10 per cent
12.	Average Rainfall	134 cms
13.	Area under irrigation	150
14.	Sources of Irrigation	Canals, Tube well, Well

Basic Details of the Stop Dams in village Bhamuria

Village Bhamuria falls under tail of canals so water comes in the canals either at the beginning or at the ends of canal opening. National Fertilizers Limited constructed two stop dams in the village *i.e.* Shiv Shankar Stop dam samooh and Umashankar Stop dam samooh. The length of these stop dams is 15 meter and 22 meter and height is 2 meter (Table 11.6). This stop dam is built on the small rivulets to harvest the rain water during monsoons, primarily for the reason of irrigation during Rabi season and groundwater recharge with supplementary use like drinking water for cattle *etc.* Farmers having agriculture land nearby the stop dams have made their own arrangement for lifting water from the dam for irrigation by installing diesel and electric pumps for irrigation. The water storage capacity of the stop dams came out to 16424 cubic meter and 24069 cubic meter.

Table 11.5: Major Crops and their Productivity (Qtls/Ha.) of the Project Area

Sl.No.	*Crop*	*Per cent of the Crop Sown to the Net Sown Area*	*Av. Productivity Project Area*	*Av. Productivity National*
1.	Maize	As mixed crop 2-3	—	15.01
2.	Black gram	Very less	—	9.50
3.	Soybean	65	8-10	13.50
4.	Paddy	25-30	12 -15	22.00
5.	Cotton	Not grown	—	—
6.	Wheat	70-75	22-24	27.00
7.	Gram	25	5-6	7.80

Table 11.6: Basic Details of the Stop Dams in Village Bhamuria, Distt Hoshangabad

Sl.No.	*Parameters*	*Stop Dam (1)*	*Stop Dam (2)*
1.	Name of District	Hoshangabad	Hoshangabad
2.	Name of Stop Dam	Shiv-shankar stop dam samooh	Umashankar stop dam samooh
3.	Name of Department	National Fertilizers Limited, Bhopal	National Fertilizers Limited, Bhopal
4.	Year of Construction	2011	2011
5.	Water Storage Capacity	16424 cubic Meter	24069 cubic Meter
6.	Length	15 Meter	22 Meter
7.	Height	2 Meter	2 Meter
8.	Back water	800 Meter	900 Meter
9.	Openings	4	6
10.	Total Cost (Rs in Lakhs)	8	8

Visible Impacts of Stop Dams

Direct and Indirect Benefits of Lalpura Stop Dam

Before the construction of Lalpura stop dam 238 acres area was under irrigation as the farmers' were drawing water from the flowing rivulets/nullahs or from the places of natural pondage or from their own sources like dug wells and bore wells. However, the availability was always scarce, resulting in poor crop productivity and failure of crop. Hence, the post construction of stop dam has made significant contribution in supplementing irrigation during rabi crop and thus provided an assured crop. Farmers got benefited as their crop production increased 10-15 per cent and estimated profit is Rs. 3600/acre. Therefore, the estimated net gain to the farmers in rabi is Rs. 8.6 Lakhs and an additional benefit of Rs. 5 lakh by vegetable production. Undoubtedly the stop dam has contributed to the family's total income by providing additional irrigation benefits fully or partially.

Figure 11.2: Shiv Shankar and Umashankar Stop Dam at village Bhamuria, Distt Hoshangabad, MP

An additional 105 acres of cultivable land has come under direct irrigation because of the Lalpura stop dam 25-acre additional area in Kharif and 80 acres additional area during rabi (Table 11.7). The total number of beneficiaries under direct irrigation is 12. Importantly, 80 acres was Rabi fallow earlier and now the farmers have started growing short matured variety of wheat (Lok-1) and in majority (84 per cent) of the cases growing gram for which two irrigations is more than adequate.

Table 11.7: Impact of Stop Dams on Direct Irrigation

Sl.No.	*No. of Stop Dam*	*Status*	*Direct Irrigation Benefits Post Construction*				
			Increased Area in Kharif (Acre)	*Increased Area in Rabi (Acre)*	*Increased Area in Summer (Acre)*	*Total*	*No. of Direct Irrigation Beneficiary*
1.	Lalpura Baigram	Complete	25	80	0	105	12
2.	Bhamuria Stop Dam (1)	Complete	15	25	5	45	05
3.	Bhamuria Stop Dam (2)	Complete	20	30	5	55	06
	Total (2+3)*		35	55	10	100	11

* The construction stop dams in Bhamuria village of Hoshangabad district have been completed in year *i.e.* 2011; therefore estimated irrigation benefits are shown in Table 11.7 and Table 11.8. Project data for Kharif are yet to be collected.

Table 11.8: Impact of Stop Dams on In Direct Irrigation

Sl.No.	*No. of Stop Dam*	*Status*	*In Direct Irrigation Benefits Due to Water Recharge in the Dug Wells and Bore Wells*				
			Increased Area in Kharif (Acre)	*Increased Area in Rabi (Acre)*	*Increased Area in Summer (Acre)*	*Total*	*No. of Indirect Irrigation Beneficiary (Recharged Wells)*
1.	Lalpura Baigram	Complete	10	200	0	210	22
2.	Bhamuria Stop Dam (1)	Complete	20	30	10	60	25
3.	Bhamuria Stop Dam (2)	Complete	30	40	10	80	25
	Total (2+3)*		50	70	20	140	50

There are many other water bodies such as dug wells and bore wells that have been recharged and helped in making water available for the crops. There has been irrigation benefits accrued indirectly also through the sub-surface and groundwater recharge to the dug wells and bore wells in the downstream of the stop dam to 210 acres of land in the nearby villages. As per farmers from the nearby villages Bherughat, Lalpura and Bigram, the water level in the dug wells and bore wells has increased significantly in the post construction of the stop dam.

Direct and Indirect Benefits of Shiv-Shankar Stop Dam and Umashankar Stop Dam

The irrigation related issue with Bhamuria was that the water was not available as per the crop requirement leading to poor crop yields. Rabi sowing often gets delayed due to late availability of canal water. After the construction of these stop dams, farmers of Bhamuria village are ensure the timely sowing of Rabi crops. A total of 11 farmers spread over 100 acres started reaping direct benefits (Table 11.7). These farmers, who had just about sufficient water available for only two irrigation in Rabi for wheat, will now have four irrigation available, and it is estimated that the production would be increased. The farmers would be benefited by raising summer fodder crops for cattle feeding as water recharges in wells and the availability of water may go till next rainy season. Moreover, 50 other farmers would be benefited indirectly and 140 acres additional area would be irrigated with enhanced water level in their tubewells and wells.

Construction of Lalpura stop dam increased water retention capacity of catchment area from 8000 cubic m to 38000 cubic meters. After the construction of dam about 80 percent rain water stored in catchment area, which ultimately increased the water level in water bodies of the nearby villages. Now farmers are taking double crops due to increased availability of irrigation water. Some farmers are also taking vegetable crops that have improved their economic condition.

Due to construction of Lalpura stop dam, drinking water now available around the year for villagers and cattles. After the construction of Lalpura stop dam, during the year 2010 and 2011 water lasted till late January and as a result of which these 11 families started taking Rabi crops for the first time. They took Lok-1 variety of wheat on their 15 acres of the command land through 3 to 4 irrigation, when not even a single irrigation was available to them earlier. The results were very encouraging as their crop production increased 5-7 per cent. The food sufficiency after the construction of stop dam has risen from 3 months to 5 months (Table 11.9).

Table 11.9: Impact of Stop Dam on Food Security

Total Surveyed Households	*Households Food Secured Before Construction of Stop Dam*	*Households Reported Achieving Food Security After Construction of Stop Dam*	*No. of Months Increased (Average)*
50	20	30	5

The immediate result of the Lalpura stop dam's construction showed 50 per cent drop in migration of both men and women as well as the number of days of migration per year (from 100 to less than 60) in Gajinda, Lalpura and Biagram villages (Table 11.10). This drop in migration could only be possible by the availability of job in agriculture due to increase in irrigation and availability of water round the year.

Table 11.10: Impact of Stop Dam on Migration

Total Surveyed Household	*Household Migrating before Stop Dam Construction*	*Household Reported Reduction in Migration due to Stop Dam*	*Household Reported No Change in Migration*
50	20	10	50 per cent

With the construction of Bhamuria stop dams now the number of irrigation and time of irrigation would be in the control of the farmers. This will lead to better crop production and increased food security. The water will not only be available for the animals but also for drinking water and other domestic purposes in summer by recharge of those dug wells and tube wells upon which villagers are depend for drinking water.

Conclusion

National fertilizers Limited under aids Corporate Social Responsibility is engaged in undertaking various activities of socioeconomic nature for the upliftment of village communities and farming communities realizing the water as one of the essential element for drinking and domestic uses, increasing the crop yields and harvesting rich dividends for the village community has taken up construction of water harvesting infrastructures such as stop dams.

The stop dams have resulted in making available water around the year both for the farmers and animals. Apart from the water harvesting structures the company is also engaged in various other activities such as children education, women empowerment, human and animal health camps as well as propagation of non-conventional sources of energy. The holistic approach adopted by the company is producing desired, consistent and sustainable results.

2013, Sustainable Approaches for Environmental Conservation *Pages* ***79–89***
Editors: **D.R. Khanna, A.K. Chopra, R. Bhutiani, Gagan Matta & Vikas Singh**
Published by: **BIOTECH BOOKS, NEW DELHI**

Chapter 12

Loss of Freshwater Biodiversity by Blue-Green Algae and its Management

Rakesh Kumar Dwivedi

Department of Botany, S.R.T. Campus,
H.N.B. Garhwal University, Badshahi Thaul, Tehri Garhwal, Uttarakhand

Freshwater ecosystem is the home of some precious biota of the global biodiversity which contribute to the significant amount to the food bowl of humankind. The loss of freshwater biodiversity due to autogenic and allogenic reasons is one of the serious concerns of the day for the range of academicians. In freshwater habitats, the blue-green algae constitute the major chunk of the aquatic flora and contribute to major part of the aquatic primary productivity and food web. But the other aspect of the same coin is its significant role in promoting autogenic succession detrimental to the momentous loss of biodiversity under certain circumstances. Understanding of the factors fuelling the overgrowth of blue-green algae in freshwater habitats, bloom formation and methods of its management are very important for the conservation of the freshwater ecosystem and preventing colossal economic loss to the aqua-culturists.

Keywords: *Rainwater, Blue-green algae, Biodiversity, Primary productivity.*

Introduction

Freshwater is the elixir of life for the terrestrial biota which has been used for the habitat, breeding, spawning, pollination, fertilization and consumption as drinking water to satiate the metabolic need. The freshwater is generally characterized by having low concentrations of the dissolved salts which separates it from the categories of marine and brackish water. Limit separating 'freshwater' from 'saline' water is commonly set at 3g/l (Williams, 1981). Naturally, the freshwater on the surface of the Earth is present in form of rivers, streams, lakes, ponds, bogs, underground aquifers and as water vapours. Precipitation in the form of rain, snow and mist is the source of all the freshwater which is governed by the water cycle of the biogeochemical cycle; an important function of the ecosystem.

The freshwater ecosystem has been categorized into lotic system which are running water forms like rivers and lentic system which are still water forms like lakes. These ecosystems are the habitat of numerous known and unknown species contributing to the dynamics to these systems. Among various aquatic plants the algae are chief source of primary productivity due to its ubiquitous distribution and make the fulcrum of the aquatic food-web. Besides, one of the fundamental metabolic processes of BGA like *Anabaena, Nostoc, Aphanizomenon* is dinitrogen fixation.

But the other aspects of the algae are seldom if ever recognized due to less familiarity with these tiny or microscopic plants having the simplest thallus organization and reproductive methods.

Biodiversity in Freshwaters

About 29,000 species have been described in freshwater ecosystems, including about 12,000 species of fish and 17,000 other species from diverse groups such as microbes, algae, nematodes, rotifers, insects, crustaceans, annelids and mollusks (Abell *et al.*, 2000). Freshwater ecosystems- rivers, lakes, aquifers and wetlands provide pivotal ecosystem services and support important fisheries and consumption for the terrestrial biota. The maintenance of biodiversity is one of the important aspects for the sustainability of the ecosystem services (Palmer *et al.*, 2000). Diverse assemblage of speies may be able to use resources more efficiently and thus generates more productive ecosystems. They may also offer increased resistance against ecosystem collapse in the face of disturbance (Loreau *et al.*, 2001, Cardinale *et al.*, 2002).

Algal Biodiversity in Freshwater

In terms of classification, the freshwater microalgae investigated by phycologists under International Code of Botanical Nomenclature (ICBN) include organisms of both eukaryotic and prokaryotic cell types. It comprises members of Cyanophyceae or Blue-Green Algae (BGA) (Geitler, 1932) or Cyanobacteria (Waterbury, 1989) both are compatible systematic terms due their prokaryotic unicellular or multicellular cells but possess chlorophyll *a* and perform oxygenic photosynthesis associated with photosystem I and II (Whitton and Potts, 2000) but still it is traditionally included in algae and other eukaryotic members like Chlorophytes, Charophytes both are included in Green Algae, Chrysophytes (Golden-brown and yellow-green algae) diatoms and sporadic representation of other groups like Rhodophytes (Red algae), Phaeophytes (Brown algae). The algal biodiversity is usually considered at the level of richness of species and higher taxonomic ranks and there is no reliable answer about their exact number discovered so far. The identification of algae is very difficult for they display few characters and due to their short generation time rate of evolution is higher resulting in several sibling species. There are about 30,000-40,000 species of algae (Andersen, 1992; John 1994) and BGA constitute 150 genera and 2000 species (Hoek *et al.*, 1995).

Blue-Green Algae Decreasing Freshwater Biodiversity

Human activities like agricultural runoff, inadequate sewage treatment, runoff from roads have led to excessive fertilization of many water bodies which is more commonly termed as: Eutrophication. Eutrophication promotes the excessive proliferation of the BGA and inter alia the Biologically Available Phosphorus (BAP) is the chief factor (Schindler, 1977). The factors decreasing freshwater biodiversity caused by some BGA, which is paradoxically the lifeline of the same ecosystem, are mainly of two types: the toxins produced by some microalgae and production of high biomass or bloom formation.

The problems associated with biomass blooms and toxic events are different. The dynamics of bloom formation varies from place to place, not depending alone of the hydrographic and topographical conditions, but also on ecological and biological characteristic of the causative organism. Increase in algal cell numbers are affected by season, temperature, amount of sunlight penetrating the water column, amount of available inorganic nutrients, competition from other algae and aquatic plants. High biomass blooms may cause significant ecological problems and harmful effects on the biota of the region like anoxia, community and food-web changes. But the toxin events may result from very low concentrations of the causative organisms and in the case of toxin events occurring with high biomass production, the toxicity level shoots to dangerous limits before the bloom is conspicuous. BGA have both the features.

Some of the main effects due to cyanobacteria blooms comprise a decrease in water transparency, heavy fluctuation of oxygen levels and release of toxins. As stated, cyanobacteria blooms are responsible for the decrease in water transparency and this can pose ecological and safety problems. During a heavy bloom, water transparency can be as low as 1- 2 cm so this disturbs the whole ecology of the ecosystem by preventing light from reaching higher water depths. Only species able to migrate in the water column such as cyanobacteria can be successful in this type of environment.

On the other side, animals that use sight to move, locate food, or find partners to mate will also be severely affected during these cyanobacteria blooms. The huge cyanobacteria biomass that composes these blooms may produce high amounts of oxygen during daytime leading to over saturation. The high respiration of all aquatic organisms during nighttime may lead then to a drop in oxygen concentration during night. Especially in areas close to the bottom or the thermocline, the oxygen levels at night may be low enough to produce the death of the more sensitive species such as some fish. The variation in pH, with high values over nine during daytime and low levels at night may also stress the environment and cause changes in the biogeochemistry that can increase the negative effects stressed before (Vasconcelos, 2001).

Unmanaged BGA growth in aquaculture ponds causes poor water quality following algal degradation. When algae reach their maximum growth phase, they flourish for a period and then die. This is known as an algal "crash". After a crash or periodical collapse of algal populations, the decomposition of these dead algae utilizes a large amount of oxygen and can cause oxygen deficits and increased concentration of toxic ammonia. Insufficient oxygen and high ammonia concentrations may, in turn kill aquaculture species; promote disease; and/or temporarily reduce the feeding and growth rates of fish or prawns.

BGA and Bloom Formation

A bloom is considered to be a sudden increase in the population of a microalgae that has encountered suitable conditions for growth, that, together with their adaptive strategies (*i.e.* migration, active swimming) and the appropriate physical conditions, can reach concentrations of 104–105 cell l^{-1} during certain period of time commonly 1–3 weeks (Maso and Garces, 2006). Although many species of freshwater algae proliferate quite intensively in eutrophic waters, they do not accumulate to form blooms of extremely high cell density, as do some BGA. In freshwater, BGA have been reported from sea level to high altitudes. They have a number of special properties which determine their relative importance in phytoplankton communities and have competitive dominance over other groups.

BGA, along with chlorophyll *a* as a major pigment, also contain other pigments such as the phycobiliproteins which include allophycocyanin (blue), phycocyanin (blue) and sometimes phycoerythrine (red). These pigments harvest light in the green, yellow and orange part of the spectrum

(500-650 nm) which is hardly used by other phytoplankton species. The phycobiliproteins, together with chlorophyll *a*, enable cyanobacteria to harvest light energy efficiently and to live in an environment with only green light. The accessory pigment c-phycocyanin allows growth to occur at low irradiance (Scheffer *et al.*, 1997). Their cellular gas-vacuoles provide buoyancy and allow certain taxa to bloom at the water surface (Reynolds *et al.*, 1987). The optimum TN (Total Nitrogen): TP (Total Phosphorus) ratios for eukaryotic algae (16-23 molecules N: 1 molecule of P) with the optimum rates for bloom-forming cyanobacteria (10-16 molecules N: 1 molecule P), showed that the ratio is lower for cyanobacteria (Smith and Bennett 1999).

Other potential causal factors are high pH and scarcity of free CO_2 (Dokulil and Teubner, 2000). During intense blooms photosynthetic activity depletes free CO_2 from water and pH rises especially in lakes. The low CO_2 stimulates the dominance of BGA which can move to the air-water interface where CO_2 is most available, shading other algae in the process (Paerl and Ustach, 1982). Dying and lysing cells release their contents into the water, where pigments may adopt a copper-blue colour. Bacterial decomposition leads to rapid putrefaction of the material.

Different species have different temperature optima for growth which influences the outcome of competition and its effect on dominance (Chu *et al.*, 2007). The warmer water favours the toxic BGA over other forms of algae (Coles and Jones, 2000). In temperate lakes and ponds, the BGA generated blooms often occurs when the temperature rises above 20°C when there is depletion of dissolved inorganic N and free CO_2 from the water and this pattern also occurs in subtropical waters (Havens 2008). With increase in temperature of water bodies above the climatic range of the local ecosystem, dominance of algal classes shifts from diatoms to green algae to BGA. Concomitantly, biodiversity and biomass increases to an upper limit at around 20-22°C, both decreasing at further higher temperatures.

BGA and their Toxins

Freshwater BGA are known to produce secondary metabolites with toxic properties to mammals but also to aquatic invertebrates. Cyanobacteria toxins may be divided into three groups according to their effects on mammals: irritants, neurotoxins and hepatotoxins.

Irritants are the toxins that may be considered least harmful to humans, with the exception made of toxins produced by *Lyngbya majuscula* in tropical waters, as these may cause severe dermatitis and are tumour promoters (Moikeha and Chu, 1971). Neurotoxins compose a group of toxins with a diverse chemical composition, although they produce similar symptoms, examples are anatoxin- a, anatoxin-a(s), saxitoxins and neosaxitoxins. The biochemical effects and the chemical structure differ among the three major groups. The anatoxin-a is an alkaloid, anatoxin-a(s) is a natural organophosphate and it is the only naturally produced organophosphate known date. Saxitoxins are alkaloids that may vary in their composition according to changes in radicals (Sivonen and Jones, 1999). These toxins inhibit the impulse generation in peripheral nerves and skeletal muscles. Animals die of respiratory arrest (Kuiper-Goodman *et al.*, 1999).

Hepatotoxins may be divided in three groups: microcystins, nodularins and cylindrospermopsins, according to their chemical structure. The first two groups are peptides while cylindrospermopsins are alkaloids. Microcystins are cyclic heptapeptides composed of common aminoacids such as alanine, arginine, tryptophane, leucine and of a special amino acid only found in these toxins and in nodularins called ADDA. ADDA is a (2S,3S,8S,9S)-3-amino-9-methoxy-2,6,8-trimethyl- 10-phenyldeca-4,6-dienoic acid and it is the structure responsible for the toxic properties of microcystins and nodularins.

Nodularins have a structure similar to that of microcystins except for the number of amino acids. Nodularins are composed of five amino acids, being ADDA one of them.

Cylindrospermopsins are alkaloids produced by *Cylindrospermopsis, Umezakia* or *Aphanizomenon* (Sivonen and Jones, 1999). This toxin induces pathological changes not only in the liver but also in kidneys, spleen, thymus and heart (Hawkins *et al.*, 1985, 1997). These toxins inhibit glutathione synthesis and protein synthesis (Runnegar *et al.*, 1994; Terao *et al.*, 1994).

Effects of Cyanotoxins on Freshwater Aquatic Community

Cyanobacteria toxins may have lethal effects on many aquatic or terrestrial organisms but others are less sensitive to intoxication. The growth of bacterial species like *Bacillus cereus, Mycobacterium smegmatis, M. phley, M. balnei, Gaffkya tetragena, Sarcina lutea* and *Staphylococcus aureus* are inhibited by extracts of BGA *Lyngbya majuscula* (Moikeha and Chu 1971). However, the extracts of *Microcystis aeruginosa* stimulate the growth of *Escherichia coli* and *Streptococcus faecalis* at the concentrations that cause the cytolethic effects on mammals (Grabow *et al.*, 1982) and it is non-toxic to *Bacillus subtilis, Staphylococcus aureus, E. coli* and *Pseudomonas hydrophila* (Foxall and Sasner 1988). These data suggest that some aquatic bacteria are able to degrade cyanotoxins.

The toxic metabolites of the freshwater BGA *Lyngbya majuscula* inhibit the development of fungi like *Penicillium, Candiada albicans* and *Cryptococcus neoformans* (Welch, 1962). Similarly, the extracts of *Microcystis aeruginosa* inhibit the growth of *Candida* species.

The BGA blooms in freshwater systems are known to cause the disappearance of phytoplankton groups other than the BGA or attain only very low densities which suggest that BGA toxins might inhibit other phytoplankton. The reproduction of alga *Cryptomonas* is inhibited by *Oscillatoria* sp. There are scant conclusive work in this field regarding such studies. The toxins produced by the BGA are known to produce the deleterious effects on zooplankton (Infante and Riehl, 1984). The toxins of *Gloeotrichia echinulata* and of *Fisherella epiphytica* are lethal to the protozoan *Paramoecium caudatum* (Ransom *et al.*, 1978). It has been observed that during blooms of *M. aeruginosa*, rotifers were usually at high density (Fulton and Pearl, 1987).

On Invertebrates

The impact of cyanobacteria on aquatic macroinvertebrates has not been studied very intensively. Eriksson *et al.* (1989) showed that freshwater mussel *Anodunta cygnea* may accumulate high levels of microcystins from *Oscillatoria agardhii* without suffering adverse effects. The mussel *Mytilus galloprovincialis* and crayfish are quite resistant to microsystin (Vasconcelos, 1995; Vasconcelos *et al.*, 2001).

On Fish

Mass mortality of fish is known due anoxia and toxins released by BGA during and after the bloom collapse (Ayles *et al.*, 1986). The chief species of BGA causing fish mortality due to their toxins are *Aphanizomenon flos -aquae, Anabaena flos-aquae* and *Microcystis aeruginosa.* The microcystin toxins also affect the development of fish embryos (Oberemm *et al.*, 1997) and gets bioaccumulated into muscle of carps, mullet or barbells (Vasconcelos, 1999). The amount of toxins detected in fish flesh is not very significant in terms of human health except for the populations with a diet based on fish. Levels of microcystins reached 250 mg/g in the edible parts while in the viscera values may be ten to hundred fold higher.

On Birds

The BGA toxins are known to cause the mortality of birds also (Bossenmaier *et al.*, 1959, Senior 1960). Researches indicate that the microsystins and anatoxin-a are the major agents contributing to the death of flamingos (Andicoberry *et al.*, 2002). Though the flamingos feed on diatoms and BGA like *Arthrospira*, but shift from BGA bloom composition due to human intervention towards toxigenic genera like *Anabaenopsis* and *Anabaena* sp. present the poisoning hazards. The toxic BGA may also enter the birds' body through drinking of water infused with mats of toxin BGA composed of *Oscillatoria* and *Phormidium* sp.

Mammals

There have been many reports of cattle mortality due to toxic BGA (Francis, 1878; Siegelman *et al.*, 1984). The toxic symptoms include loss of equilibrium, muscle trembling, weakness, diarrhoea, convulsions and finally death (McDonald, 1960; Konst *et al.*, 1965; Aziz, 1974). The toxic Anabaena flos-aquae and Aphanizomenon flos-aquae may cause muscle paralysis and respiratory arrest in a short period of time (Mc Leod and Bondar, 1952). There are reports of death of insenctivorous bats in Canada due to antoxin – a; and rhinoceros in African Game Reserve due to consumption of water infested with blooms of *Microcystis aeruginosa*. Mass wildlife mortality due to BGA have been reported in Donana National Park Spain (Lopez-Rodas *et al.*, 2008).

Management Strategies

Physical Methods

1. Artificial mixing techniques have been widely used in protection of water quality in reservoirs (Kirke, 2000). The mixing of water (with help of lift pumps, bubble plumes *etc.*) take the algae beyond the conventional photic depth and dilute their exposure to light field.
2. The barley straw has been shown to be reliably effective in preventing the growth of nuisance BGA (Barrett *et al.*, 1999; Wisniewski, 2002). It is the phenolic and ester compounds released during the rotting of the barley straw are the chief components which produce the allelopathic effect on the growth and reproduction of the BGA (Choe and Jung, 2002).
3. Generally, 450 pounds per acre is recommended for muddy lakes or lakes with extreme algae problem.

Chemical Methods

1. The treatment of water bodies with algicide like copper sulphate. But, it kills the most species of phytoplankton and crustacean zooplankton. Increased aeration is required in copper treated ponds or aquaculture ponds. But the instances of copper poisoning and potential threats of long term copper application has led to banning of its use in many countries but it is still being practiced in India.
2. Alum is also used for treatment of bloom infested ponds to reduce the availability of dissolved phosphorus and organic matter. For a lake having dimension of 17 ha (hectare) and initial total phosphate content up to 0.065 ppm the required alum for the treatment should be 5.7 mg per l (Lewandowski *et al.*, 2003).
3. By reducing phosphorus load in water bodies by application of 'phosphorus' binding clays like 'Phoslock' which is modified bentonite clay or by clay minerals based on lanthanum (*e.g.* Lanthanum chloride) (Douglas *et al.*, 1999). When phoslock is applied as a slurry, it

moves down through the water column and up to 95 per cent of the Filterable Reactive Phosphate is rapidly removed and adsorbed onto the surface forming an insoluble complex within the clay structure which is no longer available for use by algae for assimilation.

Biological Methods

1. The "Top-down" control through biomanipulation was seen as an alternative to the traditional "Bottom-up" approach through reductions in nutrient inputs (Shapiro and Wright, 1984). The bloom of BGA fuelled by extra nutrients may be minimized by zooplankton grazers to control phytoplankton. It has been shown that *Daphnia carinata*, a cladocern, can effectively graze the phytoplankton and it is quite resistant to the certain BGA toxins (Matveev *et al.*, 1994; Matveeva and Matveeva, 1997). This technique of biomanipulation has been most successful in small, shallow lakes and ponds where planktivorous fish are absent or occur in low number and where phytoplankton production is sufficiently high to sustain large *D. carinata* populations. But this method is least effective so in larger, deeper lakes (Reynolds, 1994).
2. There have been relatively few studies of the effects of nutrients, nitrogen and phosphorus on macrophytes and uptake by them. In Australia, Dudley and Walker (2000) have highlighted the importance of *Ruppia megacarpa* in maintaining low nutrient concentration. The study showed that *Ruppia* is able to take up nutrients through its leaves. Thus, the macrophyte beds represent significant competitors for phytoplankton and healthy macrophytes decrease the likelihood of phytoplankton blooms.

Conclusion

It is obvious fact that the BGA blooms are formed by eutrophication of the water bodies which have multipronged lethal effects on biodiversity not only inside the water but also outside the water using the infested waterbody for different purposes. Detailed ecological studies based on bloom forming factors and its effect on aquatic biota have been conducted in various countries but in India still such type of studies based on the similar pattern in Indian water bodies are lacking. There are huge knowledge gaps about the status of the Indian freshwater bodies *vis-à-vis* aquatic biodiversity, eutrophication, bloom incidents and their effect on biodiversity. There are various models which have been designed to manage the nuisance of algal blooms in western countries but such types of modeling is required to be done in Indian perspective to be effectively materialized. We need to make the ecofriendly, long lasting techniques for managing waterbodies like destratificaiton and biomanipulation and to develop and trial novel management solutions to mitigate bloom effect of algae.

References

Abell, R. A., Olson, D. M., Dinerstein, E., Hurley, P. T., Diggs, J. T., Eichbaum, W., Walters, W., Wettengel, W., Allnutt, T., and Loucks, C. J. 2000. *Freshwater Ecoregions of North America: A Conservation Assessment*. Washington DC andCovelo, California, Island Press.

Alonso-Andicoberry, C., Garccia-Villada, L., Lopez-rodas, V., and Costas E., 2002. Catastrophic mortality of flamingos in a Spanish national park caused by cyanobacteria. *Veterinary Record*. 151, pp. 706-707.

Andersen R.A. 1992. Diversity of eukaryotic algae. *Biodiversity and Conservation*. 1: 267-292.

Ayles, G.B., Lark, J.G.I., Barica, J. and Kling, H. 1986. Seasonal mortality in rainbow trout (*Salmo gairdneri*) planted in small eutrophic lakes of Central Canada. *J. Fish. Res. Board Can.* 37: pp. 906-919.

Aziz, K.M.S. 1974. Diarrhoea toxin obtained from a water bloom-producing species, *Microcystis aeruginosa* Kutzing. *Science,* 183: pp. 1206-1207.

Barrett, P.R.F., Littlejohn, J.W., and Curnow, J. 1999. Long-term algal control in a reservoir using barley straw. *Hydrobiologia.* 415: pp. 309–313.

Bossenmaier, E., Olson T.A., Rueger M.E. and Marshall, W.H. 1959. Some field and laboratory aspects of duck sickness at White water Lake, Manitoba. *Nineteenth North American Wildlife* Conference: pp.163-175.

Cardinale, B. J., Palmer, M. A., and Collins, S.L. 2002. Species diversity enhances ecosystem functioning through interspecific facilitation. *Nature.* 415. pp. 9-426.

Choe, S. and I. H. Jung. 2002. Growth inhibition of freshwater algae by ester compounds released from rotted plants. *Journal of Industrial and Engineering Chemistry.* 8 (4): pp. 297-304

Chu, Z., Jin, X., Iwami, N., and Inamori, Y. 2007. The effect of temperature on growth characteristics and competitions of *Microcystis aeruginosa* and *Oscillatoria mougeotii* in a shallow, eutrophic lake simulator system. *Hydrobiologia.* 581: pp. 217–230.

Coles, J. F. and Jones, R. C. 2000. Effect of temperature on photosynthesis-light response and growth of four phytoplankton species isolated from a tidal freshwater river. *Journal of Phycology.* 36: pp. 7–16.

Desikachary, T.V. 1959. Cyanophyta. Indian Council of Agricultural Research, New Delhi. pp. 686.

Dokulil M.T., and Teubner K. 2000. Cyanobacterial dominance in lakes. *Hydrobiologia.* 438: 1–12.

Douglas, G. B., Adeny J. A. and Robb, M. 1999. A novel technique for reducing bio-available phosphorus in water and sediments. In Barber, C., B. Humphries and J. Dixon (eds), International Conference On Diffuse Pollution. CSIRO Land and Water, Perth, Australia, 517–523.

Dudley, B. J. and Walker, D. I. 2000. Plant/water nutrient dynamics in a southwestern Australian estuary. *Biologia Marina Mediterranea.* 7: pp.43–46.

Eriksson, J.E., Meriluoto J.A. and Lindholm, T. 1989. Accumulation of peptide toxin from the cyanobacterium *Oscillatoria agardhii* in the freshwater mussel *Anodonta cygnea. Hydrobiologia.,* 183: pp. 211-216.

Francis, G. 1878. Poisonous Australian lake. *Nature,* 18: pp.11-12.

Foxall, T.L. and Sasner, J.J., 1988. Effect of a hepatic toxin from the cyanophyte *Microcystis aeruginosa.* In: *The Water Environment. Algal Toxins and Health* W.W. Carmichael (ed). Plennun Press, New York.

Fulton, R.S. and Pearl, H.W. 1987. Toxic and inhibitory effects of the blue-green alga *Microcystis aeruginosa* on herbivorous zooplankton. *J. Plankton Research.* 9: pp. 837-855.

Geitler, L. 1932. Cyanophyceae in Rabenhorst's Kryptogamen flora, Leipzig 14. 1196p.

Grabow, W.O., Du Randtprozesky W.C. and Scott, W. E. 1982. *Microcystis aeruginosa* toxin: cell culture toxicity, hemolysis and mutagenicity assays. *Applied and Environmental Microbiology.* 43: 1425-1433.

Havens, K.E. 2008. Cyanobacteria blooms: effects on aquatic ecosystems. *Adv. Exp. Med. Biol.* 619, pp.733-747.

Hawkins, P.R., M.T.C. Runnegar, A.R.B. Jackson and I.R. Falconer. 1985. Severe hepatotoxicity by the tropical cyanobacterium (blue-green algae) *Cqilindrospermopsis racihorskii* (Woloszynska) Seenaya and Subba Raju isolated from a domestic water supply reservoir. *Appl. Environ. Microbiol.*, 50: 1292- 1295.

Hawkins, P.R., Chandrasena N.R., Jones G.J., Humpage A.R. and Falconer I.R. 1997. Isolation and toxicity of *Cylindrospermopsis raciborskii* from an ornamental lake. *Toxicon.*, 35: pp. 341-346.

Infante A. and Riehl, W. 1984. The effect of Cyanophyta upon zooplankton in a eutrophic tropical lake (Lake Valencia, Venezuela). *Hydrobiologia.* 113: pp. 293-298.

John D.M. 1994. Biodiversity and conservation: An algal perspective. *The Phycologist.* 38: pp. 3-15.

Kirke, B.K. 2000. Circulation, Destratification, Mixing and Aeration: Why and How? *J. Aust. Water Asso.* 27: (4). pp. 24-30.

Konst, H., Mckercher P.D., Gorham, P.R., Robertson A. and Howell J.1965. Symptoms and pathology produced by toxic *Microcystis aeruginosa* NRC- 1 in laboratory and domestic animals. *Can. J. Comp. Med. Vet. Sci.*, 29: pp. 221-228.

Kuiper-Goodman, T., Falconer, I. and Fitzgerald, J., 1999. Human Health Aspects. In: *Toxic Cyunohacteria in Water.* I. Chorus and J. Bartram (eds.): 1 13- 154, Who and E and FN Spon, London.

Lewandowski, I., Schauser, I. and Hupfer, M. 2003.Long term effects of phosphorus precipitations with alum in hypereutrophic Lake Susser See (Germany). *Water Research.* 37(13): pp. 3194-3204.

Lopez-Rodas, V., Maneiro, E., Lanzarot, M.P., Perdigones, N., and Costas, E. 2008. Mass wildlife mortality due to cyanobacteria in Donana National Park, Spain. *Veterinary Record.* 162. pp. 311-315.

Loreau, M., Naeem, S., Inchausti, P., Bengtsson, J., Grime, J. P., Hector, A., Hooper, D. U., Huston, M. A., Raffaelli, D., Schmid, B., Tilman, D., and Wardle, D. A. 2001. *Biodiversity and ecosystem functioning: current knowledge and challenges.*: Stanford, California, High Wire Press (Stanford University Libraries) pp.806.

Maso, M. and Garces, E. 2006. Harmful microalgae blooms (HAB); problematic and conditions that induce them. *Marine Pollution Bulletin.* 53: 620-630.

Matveeva, V., Matveeva, L. and Jones, G. J. 1994. Study of the ability of *Daphnia carinata* King to control phytoplankton and resist cyanobacterial toxicity: implications for biomanipulation in Australia. *Australian Journal of Marine and Freshwater Research.* 45: pp. 889–904.

Matveeva, V. and Matveeva, L. 1997. Grazer control and nutrient limitation of phytoplankton biomass in two Australian reservoirs. *Freshwater Biology.* 38: pp. 49–65.

Mcdonald, D.W. 1960. Algal poisoning in beef cattle. *Can. Vet. J.*, 1: pp. 108-110.

Mcleod J. A. and Bondar G.F. 1952. A case of suspected algal poisoning in Manitoba. *Can. J. Pub. Health.* 43: pp. 347-350.

Moikeha, Chu, S.N. and Chu G.W. 1971. Dermatitis-producing alga *Lyngbya majuscula* Gomont in Hawaii. 11. Biological properties of the toxin factor. *J. Phycol.*, 7: pp. 8-13.

Oberemm, A., Fastner, J. and Steinberg, C. 1997. Effects of microcystin-LR and cyanobacterial crude extracts on embryo-larval development of zebrafish (*Danio rerio*) *Wat. Res.*, 31: pp. 2918-2921.

Paerl H.W. and Ustach, J.F. 1982. Blue–green algal scums: an explanation for their occurrence during freshwater blooms. *Limnology and Oceanography*. 27: pp. 212–217.

Ransom, R.E., Nevad, T.A. and Meier, P.G. 1978. Acute toxicity of the blue-green algae to the protozoan *Puramecium caudatum. J. Phycol.*, 14: pp. 114-116.

Reynolds C.S., Oliver R.L., and Walsby A.E. 1987. Cyanobacterial dominance: the role of buoyancy regulation in dynamic lake environments. *New Zealand Journal of Marine and Freshwater Research*. 21: pp. 379–390.

Reynolds, C. S., 1994. The ecological basis for successful biomanipulation of aquatic communities. *Arch. Hydrobiol.* 130: pp. 1–33.

Runnegar, M.T.C., Kong, S.M., Zhong, Y.Z., Ge, J.L. and Lu, S.C. 1994. Inhibition of reduced glutathione in the toxicity of a novel alkaloid cylindrospermopsin in cultured rat hepatocytes. *Biochem. Biophys. Res. Commun.*, 201: pp. 235-241.

Scheffer M, Rinaldi S, Gragnani A., Mur L.R., and van Nes E.H. 1997. On the dominance of filamentous cyanobacteria in shallow turbid lakes. *Ecology*. 78: pp. 272–282.

Schindler, D.W. 1977. Evolution of phosphorus limitation in lakes. *Science*. 196. pp. 2-26.

Senior, V.E. 1960. Algal poisoning in Saskatchewan. *Can. J. Comp. Med.*, 24: pp. 36-40.

Siegelman, H.W., Adams, W.H., Stoner, R.D. and Slatkin, D.N. 1984. Toxins of *Microcystis aeruginosa* and their haematological and histopathological effects. *American Chemical Society Symposium on Seafood toxins*, pp. 407-413.

Sivonen K. and Jones, G. 1999. Cyanobacterial toxins. In *Toxic Cyunobacteria in Water:* (I. Chorus and J. Bartram, eds.): pp. 41-111, Who and E and FN Spon, London.

Shapiro, J. V. and Wright, D. I. 1984. Lake restoration by biomanipulation: Round Lake, Minnesota, the first two years. *Freshwater Biology*.14: pp. 371–383.

Smith V.H. and Bennet S.J. (1999). Nitrogen Phosphorus supply ratios and phytoplankton community structure in lakes. *Archiv fur Hydrobiologie*. 146: pp. 37–53.

Terao, K.S., Ohmori, K., Igarashi, I., Ohtani, M., Watanabe, K., Harada, E. and Watanabe, M. 1994. Electron microscopic studies on experimental poisoning in mice induced by cylindrospermopsin isolated from blue-green alga *Umezakia natuns. Toxicon.*, 32: pp. 833-843.

Van den Hoek, C., D. G. Mann and H. M. Johns. 1995. *Algae: an Introduction to Phycology*. Cambridge University Press, Cambridge, pp. 623

Vasconcelos, V.M. 1995. Uptake and depuration of the peptide toxin microcystin-LR in the mussel *Mytilus galloprovinciallis. Aquatic Toxicology*, 32: pp. 227-237.

Vasconcelos, V.M. 1999. Cyanobacteria toxins in Portugal: effects on aquatic animals and risk for human health. *Brazilian Journal of Medical and Biological Research*, 32: pp. 249-254.

Vasconcelos, V.M. 2001. Cyanobacteria toxins: diversity and ecological effects. *Limetica*. 20 (1): pp. 45-58.

Waterbury, J.B., Watson, S.W., Guillard, R.R.L. and Brand, L.E., 1989. Widespread occurrence of a unicellular marine, planktonic cyanobacterium. *Nature*. 277: 293-294.

Welch, A.H. 1962. Preliminary survey of fungistatic properties of marine algae. *J. Bacteriology*. 83: pp. 97-99.

Whitton, B. A., and Potts, M. 2000. Introduction to the cyanobacteria. In *The ecology of cyanobacteria: Their diversity in time and space*. B. A. Whitton, and M. Potts (eds). Dordrecht, Kluwer Academic Publishers, pp.1-11.

Williams, W.D. 1981. Inland salt lakes: An Introduction. *Hydrobiologia*, 81: 1-14.

Wisniewski, R. 2002. Attempts to eliminate cyanobacterial blooms in Lake Leasinskie. Environment Protection Engineering. 1: pp. 15-26.

2013, Sustainable Approaches for Environmental Conservation *Pages* **91–96**
Editors: **D.R. Khanna, A.K. Chopra, R. Bhutiani, Gagan Matta & Vikas Singh**
Published by: **BIOTECH BOOKS, NEW DELHI**

Chapter 13

Moulds Intercepted on Seed and Dal of Lathyrus Collected from Severe Flooded Areas of Kalamna Grain Market of Nagpur

Gajanan Mate[1], *Sanjeev Charjan*[1], *Umesh Kakade*[2], *Rajesh Gadewar*[3] *and Ashish Lambat*[3]

[1]*Dr. P.D.K.V's College of Agriculture, Nagpur, M.S.*
[2]*Ismail Yusuf College, Mumbai, M.S.*
[3]*Sevadal Mahila Mahavidyalaya, Nagpur, M.S.*

The study was conducted to determine the occurrence of moulds on seed and dal of lathyrus collected from severe flooded area of Kalamna grain market of Nagpur. Seed and dal samples of lathyrus collected in polyethylene bags. In all seed and dal samples, 30 fungal organisms were isolated belonging to 2 genera of fungi. Out of them *Aspergillus flavus* and *A. stellatus* were abundant and *A. niger* and *Rhizopus* sp. were frequently observed while *A. amestelademi*, *A. tamari* and *Drechslera halodes* were less frequent. Rest of the fungi showed their occurrence. The *A. flavus* dominated over other fungi. Among *Penicillia*, *P. islandium* was most commonly recorded.

The presence of *Aspergilli*, *Penicillia* and *Fusarium* showed that there were higher moisture contents in seed and dal as these fungi are moisture loving. All these fungi are mycotoxin producing fungi. These fungi contaminate all seed and dal lots in godown and spread and create the infection which causes health hazard problem in human being and animals. Therefore, the consumption of infected seed and dal of lathyrus has to be avoided for consumption purposes.

Keywords: Moulds, Dal of lathyrus, Mycotoxin, Godown.

Introduction

Nagpur is one of the most famous centres for orange and food grains. The lathyrus is valuable pulse crop important for its high protein content, starch and vitamins and considered of a great nutritive value which is used in our daily diet as a dal. The moisture content plays an important and decisive role for the attack of moulds which causes deterioration and degradation in grain quality and storability. The storage moulds impose health hazard problem in human beings and animals (Christensen and Kafman, 1969; Betina, 1989, Charjan *et al.*, 2006, Charjan *et al.*, 2010, Lambat *et al.*, 2011, Gadewar *et al.*, 2011, Charjan *et al.*, 2011). The present study deals with the moulds intercepted on seed and dal. Samples of lathyrus collected from severe flooded areas in Nagpur grain market, Kalamna during the month of August, 2010, when almost area of Nagpur City was submerged under water of the Nag river.

Materials and Method

Ten seed and dal samples of lathyrus were collected separately in aerated polythene bags from severe flooded areas of Nagpur grain market, Kalamna. Four hundreds seeds and dals were plated out of equidistance and examined for the occurrence of moulds following standard moist blotter techniques (Anonymous, 1985). The antibiotic antibacterial substance like streptomycin was used in medium for the inhibition of bacterial colonies. The plates were incubated at 28 ± 1°C for 7 days. The fungi developed on seeds were identified with the help of standard monographs (Barnett and Hunter, 1972, Ellis, 1971; Domesch *et al.*, 1980).

Results and Discussion

In all 44 fungal isolates were recorded and categorized under 16 genera of fungi (Table 13.1). Out of them, 10 sp. of *Aspergillus* (*A. flavus, A. fumigatus, A. japonicus, A. nidulans, A. niger, A. ostianus, A. stellatus, A. tamari, A. tenuis* and *A. ustus*); 2 sp. *Alternaria* (*A. alternata* and *A.lunata*); 3 sp. of *Cladosporium* (*C. cladosporioides, C. sphacrosphermum* and *C. tenuissimum*); 3 sp. of *Drechslera* (*D. gustraliensis, D. hawaiensis* and *D. specifer*), 4 sp. of *Fusarium* (*F. equiseti, F. moniliforme, F. oxysporum* and *F. semitectum*), 8 sp. of *Penicillium* (*P. chrustosum, P. citrunum, P. chrysoginum, P. cyclopium, P. griseofulvum, P. islandium, P. lividium* and *P. simplicisssimum*) and 3 sp. of *Rhizopus* (*R. nigricans, R. stolonifer* and *R. arrhizus*). Other fungi found were *Cephalesporium* sp., *Chaetomium globosum, Curvularia lunata, Helminthosporium tetramere, Macrophomina phaseolina, Mucor* sp., *Neurospora sitophila, Rhizoctonia bataticola* and *Verticillium* sp.

The percentage occurrence of each fungus was calculated and finally the results are presented in Table 13.1. In the present study the *Aspergilli* were dominant over other fungi, interestingly in *Aspergilli, Aspergillus flavus* and *Aspergillus niger* was present in all samples of seed and dals with maximum average value while *Aspergillus fumigatus* was found minimum. Dominance of *Aspergilli* on stored grains has been reported earlier by Deo *et al.* (1980), Charjan and Tarar (1992), Charjan and Tarar (1993), Charjan and Tarar (1994), Charjan and Gupta (1996), Charjan *et al.* (2006), Charjan *et al.* (2010), Charjan *et al.* (2011)

Moreover, *Penicillia, Rhizopus, Fusarium, Drechslera* and *Cladosporium* were next to *Aspergilli. Cephalesporium* sp., *Colletotrichum truncatum, Curvularia lunata, Macrophomina phaseolina* was present in only one sample. Similar types of fungi were also isolated on different crop seeds by Giridhar and Reddy (2001), Reddy (2006), Sinha (2007), Agrawal (2007), Mohd. Shaker (2010), Charjan *et al.* (2011) on raisins, gram, blackgram, greengram, some pulses and soybean seeds respectively.

Table 13.1: Incidence per cent of Fungi on Seeds and Dals of Lathyrus Collected from Flooded Area Kalamna Grain Market of Nagpur.

Sl.No.	Name of Fungi	Average per cent	T.P.I.	S.L.I.	P.S.L.I.
A1	*Aspergillus flavus*	36.8	368.0	10	100.00
A 2	*A. fumigatus*	0.40	4.0	2	20.00
A 3	*A. japonicus*	2.00	20.0	4	40.00
A 4	*A. niduians*	1.00	10.0	2	20.00
A 5	*A. niger*	28.20	282.0	10	100.00
A 6	*A. ostianus*	2.80	28.0	3	30.00
A 7	*A. stellatus*	7.90	79.0	8	80.00
A 8	*A. tamari*	5.00	50.0	5	50.00
A 9	*A. tenuis*	1.70	17.0	2	20.00
A 10	*A. ustus*	2.50	25.0	3	30.00
A 11	*Alteranaria alternata*	1.00	10.0	2	20.00
A 12	*A. lunata*	1.70	17.0	2	20.00
B 13	*Cephalesporium sp.*	0.80	8.0	1	10.00
C 14	*Chaetomium globosum*	1.20	12.0	2	20.00
D 15	*Cladosporium cladosporioides*	0.50	5.0	2	20.00
D 16	*C. sphacrosphermum*	0.70	7.0	2	20.00
D 17	*C. tenuissimum*	0.80	8.0	2	20.00
E 18	*Colletotrichum truncatum*	0.50	5.0	1	10.00
E 19	*C. truncatum*				
F 20	*Curvularia lunata*	0.60	6.0	1	10.00
G 21	*Drechslera australiensis*	0.40	4.0	2	20.00
G 22	*D. hawaiensis*	1.20	12.0	3	30.00
G 23	*D. specifer*	0.10	1.0	2	20.00
H 24	*Fusarium equiseti*	0.40	4.0	2	20.00
H 25	*F. moniliforme*	0.50	5.0	2	20.00
H 26	*F. oxysporum*	0.50	5.0	2	20.00
H 27	*F. semitectum*	0.30	3.0	3	30.00
H 28	*Helminthosporium tetramera*	1.30	13.0	2	20.00
J 29	*Macrophomina phaseolina*	0.10	1.0	1	10.00
K 30	*Mucor sp.*	0.60	6.0	2	20.00
L 31	*Neurospora sitophila*	1.00	10.0	2	20.00
M 32	*Penicillium chrustosum*	0.80	8.0	2	20.00
M 33	*P. citrunum*	0.40	4.0	5	50.00
M 34	*P. chrysoginum*	0.50	5.0	3	30.00
M 35	*P. cyclopium*	1.00	10.0	2	20.00
M 36	*P. griseofulvum*	0.40	4.0	2	20.00

Contd...

Table 13.1–Contd...

Sl.No.	Name of Fungi	Average per cent	T.P.I.	S.L.I.	P.S.L.I.
M 37	*P. islandium*	0.30	3.0	2	20.00
M 38	*P. lividium*	0.20	2.0	2	20.00
M 39	*P. simplicissimum*	0.30	3.0	4	40.00
N 40	*Rhizoctonia bataticola*	1.00	10.0	2	20.00
O 41	*Rhizopus nigricans*	2.10	21.0	4	40.00
O 42	*R. stolonifer*	5.0	50.0	6	60.00
O 43	*R. arrhizus*	5.50	55.0	5	50.00
P 44	*Verticillium sp.*	2.0	20.0	2	20.00

T.P.I.: Total Percentage Incidence; S.L.I.: No. of Seed Lots Infested; P.S.L.I.: Percentage Seed Lots Infested; 75-100 per cent: Abundant; 50-74 per cent: Frequent; 25-49 per cent: Moderate and below 25 per cent: Poor.

The fungal percentage seed lot infected was also considered and found that *Aspergillus flavus, Aspergillus niger* and *Aspergillus stellatus* were abundant, and *Aspergillus tamari, Rhizopus arrhizus, Rhizopus stolonifer* and *Penicillium citrunum* were frequently present in the seed and dal samples, while *Aspergillus japonicus, A. ostianus, A. ustus, Drechslera hawaiensis, Fusarium semitectum, Penicillium chrysogenum, P. simplicissimum* and *Rhizopus nigricans* were moderately found and rest of fungi were poorly observed.

Further it was observed from the experiment that the presence of *Aspergilli, Penicillia, Rhizopus, Fusarium, Drechslera* and *Colletotrichum* showed that there were higher moisture contents in grains and dals as these fungi are moisture loving. All the above said are mycotoxin producing fungi. Natural contamination and storage deterioration of starchy, oily and proteinaceous food commodities by mycotoxins has been reported earlier by several workers (Prasad *et al.*, 1984, Giridhar and Reddy, 2001). These fungi contaminated all the seed lots in godowns and spread and create the infection in other agricultural commodities. Association of the moisture loving fungi with seeds are found to be harmful to the seed health and seed content, such mouldy seeds are known as spoiled or deteriorated and the process is termed as biodeterioration. The fungal attack on the pulses seeds lead to hydrolytic breakdown of carbohydrate, protein and lipid content of the seed at different levels. The hydrolytic cleavage is correlated with potentiality to the associated seed borne fungi for their extracellular enzyme make up.

The results suggested that it is desirable to avoid lathyrus seed lot with higher percentage of moisture (flood affected) for consumption purposes. Because above moisture loving fungi are mycotoxin producer causes health hazard problem in human beings and animals.

References

Anonymous (1985). International rules for seed testing. *Seed Sci. and Technol.* 13: 299-513.

Agrawal V. K. (2007). Techniques for the detection of seed borne fungi. *Seed Res.* 4:24-31.

Barnett, H. L. and Hunter, B. B. (1972). *Illustrated genera of imperfect fungi*, Burgess Publishing Company, USA.

Betina, V. (1989). *Mycotoxins chemical, biological and environmental aspects.* Elsevier, Amsterdam, Oxford, New York.

Christensen, C.M. and Kafmann, M. H. (1969). Deterioration of Stored grains by fungi. *Ann. Rev. Phytopath.* 3: 69-84.

Charjan, S.K.U. and Tarar, J. L. (1992). Effect of storage on germination and microflora of soybean seed. *Indian J. Agric. Sci.* 62(7): 500-502.

Charjan, S.K.U. and Tarar, J. L. (1993). Studies on seed quality parameters and mycoflora associated with bold and shriveled seeds of greengram. *Biojournal*, 5:71-72.

Charjan, S.K.U. and Tarar, J. L. (1994). Viability, Vigour and mycoflora changes in pulse beetle damage seeds of greengram. *Proc. Nat. Acad. Sci.* 64(B)I: 95-98.

Charjan, S.K.U. and Gupta, V. R. (1996). Impact of storage condition on fungal flora and germinability of gram seeds. *J. Soils and Crops.* 6(2): 136-138.

Charjan, S.K.U., Wankhede, S. R., Tayade, K. G. (2006). Viability, vigour and mycoflora changes in pulse beetle damaged seeds of cowpea. In *proceedings of National Seminar on economics of sustainability of dry land agriculture*, India, 5 to 6 January 2006: 5-6.

Charjan, S.K.U., Patil, A.E. and Udasi, R. N. (2010). Seed quality and mycoflora associated with bold and wrinkled seed of cowpea. *J. Phytol. Res.* 23(2):379-380.

Charjan, Sanjiv, Mohod Vandan, Rokade Shashikant, Gadewar Rajesh, Lambat Ashish and Charde Pravin (2011). Effects of seed size and storage time on soybean seed qualities. In *Proc. International Conference in Agricultural Engineering (Post Harvest Technology)*, Thailand, 31 March-1st April, 2011: 1-6.

Deo, P. P., Gupta, J. S. and Gupta Pushpa (1980). Moulds intercepted on seed and dal of lentil collected from severe flooded areas in Agra mandi. *Seed and farms* Nov. 1980: 11-12.

Domesch, K., Garms, W. and Anderson, T. H. (1980). *Compendium of soil fungi.* Academic press, New York.

Ellis, M. B. (1971). *Dematiaceous Hyphomycetes.* Commonwealth Mycological Institute, Kew, Surrey, England.

Gadewar, Rajesh, Lambat, Ashish, Charjan, Sanjiv, Charde, Pravin, Cherian, Konglath and Lambat Prachi (2011). Post harvest losses in qualities of pegionpea seed due to pulse beetle infestation. In: *Proc. International Conference in agricultural engineering*, Thailand, 31 March – 1st April 2011: 14-20.

Giridhar, P. and Reddy, S. M. (2001). Incidence of mycotoxigenic fungi on raisins. *Ad. Plant Sci.* 14(1): 291-294.

Lambat, Ashish, Gadewar, Rajesh, Charde, Pravin, Charjan, Sanjiv, Cherian, Konglath and Lambat, Prachi (2011). Effect of post harvest operations on seed quality of urdbean. In *Proceeding of the International Conference in Agricultural Engineering (Post Harvest Technology)*, Thailand, 31 March-1 April, 2011. PP.14-20.

Mohd. Shaker, Momin, R.K. and Hashmi Seema (2010). Isolation and identification of some pulse mycoflora. *Bionano Frontier.* 2: 321-324.

Prasad, T., Bilgrami, K.S., Thakul, M. K. and Singh, A. (1984). Aflatoxin problem in some common spices. *J. Ind. Bot. Soc.* 63: 171-173.

Reddy, B. N. (2006). Evaluation of fungicides against seed borne fungi of blackgram. *Seed and Farms*, February, 2006: 162-163.

Sinha, M. K. (2007). Changes in the starch content of some fungal infected pulses. *Indian Phytopath* 34: 269-271.

2013, Sustainable Approaches for Environmental Conservation *Pages* **97–101**
Editors: **D.R. Khanna, A.K. Chopra, R. Bhutiani, Gagan Matta & Vikas Singh**
Published by: **BIOTECH BOOKS, NEW DELHI**

Chapter 14

Green Building: Concept to Practice in Indian Scenario

***Priya Gautam*[1], *G.C. Mishra*[1] *and K.R. Nisha*[2]**
[1]*Department of Civil Engineering, Lingaya's University, Faridabad, Haryana*
[2]*Department of Mathematics,*
Rawal Institute of Engineering and Technology, Faridabad, Haryana

Green construction refers to a structure and process that is environmentally responsible and resource-efficient throughout a building's life-cycle: from sitting to design, construction, operation, maintenance, renovation and demolition. This practice expands and complements the classical building design concerns of economy, utility, durability and comfort. Although new technologies are constantly being developed to complement current practices in creating greener structures, the common objective is that green buildings are designed to reduce the overall impact of the built environment on human health and the natural environment. Apart from above parameters few more considerations are taken into account for green building such as: Life cycle assessment (LCA), Structure design efficiency, Energy efficiency, Water efficiency, Materials efficiency, Indoor environmental quality enhancement, Operations and maintenance optimization, Waste reduction, Cost and payoff, Regulation and operation. Although these terms have different meaning and are poorly understood, several countries have adopted this vision as a long-term goal of their building energy policies. What is missing is a clear definition and international agreement on the measures of building performance that could inform "zero energy" building policies, programs and industry adoption.

The objective of this research paper is to develop a common understanding and to set up the basis for an international definition framework of Net Zero Energy Buildings (Net ZEBs) and to describe and evaluate existing ZEB definitions and also foresee the scope of such concepts in Indian scenario in terms of its feasibility from govt. policy point of view as well as economic viability by seeing the affordability of masses.

Keywords: *Green building, Zero energy buildings, Architects.*

Introduction

Green building development has gained momentum in India despite global economic slowdown, with its supply rising manifold in the last few years. Green buildings are designed to have a longer life-cycle and help conserve natural resources like water, while consuming minimal power and energy. "The concept of green building development in the country has witnessed a sustained momentum despite the overall weakening macroeconomic environment, witnessed in the last few years. The supply of green space has been contributed by various industrial segments but IT sector contributed the maximum at 58 per cent of the total supply in the country.

The Turbo Energy R&D and administrative block in Paiyanur, Chennai, has been certified by globally renowned Leadership in Energy and Environmental Design (LEED) as the greenest building in India. Other prominent green projects include ITC Green Centre (Gurgaon), CII-Godrej Green Business Centre (Hyderabad) and Kalpataru Square (Mumbai). Initiatives undertaken by occupiers and developers, as well as those by state and central governments, have driven this progress. Floor area in green buildings grew by over 50 per cent in 2009 despite an overall sluggish macro-economic environment.

Indian Green Building Council

The council is represented by all stakeholders of construction industry comprising of: Corporate, Government, and Nodal agencies, Architects, Product manufacturers, Institutions *etc.* The council operates on a consensus based approach and is member-driven. The vision of the council is to usher in a green building revolution and facilitate in India emerging as one of the world leaders in green buildings in coming years.

Leadership in Energy and Environmental Design (LEED)

It is an internationally recognized green building certification system, providing third-party verification that a building or community was designed and built using strategies intended to improve performance in metrics such as energy quality and stewardship of resources and sensitivity to their impacts. Savings, water efficiency, CO_2 emissions reduction, improved indoor environment. The LEED rating system has been popular in both public and private sector because of its consensus based, flexible approach of defining and measuring green building and development.

India's Green Building Code, Still Evolving

The code for green building are more voluntarily than specific, it is yet to evolve.

The recent acclivity witnessed in energy costs, has had an increasing number of countries, states and cities adopting policies and implementing laws that encourage or require new construction and existing buildings to be energy-efficient. Many of these make it mandatory to meet the terms set down by their green building councils. India is one of the fastest growing countries in terms of demand and growth for construction industry and infrastructural developments; with a mere 20,000 square feet in 2003 to more than 20 million square feet of green buildings in the recent times. However, it is important to note that the Green Building Code in India is a medley of codes, standards established by the State by-laws, National Building Code, the Energy Conservation Building Code (ECBC) in combination with the norms set by ratings programmes such as Leadership in Energy and Environmental Design-India (LEED-India), the standards and guidelines put down for the Residential Sector by the Indian Green Building Council (IGBC), TERI-GRIHA. The National Building Code (NBC), designed by the

Bureau of Indian Standards (BIS) is limited to offering basic and general guidelines for efficient energy usage.

Voluntary Laws

While it is true that the green building laws and codes established in India are voluntary in the absence of an explicit nationalized green building code LEED-India in fact sets down standards that have been customized according to Indian conditions in terms of the design, construction and operation.

LEED-India

LEED in India has also been at the receiving end of some faultfinding. It has been criticized for including concepts that are specifically western, especially in terms of air-conditioning usage and temperature control. This in turn has resulted in several changes being incorporated in the LEED India rating system, which is constantly being amended to address Indian environment and climate conditions. IGBC is attempting to indigenize LEED according to the environmental conditions prevalent in India. Together with IGBC Green Homes rating, which is perhaps the first rating programme exclusively created for Indian homes.

Rediscovery of the Indian Ethos

5 elements of Nature (Panchabhutas):

1. Prithvi (Earth) - Sustainable Sites
2. Jal (Water) -Water Efficiency
3. Agni (Fire) -Energy and Atmosphere
4. Vayu (Air) -Indoor Environmental Quality
5. Akash (Sky) -Daylight, Night Sky Pollution

India Surging Ahead

Green building movement, though in its nascent stage in India, has already put the country ahead. Presently LEED-India and IGBC offer certifications for two categories. The first includes the builder or the construction company that is developing a particular green site/building for a client, which will be leased out. The second category belongs to those who are undertaking a green initiative for their own personal use and this could either be a commercial space or corporate office. Standards vary for both.

Varied Perception

Ironically, India houses individual business units with contradicting mindsets. We have brand and environment-conscious firms that will leave no blueprint unchecked to ensure strictest green designs and at the same time there are an equal number of developers, architects and designers who have not yet realized the importance of sustainable construction and green buildings. Unfortunately, most developers are under the misconception that green buildings are costlier and require complete air-conditioning. They couldn't be more wrong because the truth is otherwise. Green buildings, while requiring a very small (3 to 4 per cent) additional amount of initial investment, offer long-term cost savings, which are likely to have payback of 2-3 years.

Advantage India

Being a tropical country, we have the advantage of abundant day lighting and can operate an entire building throughout the day without having to switch on an artificial light. The Indian government has been implementing energy policies and amending them as per the demand route undertaken by sustainable development. The Integrated Energy Policy (IEP) brought out in 2006, has been designed to address the climate change issues and support sustainable development.

Conclusion

In view of above facts, the vision and mission of this work is to make India amongst the world leaders in sustainable developments, through the training of professionals, sensitizing users, encouraging builders and developers to adopt Green Building measures and making children our human resources for preserving the planet.

Green building materials offer specific benefits to the building owner and building occupants:

- ✩ Reduced maintenance/replacement costs over the life of the building.
- ✩ Energy conservation.
- ✩ Improved occupant health and productivity.
- ✩ Lower costs associated with changing space configurations.
- ✩ Greater design flexibility.

Building and construction activities worldwide consume 3 billion tons of raw materials each year or 40 per cent of total global use. Using green building materials and products promotes conservation of dwindling nonrenewable resources internationally. In addition, integrating green building materials into building projects can help reduce the environmental impacts associated with the extraction, transport, processing, fabrication, installation, reuse, recycling and disposal of these building industry source materials. Green building materials are composed of renewable, rather than nonrenewable resources. Green materials are environmentally responsible because impacts are considered over the life of the product. Depending upon project-specific goals, an assessment of green materials may involve an evaluation of one or more of the criteria listed below. Through green building incentives local government can improve overall quality of the built environment help increasingly to preserve natural resources and help combat global climate change.

References

Abbaszadeh, S., Zagreus, L., Lehrer, D., Huizenga, C. 2006. Occupant satisfaction with indoor environmental quality in green buildings. Proceedings of Healthy Buildings, Lisbon. Volume III:365370.

Arnold, J, Cochrane, L. 2009. Future opportunities for the energy ecosystem. Power Engineering July 2009. 2pages.

Baum, M. 2007. Green Building Research Funding: An Assessment of Current Activity in the United States. U.S. Green Building Council. 37 pages.

Boulos, M.N.K., Wheeler, S. 2007. The emerging Web 2.0 social software: an enabling suite of sociable technologies in health and health care education. *Health Information and Libraries Journal*. 24(1):2 23.

Brager, G.S., Paliaga, G., and de Dear, R. 2004. Operable windows, personal control and occupant comfort. *ASHRAE Transactions: Research*. RP1161:1735.

Browning, W.D., Romm, J.J. 1994. Greening the Building and the Bottom Line. 15 pages.

Dannenberg, A.L., Jackson, R.J., Frumkin, H., Scieber, R.A., Pratt, M. Kochtitzky, C., Tilson, H.H. 2003. The impact of community design and landuse choices on public health: a scientific research agenda. *American Journal of Public Health* 93(9):1500-1508.

Dearry, A. 2004. Impacts of our built environment on public health. Environmental Health Perspectives 112(11):A600-A601.

2013, Sustainable Approaches for Environmental Conservation *Pages* **103–113**
Editors: **D.R. Khanna, A.K. Chopra, R. Bhutiani, Gagan Matta & Vikas Singh**
Published by: **BIOTECH BOOKS, NEW DELHI**

Chapter 15

Man and Nature: For Peaceful Coexistence (Based on the *Vedic Samhitas* and the Works of Kalidasa)

Vakil Ahamad

Special Centre for Sanskrit Studies,
Jawaharlal Nehru University, New Delhi

Human being is the only earthly species which have the capacity to investigate the nature of the universe which it inhabits and manipulates its powers. This heightened the problem of the relationship between man and nature.

During the recent years of the growth of the physical sciences and the consequent advances in technology and industrialization, the natural balance is being upset. For the first time in the long history of mankind we have given rise to a huge technological civilization which though apparently glorious, has sown the seeds of disintegration of the natural order. It has given rise to new problems. Thus man has become inimical to man in spite of his advancement of knowledge. But more than that man has become inimical to nature itself.

The relationship of the man with nature will be discussed in the following sub-themes:

1. Based on the Cartesian dualism, and
2. Based on wholistic or pragmatic approach to studying the Man and Nature.

Keywords: *Cartesian dualism, Vedic Samhitas, Wholistic or pragmatic approach, Zoological species, Homo sapiens, Man and nature, Rita, Hiraöyagarbha, The primeval water.*

The Man (*Homo sapiens*)

In 1739 the Swedish naturalist Carolus Linnaeus classified man – *Homo sapiens*- in his *Systema Naturae* as a primate. Ever since that time, there has never been any scientific doubt that this is where man belongs in the zoological system that embraces all living forms in a unified classificatory relationship based primarily on anatomical structure (Collier's, 1987). The man we are talking about here is the zoological species to which we belong and of which we are so proud. May be once again Linnaeus selected the correct epithet when he gave us the specific name *sapiens*, but still this does not imply that man is generally wise. Because man needs freshness, purity, wholesomeness in his environments. But he is now crying for peaceful life, pure air, wholesome heat and light emerges, pure water and fragrant soil and tasty and hygienic food. But these things are becoming rarer and rarer on account of his unwise use of his technological knowledge. Infact he alone is largely responsible for digging his own grave in the very place where nature has made his earthly home. The assumption being that nature exists solely for man's pleasure or that he can only come into his own, be liberated, by doing battle with and subduing the forces of nature. And during the recent years of the growth of the physical sciences and the consequent advances in technology and industrialization, along with the geometric growth of the human population, the Nature's balance is being upset. For the first time in the long history of mankind we have given rise to a huge technological civilization which though apparently glorious, has sown the seeds of disintegration of the natural order. Thus man has become inimical to man in spite of his advancement of knowledge. But more than that man has become inimical to nature itself.

In the beginning of the human history, owing to the vast ignorance of man regarding the forces of the nature, human being must have suffered at the hands of these overpowering forces. The nature appears to have been all- powerful and man very weak and insignificant. But gradually man comes to understand the laws of the behaviour of the nature, which knowledge becomes instrumental to regaining self-confidence in man. Man is the only earthly species which have capability to investigate the nature of the universe which it inhabits and manipulate its powers. He utilised this knowledge for his own advantage. His confidence grew so much that he thought in terms of conquering nature. This heightened the problem of the relationship between the man and the nature. After all, man remained undoubtedly the most intelligent and powerful of the earth's inhabitants. There has been nearly unanimity on the need for harnessing these instances of nature as and when possible, to the use of man, the assumption being that nature exists solely for man's pleasure, or that he can only come into his own, be liberated, by doing battle with and subduing the forces of nature. This has further intensified, beyond the increase in power to utilise and understand the workings of nature through the invention of tools and machines, into an increasingly aggressive approach towards the nature climaxing in modern experimental science. This might be call the turning point in the history of mankind. While formerly, nature was power, now man's knowledge of nature is his power.

Challenges

Human knowledge has mainly resulted in artificial intervention in the process of nature. It has created impediments in the smooth working of natural courses. Many artificial synthetic products are invented, which when useless do not get assimilated in the circulation of nature. Thus the naturally recurring circle of nature is broken because of human industry. It is the ever enlarging human activity of the intervention of nature that has resulted into the pollution of the elements. Through the various kind of jarring noises the ethereal peace is disturbed. Through the nuclear energy poisonous radiations are let loose. They dangerously affect atmosphere, water and earthly existence, living and non-living. Through the unchecked growth of urbanization; forest life is in danger of extinction, which has an important part to play in maintaining the balance of nature. Through the artificial fertilizers the immediate gains in agricultural products are made at the cost of the natural quality of the soil.

Hypothesis

Man, the dominant organism on earth, has stood apart from the interaction of living things; yet, as he is discovering, he is inextricably tied to his environment. The air he breaths, the water he drinks, the food he consumes and the products he uses and throws away bind him to the functions of local and global ecosystems. He is a part and parcel of nature. Man and nature are truly members of one another. They are rather cooperative members in the big scheme. They form each other's environment which with friendly and harmonious relationship must be unegoistically understood by man. According to the Vedas *Rita* was the observed order and law that governs the physical universe or cosmos. The basis of the Vedic structures is based on philosophically wholistic vision of life. It is indeed time for him to realize that his ideology of conquering nature is ill-conceived. Actually nature's bounties are on him if he truly understands her ways. Man is the child of nature. He feeds and grows upon nature. Earth is verily his mother; sun his father as the primary source of all energy on earth. Thus neither of them is inimical to other. But while nature spontaneously follows its own course, it's working are to be consciously understood by man and accepted ungrudingly and disinterestedly. Nature which has given birth to man does not want to kill him. But man may behave want only and invite death. As is evident from his unmindful violence to nature, as well as to brother man, by piling hosts of nuclear weapons. If used against the imagined enemy, he would invite collective death of mankind and possibly of all life.

Methodology

Research and investigation in any subject, be it science, arts, philosophy or theology requires certain methods to be followed. Ultimate analysis of these methods shows that there are certain basic principles, which are essentially the same for every subject of study. These basic principles are sought, investigated and formulated. Vedic culture was pragmatic and could not have been otherwise, the philosophical speculations of the philosophers pursued at the same time two main streams, the abstract or speculative, and the pragmatic or the applications of knowledge and understanding to everyday life. Man and Nature is usually studied now by taxonomists, who take great pains in collecting, identifying, documenting and describing the elements of diversity. But, the approach/ methodology usually is very materialistic, which is based on Cartesian dualism. A plant or animal is just one unit in their inventory. Plants and animals are taken to possess genes and chemicals and not spirits.

I shall apply pragmatic approach to studying the Man and Nature. An understanding of the parallelism, corollaries, deficiencies, advances and the difference in approach and methodologies in the old as compared to the new.

Man and Nature

In the history of science the views and concepts of the physical world coming to the fore, both intuitively and on logical grounds, should be understood in the wholistic light. The early Greek thinkers attempted for the first time to give a rational interpretation of the natural occurrences. Some of them tackled the problem of one in many, of finding the single principle, which could explain all the observed phenomena. Thales (Thilly, 1956) thought that water was the substance or primal matter which would remain unchanged amidst all observed modifications (Thilly, 1956). Obviously he had observed how water would change, flow without any shape or colour and undergo a cycle of movement on earth and in air. He conjectured that their was a substance (water) from which all other things came to be, it is being conserved. Heracleitus (Thilly, 1956) regarded fire as the fundamental principle in an ingenious way. Fire changes into water and then into earth, and earth changes back again into water and fire (Thilly, 1956). About the cosmology of Pythagoras in Greece, who speculated about the cause of the motion of the objects of the sky- that there must be a 'central fire' at the origin of the spherical universe, causing all of the objects of the sky, including Earth to move ! (Guthrie, 1962)

Now let us discuss the speculative enquiry of the primitive Vedic thinkers. The primeval water, mentioned in the *N asad ya s kta* of *Rigveda*, is the first element in the Indian speculation. Later in the *Brihadaraöyaka Upni·ad* (V. 5.1), as well as in the *Chandogya Upanisad* (VII, 10), it is stated: in the beginning this world was just water. *Na sad ya s kta* of *Rigveda*, throws some more light on the issue of the existence of the pre-cosmic waters.

Na asat s t no iti sat s t tad n m na s t raja no iti viyom para yat/
Kim var variti ku a kasya arman ambha kim sit gahana gabh ra×// Rigveda X-129.I)
Tama s ta tamas guhyam agre apraketa salila sarva ida ah/
tuchena bhu apihita yat s t tapasa tat mahim aj yata eka// (Rigveda X-129.3)
k ma tat agre samavartat adhi manasa reta prathama yat s ta/
sata ba dhu asati nir avindan h di prat ya kavaya man// (Rigveda X. 129.4)

Compounding the verse one, three and four, it could be inferred that from the primordial waters arises the one principle through tapas or the power of heat.

Scientifically we measure heat in terms of temperature and we admit total absence of movement only at absolute Zero temperature or minus 2370 c. The energy responsible for the dynamics of nature, implied in the notion of movement is called fire and heat, that pervades the whole universe. Fire as agni is the main causative of movement. The first verse of the *Hiraöyagarbha hymn* (X.121.1) is as follows: *Hira yagarbha samvarta-agre bh tasya j ta patireka s ta.* How did the Hira yagarbha originate? The answer is found in verse (*Rigveda* X 121.7) which is of considerable importance to us for our present discussion.

poha yadb hat rvi vam yangarbha dadh n janayat ragni/
tato dev n samavartat sureka kasmai dev ya havi a vidhema// (Rigveda X.121.7)

According to the Hira yagarbh Hymn (X-121.7) the waters *Ap* with mighty motion (*po yaû brihat h*) generate fire agni the first evolute and for this reason Agni is called the son of the waters (*ap m napt*) (Macdonell, 1960).

There are few qualifications attributed to the primordial waters in this hymn that are worthy of consideration. The word *ap* has movement and moves holding the universe. Creative potency of the dynamic waters continues to represent chaos and undistinguishability despite their ability to generate the sky principle of orders, and distinguishability. However, the *N sad ya* and *Hira yagarbha* posit waters as the primordial substance that bears and generates the first creative principle, the *tadekam* or the Agni. The significance of this cosmogony is that, there is a postulation of a material and not any spiritual principle as the primordial base. This gives rise to a naturalistic account of creation and is markedly different from the other kind of theories offered to explain the primary creation.

The *Vedic Sa hitas* has principally drawn our attention to this friendly character of the relation which nature holds to man. The origin and sustenance of all life including that of man depends upon the natural environment. Actually the natural environs have given man a cosy home in the form of earth. It is in the motherly care of the earth that man gets his nourishment. It is ecology again that has held a red signal before man about his unthoughtful utilization of his scientific and technological knowledge. His intervention into the smooth processes of nature has been, of recent, proving to be of dangerous consequences. Man's activity results in damaging the natural environs. Ultimately it proves to be a very suicidal activity. This has a lesson for man. He ought to change his motivation in dealing with nature. He ought to pursue his science and technology in such a temperate way as would not come in the way of the orderly design of nature (*Rita*). His technological civilization must be founded on his right attitudes to the natural pattern. In short his inner culture ought to be the basis for a healthy civilization.

The Intimacy of Man with Nature as Revealed in the Vedas

The intimacy of man with nature is expressed with great joy (*Rigveda.*VIII; 31,10) "we solicit the happiness afforded by the hills, the rivers, the sun..".Agni is the first god..who makes our fields prosperous". (*Rigveda.*X; 85,1-19). Vedic people were one with nature. (*Rigveda*, VIII; 58,2-3; Swami Satya Prakash Saraswati and Satyakam Vidyalankar, a VŒakhilya hymn): "One is that which manifests in all", which in contemporary ecological terms is expressed as 'everything is related to everything else". Oh king, the rivers are the veins of the Cosmic Person and the trees are the hairs of His body. The air is His breath, the ocean is His waist, the hills and mountains are the stacks of His bones and the passing ages are His movements. [Taken from *Srimad Bhgavatam*, 2.1.32-33]. In the Vedic literatures mother Earth is personified as the goddess *Bhumi*, or *Prithvi*. She is the abundant mother who showers her mercy on her children. Her beauty and profusion are vividly portrayed in the beautiful Hymn to the Earth in the Atharva Veda,"O mother, with your oceans, rivers and other bodies of water, you give us land to grow grains, on which our survival depends. Please give us much milk, fruits, water and cereals as we need to eat and drink." (verse, 3). and may you, our motherland, on whom grow wheat, rice and barley, on whom are born five races of mankind, be nourished by the cloud, and loved by the rain (Dwivedi and Tiwari, 1987). "Trees have five sorts of kindness which are their daily sacrifice. To families they give fuel; to passers by they give shade and a resting place; to birds they give shelter; with their leaves, roots and bark they give medicines." [*Vara Purana*, 162. 41-42].

The Veda probably took shape in Balkh (Dandekar, 1981) (at the time when the Aryans lived there before emigrating further towards Saptasindhu; the *Samhita* from that we have now took shape at a later stage. The subject matter of the Veda originated from a continuous process of intellectual creation and revelation that spans a long time, probably many centuries of oral tradition. Innumerable studies and scholarly text have been published on this subject, but whatever the historical details or texts may be, the Vedas as we have them, specially the *Rig Veda*, the *Sama Veda* and the *Atharva Veda* are the expression of very ancient men. Who, to my mind, were also good observers of nature. The seers/poets, who formulated the hymns lived with nature that they studied and used. A strong feeling of man's participation in the universal order pervades the Veda; man is inseparable from his cosmic environment, he is not distinct from it but imbued in it. Man becomes immortal through his descendants and through his spirit that merges with the universal spirit on the death of the body. Fire as energy is the present in môvable and immovable natural things; Agni, the original power or God is present everywhere at all times, even in stones (Atharva Veda, III; 21,1) and Agni expresses and manifests itself in many forms. Who, Agni, is thy kin of men? Who is thy worthy worshipper? On whom dependent? Who art though? [*Rig Veda*,.1: 75, 39 (translation, Griffith).]. Energy and matter as well as the essence of nature and man are one, the same and interchangeable and when bodily life ends, man like other movable and immovable things, merges with nature or, in modern ecological terms, matter is recycled and energy is one under many interchangeable forms. In fact, even human mind is part of the cosmic soul, or energy.

Vedic Samhitas have some tenet in their philosophy about the sanctity of natural wealth, for instance of water and of the merit of conserving water. Himalayan ranges with plants, especially grasses, trees, medicinal plants and grapes. Where the word Soma is in the plural. The *soma yaj–a* was an important ritual later conducted preferably on hill tops, about which there are frequent references; there are remains of stone alignments or slabs on many hilltops in the, Sacredness of mountains and hill tops is frequently mentioned (*Rig Veda*,. V; 41, 42) and still observed at present. Trees, forests, rivers, winds would be the good providers but also a constant menance when unleashed in their fury. A whole hymn is dedicated to the lady or Goddess of the forest; *Aranyani* (*Rig Veda*, X; 146). "Mother of all forest beings who tills not but has stores of food."

The wonders of the new environment are frequently referred to: "the big rivers of *Sapta Sindhu* can be forded". (*Rig Veda*,.X; 113, 10). But it is important to know the right time and places where and when they can be managed. Man continues to apply more and more pressure on the environment at higher and higher altitudes, provided as he is with the powerful means of modern technologies. Mountains slopes are deforested, fairies and *yaksas* lose their abodes and cry Loudly in the unchecked winds, demons let loose avalanches and landslides come down on their own, with no vegetation cover to keep the loose boulders and soil in place. Downhill floods become more frequent, widespread and devastating ...but that is a sad story of the present, when the conservation practice as expressed in poetic injunctions by the *rsis*, are no longer heeded.

For the Vedic man the mountain valleys are perennially great: "untouched by time, never lacking green trees and vegetation, with their voices they have caused heaven and earth to hear". [*Rig Veda*,. X; 94, 12]: "from the mountain came the pressing stones" and probably other tools as well. Stones are respected as an important and useful element. Of earth and mountains are the maximum expression of immortality: the hardness and permanence of Rocks. Stones are spared even by Agni who "eats everything else "who in his wrath shaves a hill as a barber shaves a head" [*Rig Veda*,. X; 142, 4]. Agni's fury can be terrible (Atharva Veda, VII; 70 (73), 4,5), the " Rudra that is in fire." "To that Rudra, Agni, be homage"(Atharva Veda. VII; 87(92), 1). Fire and smoke are also used to frighten away enemies (Atharva Veda, VIII; 8, 2). Agni is present even in water as dynamic energy in water flow, rivers, mountain streams, ice and snow avalanches, torrential rains that could be harnessed for different purposes; Agni becomes the associate of water for the creation of plant and animal life (*Rig Veda*,. X; 142,3,4); " Agni descending with water impregnates in the plants" (*Rig Veda*,. X; 21,8]. There are also "plants with flowers and without, with fruits and without." Plants that (soma is here in the plural) give us different things" which, in this context, means different natural products, including what probably are intoxicating or psychedelic drugs; "be glad and joyful in the plants" (*Rig Veda*,. X; 97,3) and "bringing back thy very self, O man". It also indicates that mixture of medicinal plant extracts were prepared, because they "help and lend assistance to each other." In *Ayurvedic* medicine, mixtures are used rather than purified extracts of single plants.

Microcosm and Macrocosm

How Vedic people Looks at Nature: A Vedic people residing in *Aranya i.e.* an insider in that environment acquired knowledge and familiarity, and establishes intimate personal relationship with the bioresources, as he sees and feels them, walks by them, sits near them, and presumably talks to them each day of his life. Aspects of biodiversity is invisible and inaudible to most of us. Some aspects of ancient or traditional knowledge like on Ayurveda are available in easily accessible literature. Much however is not accessible, or studied and understood.

Human culture and material uses of bioresources, that is the total human relationships with bioresources, are largely shaped by history, and by physical and social environments (Alcorn, 1995). These relationships cover a very wide canvas, from wild foods, medicines, fibres, fodders, dyes and body ornamentation, *etc.* to still more important but less understood areas of the social and religious relationships, like beliefs, faith, taboos, worship and even protection and preservation.

The Vedic man shows his intimacy with nature as he sees health as the law and disease as the breakdown of order or law, which in this case is the break down of the equilibrium among internal factors of the body, of the mind, of their physiology as well as the breakdown of the balance between body and mind on one side and the environment on the other. *Ayurveda,* the traditional Indian Medical Science, shows a close interconnection between physical nature, human nature, biology, biochemistry and technology in the form of medicine and surgery. If we go carefully through the *Bhela, Caraka, and Susruta Samhita,* it become clear that *Ayurveda* deals with not only the therapeutic technique, it also has a deep ecological, metaphysical, moral, as well as humanistic values intertwined. The scope of *Ayurveda* is not limited to physical health alone. It also seeks to promote a totality of physical, mental, and spiritual health in the context of man's interaction with his environment. Since the biological regularities of living organism display an active and intimate engagement with their environment that is categorically not different from that of inorganic matter. There is a remarkable theory in *Ayurveda* to the effect that 'man is an epitome of the universe, a microcosm of the macrocosm'. Both the universe and man are manifestations of one and the same eternal spirit. Spirit and matter are equally integrated in both. The material contents of both are constituted of the same five primal elements, endowing their specific and resultant characteristics to both. Man the microcosm was inevitably moulded by the forces of the macrocosm (Deshpande *et al.,* 1970). The concept of the 'microcosm' follow the Philosophical doctrines of the SŒkhya-Yoga school of Indian system of Philosophy. The SŒkhya doctrine of the five elements did influence a great deal in the development of certain concepts of the Ayurveda, the principal Indian medical system. For, the *Ayurveda* has adopted the principles of SŒkhya with certain modifications, bestowing special importance on the five elements and their role even in the physiological process (Sengupta and Sengupta, 1938).

The human body is considered as a combination of the five elements and soul. In the Susruta Samhita (Sengupta and Sengupta, 1938) there is a clear exposition of the way in which the five elements constitute a human body from its very conception. The cosmic elements rule over the human body and cause its functions and malfunctions, the human body being the microcosm of the universe. The theory that the same five elements, *prithivi* (earth), *as* (water), *vayu* (air), *jyoti* (fire) and *k sa* (ether) constitute the human body (microcosm) as well as the macrocosm is accepted as axiomatic truth in many passages.

An Ethics of Equilibrium

We are going to have to become more responsive to long- term consequences of our actions if man is to persist on earth and enjoy inalienable rights. To guarantee these rights for future men, we will have to develop a new ethic, an ethic of equilibrium. Acceptance of this ethic is tantamount to accepting ecological principles as facts and as commandments. Such an ethic would dictate that we consider it immoral to disturb existing equilibrium (at the very least they should not be perturbed beyond a state where equilibrium can be reached once again in a reasonable amount of time); more-over, we will have to do our best to restore equilibrium that have been upset previously by men without our vision of equilibrium. Agni and water are givers and sustainers of life, they are "affectionate mothers....Givers of all, givers of life" [*Rig Veda*,. IX; 2]. They have healing power [*Rig Veda*,. X, 137], But "most excellent are waters..most excellent is Agni..praised is the lightening cloud".[*Rig Veda*,. I; 161]. It is also made clear that the whole is the most excellent", that each part plays its role excellently and that the whole cannot function without the adequate function of the parts, in this case, fire and water. The whole, whatever it is, cannot function if Rita is not observed and followed obediently. Soul and body are the Fire and water pair that enables life to go on [*Rig Veda*,.X; 11, 9] And Varuna's magic "puts fire in the water" [*Rig Veda*,.V; 85, 2]. The global water cycle was fully understood by the ancient *rsis*, beginning with the endless flow of water into the sea that remains unaltered; then "water vapour ascends to heaven" [*Rig Veda*,. VIII; 41, 6-8; 42] And "Varuöa is the large hidden ocean who envelopes the world with his robe of humidity".

Clouds are important for people spending most of their time in the open: *Parjanya*, the cloud, is often remembered and one simile is biologically very expressive. He is requested to drop on earth the semen (*pani* or water is a word that can be used, as well as *bij* or seed, as a synonym of semen) of father sky that generates fecundity in mother earth [*Rig Veda*,. VII; 101, 102]. The famous and usually misinterpreted and misunderstood hymn to the frogs [*Rig Veda*,.VII; 103]. Is a marvellous piece of god biological and ecological insight. O Varuna..let the frogs ..croak along the water courses" [Atharva Veda, IV; 15, 12-14]. The hymns also contains a good description of the female frog spreading out the four legs for reproduction; this is the posture she assumes to receive the male's amplexus. Good monsoon rains breed many frogs as well as good pastures and contended grazers such as cattle, horses goats and sheep that will be healthy and productive. This is how plenty of frogs are a good omen for the next season.

Air is recognized as breath of life, *prana*, of all living beings, including fire that cannot exist without air [*Rig Veda*,.X; 151,4], also plants and animals cannot exist without air. The recurrent word *bhèta* indicates all beings and being alive means to have a spirit, therefore whether material or immaterial or just air, *bhètas* have a real existence. Please do not give us too much of anything, even grace. In a general way, however, the cloud bearing winds, the Maruts, are welcome [*Rig Veda*, I; 167-169]. Which is natural in monsoonal areas where all nature thirstily awaits the rains after long months of dryness. Air and water are means for the transmission of vibrations, visible or invisible, measurable or not, physical or mental [*Rig Veda*,. 1; 171].

Some further aspects of the application of the theoretical knowledge to practical aspects can be mentioned here. Health was perceived as the normal state, the maintenance of order in the body of man and animals as well as in the body of nature. Health is the law, ill-health is disruption of the laws that govern natural forces, due to ageing, a curse, witchcraft or poisoning. Health is physiological and ecological balance; disruption of order, *Rta*, causes biological and ecological imbalance, impoverishment or degradation of the patient, be it man, animal, plant or the environment. Degradation causes poor pastures, wear and disease in animals, decrease in productivity, all of them causes of migration in search of new pastures.

Agni who sustains, maintains and controls *Rita*; logically, when disorder sets in, Agni is invoked to restore order. "all the active or productive aspects of *pari ma* or evolution, at the macro-and micro-cosmic levels are due to *rajas* or energy". " Agni, according to the SŒkhya system". And Agni is the "kinetic factor at all levels of nature." Interestingly Lord Savitr is invoked to keep away disease and in this context also the mountains, the pressing stones and the herbs who shed juices are taken together

[*RigVeda*, X; 100, 8]; ecologically this makes sense, since the mountain gives the pressing stones and gives many of the plants used for different purposes, including the preservation of health. The natural elements: fire, wind, water, air, seasons, birth, growth, old age and death of plants and animals, reproduction and reproductive cycles, floods, landslides, avalanches, earthquakes and intellect are all expression of the divine and they all fall within the realm of the eternal law and order or, alternatively, temporary disruption of *Rita* or disorder.

The Intimacy of Man with Nature Vased on Works of Kalidasa

It can be better stated in the words of Ryder."even Shakespeare, for all his magical insight into natural beauty, is primarily a poet of the human heart. That can hardly be said of Kalidasa, nor can it be said that he is primarily a poet of natural beauty. The two characters unite in him, it might almost be said chemically. And he further says that the *Meghadutam* best illustrates this feature. "The former half is a description of external nature, yet interwoven with human feeling; the later half is a picture of a human heart, yet the picture is framed in natural beauty. So exquisitely is the thing done that non can say which half is superior." To Kalidasa external nature is fraught with human feelings; and man, on the other hand, is a part of nature. Illustrations of this can be found on every page of Kalidasa; and seen as above this is especially so in the *Meghdutam.*

No doubt, when the Yaksha decided to send a message through the cloud, an *acetan,* a non-sentient thing, he, as if were to justify his position, specially points out- "Never yet was lover cloud discriminate ''to life and lifeless things, in his love-blinded state.' (*kamarta hi prakritikripna-scetanascetanesu.* [*purva megha*, verse 5] and tries for a time, though ever so short, to treat the cloud as pure nature. But it is only temporary; the poet also, it appears, soon forgets that the cloud is only a collection of 'smoke, light, water, and wind' and treats him as a living entity, having human feelings, yet possessing all elemental qualities of Nature. This is often done so intricately that it is really very difficult to separate these two elements. The Citrakuta mountain and the cloud have a very sweet friendship between them; every year the cloud meets him after a long separation and then again he has to cut himself off from his friend: profuse indeed are the tears that the cloud shades at the time of this separation. Remarks the Yaksha.

kale kale bhavati bhavato yasya samyogametya /

snehavyaktiscirvirahjam muncato vaspamusöam // (*Pèrva Megha-12*)

The cloud, being Yaksa's messenger, naturally stands in the relation of friendship to the Yaksapatni, who thus becomes his *sakhi.* When the cloud actually sees her in her love-lorn condition, *dharayanti shayotsage nihitamasakriduhkhaena gatram* (Uttar Megha-verse-33), he would naturally not be able to check the flow of his tears, and the more so, because he is very kind-hearted. Hence the Yaksha tells him:

tvamapyasraa navajalamayam mocayisyatyavasayam /

prayau sarvo bhavati karuöav ittiraardrantaratma //

The more one searches the Meghadutam, the more will one find, almost everywhere, this unison of Nature and of man, which is one of the abiding elements of the Indian mind, inherited from the horay times of the Rigveda onwards. The Vedic sage considered all Nature to be replete with human, or more correctly, divine life and deified different aspects of Nature. He considered, for example, vata (wind) as a mighty being, which moves on a big car, but strangely enough his form is never seen, *ghosa idasya shrinvire na rapaa.* In the words of Saakhyas all is *prakriti* (nature), their being no such difference in reality as human and non-human. In order to explain, or to offer a sort of raison detre for this wonderful mixture of man and nature in the poetry of Kalidasa, one may say that he is under the influence of this Indian philosophical and emotional heritage of the unity of Man and nature. Only the poet offers it in a beautified form, mellowed by his imagination, and sustained in its appeal by the use of his many conceits, which present it in a lovelier form. In the wanderings of *Pururavas*, who was mad due to the sudden disappearance of *Urvasi*, the presentation of Nature in human terms almost reaches perfection. The king piteously asked every denizens of forest, the trees, mountains, swans, elephants, about his beloved *Urvasi,* and wanders on madly. About this one feels inclined to say with Ryder "It is hardly true to say he (Kalidasa) personifies rivers and mountains and trees; to him they have a conscious individuality as animals or men or gods. (Ryder: introduction, p. xix). His addresses to the different plants and animals in the forest are so pathetic and ever so replete with a feeling of unison with nature. He asked the Cuckoo, "They call thee the messenger of love, O Cuckoo, though art the best remedy for taming the pride of women; please, therefore, bring my beloved to me or carry me there where that sweet speaking one is.

tvam kamino madanadutimudaharanti

manavabhanganipunam tvamamoghamastram /

tamanaya priyatamam mama va samipam

mam va nayashu kalabhasini yatra kanta // (Vikram, iv. ii)

Hillebrandt has rightly remarked in in connexion with *Meghdutaa,*" whose heart was moved not by the cool whiff of the North, but by the Malaya-wind fragrant with sandal-dust (Hillebrandt).

Concluding Remarks

M y concluding remark is that the Vedic seers considered all nature to be replete with human, or more correctly, divine life and deified different aspects of Nature. In the words of Saakhyas all is *prakriti* (nature), their being no such difference in reality as human and non-human. In order to explain, or to offer a sort of raison d'etre for this wonderful mixture of man and nature in the poetry of Kalidasa, one may say that he is under the influence of this Indian philosophical and emotional heritage of the unity of Man and nature envisioned by the Vedic seers.

Notes and References

1. Collier's *Encyclopedia*, Vol. 9, Macmillan Educational Company New York, 1987, p. 482.
2. Thilly, Frank, 1956, *A History of Philosophy,* p. 24, Central Publishing House Allahabad.
3. Thilly, Frank, 1956, *A History of Philosophy*, Central Publishing House Allahabad, p. 337.
4. Guthrie, W.K.C., 1962, *History of Greek Philosophy*, Cambridge.
5. Macdonell, A.A.,1960, *A Vedic Reader for the Students*, Oxford University Press, p.47, *Rigveda* ii.35.
6. Dwivedi, O.P. and Tiwari, B.N., 1987. Verses from *Atharva Veda* 12.1, paraphrased from the translations quoted. *in Environmental Crisis and Hindu Religion*, Gitanjali, New Delhi, (verse, 42)
7. Dandekar, R.N., 1981, *Excercises in Indology,* Select writings, III,PP.385, Ajanta Publ. Delhi at Bhandarkar Inst. Press, Poona., p.25.
8. Alcorn, J., 1995, *The Scope and Aims of Ethnobotany in a Developing World in Ethnobotany*, Schulters R.E. and Von Reis, S. (ed.), pp. 23-39).

9. Deshpande, P.J., K.R. Sharma,and G. C.Prasad, 1970, *Contribution of Susruta to the Fundamentals of Orthopaedic Surgery*. Indian Journal of the History of Science,vol.5,no.1,pp.13-35. L.M.Sing, K.K. Thakral and P.J. Deshpande, *Susruta's contributions to the Fundamentals of Surgery*. Ibid. pp.36-50 see also J. Filliojat, 1949, *La. Doctrine Classique de la midecine indienne; ses origins et ses paralleles grecs*, imp. Nat. Paris, ch.viii.

10. *Susruta Samhita,* Translated into English by Kunjalal Bhishagratna, 1907-15 vols. 3, Calcutta,; 2nd edition, Chowkhamba Sanskrit Series Office Varanasi, 1963, edited by Nripendranath Sengupta and Balai Chandra Sengupta with the commentary Nibandhasamgraha, 2 parts, Calcutta,1938. Sutrasthana, 1, 4-6

11. ibid. 5, 2

12. *Aitareya Brahmaöa*: edited by Satya Vrata Samasrami, 1895-1907, with the commentary Vedarthaprakasha of Sayaöacarya, 4 vols. Asiatic Society Calcutta. Translated by Martin Haug 2 vols. Bombay 1863 ii,3,2.2; *Aitareya Upanisad*: edited by Vidyaranya with the commentary Bhasyam of Sankaracharya. 2vols. Poona 1889 Translated into English by S. Radhakrishnan, vide his *Thirteen Principal Upanisads.* London,1953. Iii,5.3; *Kausitaki Aranyaka;* Edited and translated by A. B. Kieth, Oxford at the Clarendon Press, 1909 vii,2.2.

13. quoted from Ryders not literal yet effective translation of the Meghduta in Kalidasa: Translations of "*Shakuntala*", and other works,' Everyman's Library no. 629 (reprinted 1928), introduction, p. xix.

14. Ryder, ibid., p. xx.

15. Uttar Megha verse-33, The Meghaduta of Kalidasa, Delhi, 1999.

16. Hillebrandt's *Critism of the Meghadutas*: translated by the author from German: *Kalidasa*, pp. 29-32.

2013, Sustainable Approaches for Environmental Conservation *Pages* **115–121**
Editors: **D.R. Khanna, A.K. Chopra, R. Bhutiani, Gagan Matta & Vikas Singh**
Published by: **BIOTECH BOOKS, NEW DELHI**

Chapter 16

Rapid Groundwater Depletion in Firozabad: A Critical Study

***K.K. Bhardwaj*[1] *and Vishal Pathak*[2]**

[1]Department of Chemistry, Govt. P.G. College Fatehabad, Agra, U.P.
[2]Department of Chemistry, Paliwal P.G. College, Shikohabad, U.P.

The present investigation has been carried out in the city Firozabad of Uttar Pradesh (India) to find out the root cause for rapid groundwater depletion in the last 10 years that has gone from 100 ft to 250 ft. The problem seems to be study of a particular area but it is of big concern for whole of the world. The main motto behind the analysis is to provide a permanent solution with very less cost and simplicity that could be executed easily and effectively.

In the study, the affect of groundwater quality on human health has also been discussed. 05 tube wells were chosen from different locations. Their groundwater properties are discussed with human health issues.

The study points out the necessity for groundwater recharging. Its methodology, execution part, essential factors have also been studied in the investigation. The approach emphasize on awakening of human being before it is too late.

Keywords: *Root cause analysis, Rain water harvesting, Groundwater, Wells, Hand pumps.*

Introduction

Water, a precious natural resource, now has become a commodity for trading. Man is responsible for this anomaly. The nature blessing that was in abundance yesterday, becomes a rare treasure today. We have ruined all of our earth. The cost of development is excessive exploitation of environment. As a result, now we are facing the environmental crisis through out the world. Usually it is said that 3rd World war will be for water.

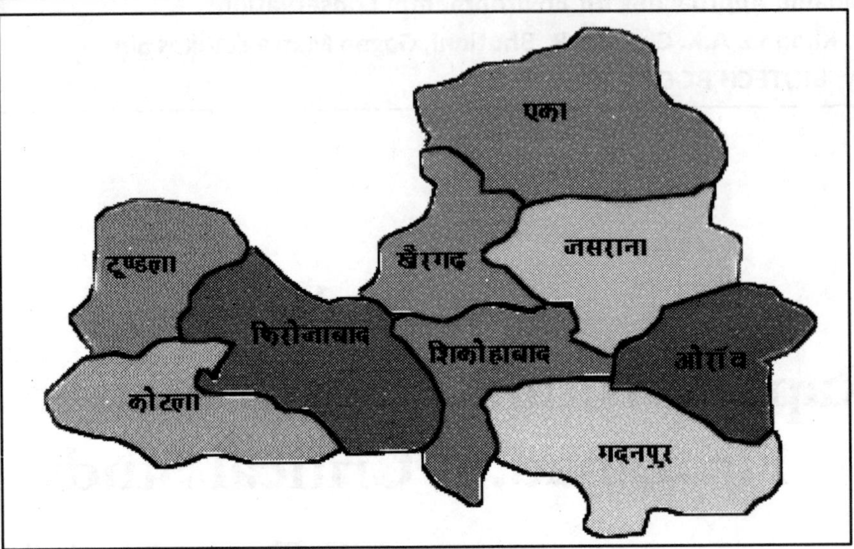

Figure 16.1: Firozabad District

Water scarcity is found in every corner of the world. Every city of the world is confronting with this problem. Groundwater level is going deeper and deeper day by day. The present study deals with this concurrent issue. The study has been carried out for root cause analysis of the rapid groundwater level depletion at Firozabad, one of the city of Uttar Pradesh (INDIA).

The District Firozabad is situated between 27º 27 º 24′ North latitude and 77 º 60′ and 70º 04′ East longitude in the south west corner of Uttar Pradesh. It is situated at about 40 km east of Agra Distt. The district comprise of an area of about 263.86 sq km. The total population of the district is 2081752 of which 1411513 reside in rural areas while 670239 in urban areas.

Firozabad well known as *'Suhag Nagari' is* famous for its glass industry through out the world. Today Firozabad is a modern town with all of the amenities. But this urbanization and modernization leads the city to a drastic situation of water scarcity.

Statement of the Problem

Today city is at a dreadful end of water. Whole of the city is striving for water. Groundwater has gone to the level of around 250 feet (for almost whole of the city). There is complete mess. The water supply is totally mismanaged. There are 16 overhead tanks and 198 tube wells. In place of supplying through OH Tank, usually there is a ill practice of supplying water directly through tube wells. There is also not a proper timing of tube well operations. The municipality is struggling with funds. Just 10 years back water level was around 100 feet but due to excessive exploitation of groundwater, level is going deeper and deeper day by day. One could imagine about the stage after 10 years. Man is not only the culprit but also the victim for this situation. It is the time to awake and react.

Experimental

Among 198 tube wells, 05 were chosen for groundwater study. These 05 tube wells are located at 05 corners of the city. This study has not done for research point of view but it has been done for the sake of implication of deeper groundwater on human health.

Parameters

Among all of the parameters (Physical and Chemical) pH, TDS, DO, COD, BOD, Fluorides, Arsenic has been analysied by following the methods of APHA (1975).

Parameter	*TW 1*	*TW2*	*TW3*	*TW4*	*TW5*
pH	6.8	6.7	7.2	6.5	6.8
TDS (mg/l)	1356	1863	1962	1298	1530
DO (mg/l)	0.8	0.6	0.4	0.9	0.7
COD (mg/l)	370	455	630	586	443
BOD (mg/l)	132	165	201	187	167
Fluoride (mg/l)	21	23	26	19	24
Arsenic (mg/l)	00	02	04	00	01

TW1: Tube Well Tilak Nagar, TW2: Tube Well Vibhav Nagar, TW3: Tube Well Indira Colony, TW4: Tube Well Ram Nagar, TW5: Tube Well Rasoolpur.

Results and Discussion

The study tells that arsenic is in excess in TW3 (Indira Colony) an alarming situation. Flouride content is also a troublesome factor. DO is very low at all the 5 tube wells. The oxygen demand (Both BOD and COD) is also very high in all the 5 tube wells. TDS is also showing larger values.

In a nut shell, we can say that TW3 is at a dreadful end but other tube wells are also not very good. People are drinking water directly that causes many problems. People are getting sick due to this unhygienic water. The common diseases seen in the area is arthritis, diarrhoea, cholera, osteoporosis *etc.*

Root Cause Analysis of the Issue: The Findings

The present study has been carried out not for the paper work but for the ethics of civilization. Man's deeds should not destroy him. He should get awakened early so that future generation will not fight 3rd world war for water.

Going through the different phases of study, we have found out that it is not a specific problem of a city or a District or a state or even a country. It is a generalized problem prevalent everywhere in the world. These are some of the main reasons that generated the issue. These are:

- ☆ Excessive installation of House Hold as well as Industrial Submersible Pumps/Tube Wells that causes 200 ltr waste for the requirement of 2 ltr.
- ☆ Human habits.
- ☆ Presence of very less Effluent Treatment Plants (ETP) at industries that can recycle the water again in the industry.
- ☆ Negligible water conservation efforts in the world.
- ☆ Water Management System is not well planned.
- ☆ Rejuvenation activities of rivers, ponds, lakes *etc.* is not seen commonly.

List can be enriched more but these are the basic things that need the attention of every one.

Figure 16.2: The OH Tank

The Solution

As earlier said that it is a ground work that provides some solution for the problem. In place of doing complex things at complex level, simple things are required to be done but with a visionary approach and in a well planned and organized manner. The solution for this problem is *Rain Water Harvesting/Groundwater Recharging.*

This is a very traditional and old approach but still very successful. The basic point we should remember that if we take some thing from the earth/environment, it is our duty to return it in equal proportion so that ecological balance is maintained.

In place of digging at 800 or 1000 ft in the earth (that cost also too much), we could do it with very less efforts and cost. It is too much effective also for enhancing the water level.

Action Plan

Water recharging is required to be done at macro + micro grounds. For it recharge points are needed. New recharge points could be generated easily while the old ones are easily identified. These could be:

- ☆ Failed Boring of Submersible pumps
- ☆ Hand Pumps that refused to work due to water level depletion
- ☆ Ponds that have disappeared due to urbanization.
- ☆ Deep disappeared Wells
- ☆ Reservoir could be constructed

All these are the best source for recharging. If serious efforts could be made in this area, not only the water level would be enriched but also earth would be a better place for living. For it a *PPP (Public Private Partnership) Model* is the best one. NGO, GOVT, People and participation of common man would be a better way.

Implementation

For Household

Water recharging is a totally practical aspect. If it is applied with a visionary approach, much of the problems related with water can be sorted out.

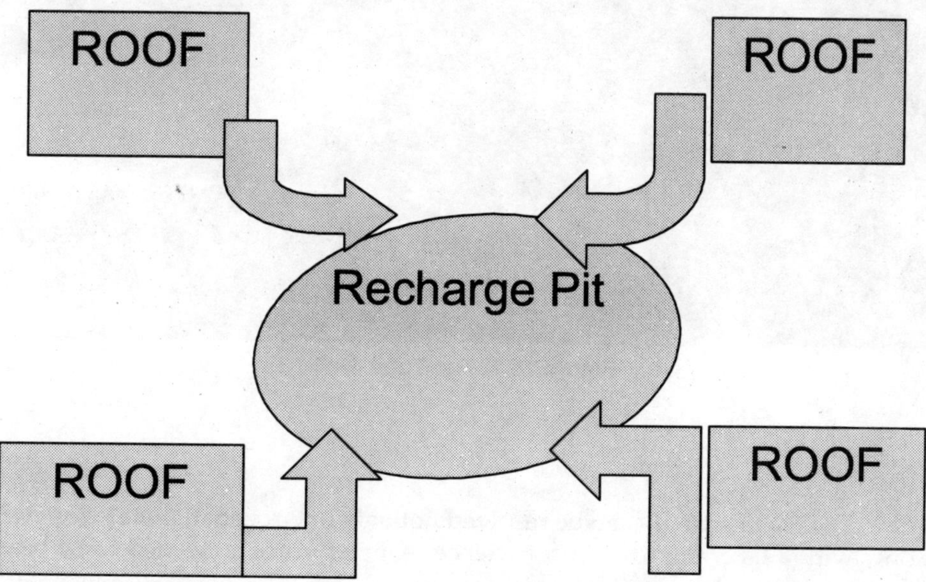

This is the most easy and basic concept for rain water harvesting in which roof water of house is attached with the recharge point. If only this is done in every street of the city, there would be no problem for generations to generations.

For Industry/At Micro Level

For industries or for Govt organizations, work is needed to be done with broad, careful and serious attitude. Either the small reservoir could be made or different approaches could be taken to preserve the colorless gold.

As a whole we could say that there is no fixed technique to conserve the water but approach is to be made accordingly with the geographical and demographical perspective.

Figure 16.3: The Tube Well

Conclusion

The problem of water scarcity can be resolved not only in Firozabad but also anywhere in this world but only with a listening to own conscience. A inner voice is needed to be heard when it murmurs "Conserve the water. Don't waste it'

Somebody did a golden deed;

Somebody proved a friend in need;

Somebody sang a beautiful song;

Somebody smiled the whole day long;

Somebody thought, "It's sweet to live;"

Somebody said, "I'm glad to give;"

Was that Somebody".... you? (You, 1996).

References

Courtesy: http://updes.up.nic.in/spatrika/engspatrika/tab7.asp?formd=21+ Firozabad +++++++++++++++++++++&formy=0910

http://updes.up.nic.inspatrikagraphicaldist21.asp

Standard methods of examination of water and wastewater 14th editions (1975) APHA, WWW, WPCF

You, 1996. Inc.: Burke Hedges, Inti Publication.

2013, Sustainable Approaches for Environmental Conservation *Pages* ***123–125***
Editors: **D.R. Khanna, A.K. Chopra, R. Bhutiani, Gagan Matta & Vikas Singh**
Published by: **BIOTECH BOOKS, NEW DELHI**

Chapter 17

Environment Education

Siddharth Kumar Pokhriyal

Uttarakhand Open University, Haldwani, Nainital, Uttarakhand

Environment is a broad concept encompassing the whole range of diverse surroundings in which we perceive experience and react to events and changes. It includes the land, water, vegetation, air and the whole gamut of the social order. It also includes the physical and ecological environment. It concerns man's ability to adapt both physically and mentally to the continuing change in environment. The environment comprises of the Abiotic (the non-living), Biotic (the living) and the human components. All the components are deeply inter-related and interwoven. They influence the life of an individual organism. They also influence the life of individual organism. They also influence men, their habit, custom and activities. Economic development is a must for the human welfare, but at the same time due care has to be taken of the environment. Maintains of its quality is equally essential otherwise the future generation of humanity will have to suffer in many ways.

Keywords: Environment education, Pollution, Earth, Development.

The word *environment* comes from the French word *"environmer"* which means which means to encircle or surround. The term environment refers to the circumstances or conditions that surround an organism or group of organism. This term also refers to the social or cultural conditions.

Environment is a broad concept encompassing the whole range of diverse surroundings in which we perceive experience and react to events and changes. It includes the land, water, vegetation, air and the whole gamut of the social order. It also includes the physical and ecological environment. It concerns man's ability to adapt both physically and mentally to the continuing change in environment.

Environment is not static. It is dynamic and the changes occur even if there is no human interference. In its natural uninterfered conditions, the Environment of any region is in the state of dynamic equilibrium. This is what is called the balance of nature.

The environment of earth is a combination of two things. One is called physical environment and the other environment. The physical environment includes the non-living elements such as land, water and air. The biological environment includes plant, animals and micro organism. Physical and biological environment are inter-dependent. For instance, excess felling trees leads to decline in the population of wildlife as well as increase in temperature.

The environment comprises of the Abiotic (the non-living), Biotic (the living) and the human components. All the components are deeply inter-related and interwoven. They influence the life of an individual organism. They also influence the life of individual organism. They also influence men, their habit, custom and activities.

Environment sets limit to all human activities. A particular environment may act as a favorable factor in the economic or social development of a region or it may prove to be an obstacle in its growth and development. Infact environment is the area within limits of which all human actions are determined.

From socio-economic context, environment can be regarded as the stock and flow of resources not created by man. These resources are exploited by man for his socio-economic benefits.

Environment is both physical and cultural. Physical environment is the product of the nature while cultural environment is the product of man.

Cultural environments are also known as social environments. These are those which have been created artificially or one man-made, *e.g.*, cancels, railway factors like popular density, cultural background of the people, the level of technological development and factors like personal, domestic, religious, educational means of transport market facilities, economic conditions, policies of the government *etc.* Social factors also play an important role to influence the nature of agriculture in a region.

Environmental Impact on Human Society and Human Impact on Environment

All the environmental forces influence the living of a man, his number, density and distribution and occupation structure. So men too, by his continued occupance of an area invoke responses from the environment. This human impact on environment depends not only on his number and density but also on his level of culture, technology and relative rate of progress. Invariably, the rate of impact of human action on environment is more rapid than the reverse process except, possibly in areas of very low density of population and of tribal food gathering societies.

The expression of the impact of human action on environment are varied. They are seen at local regional and global level, *e.g.* reclamation of saline soils, irrigation canals, creation of urban and industrial areas on virgin lands, deforestation, depletion of natural resources, retreat of wildlife, creation of polluted environment, mining landscape, with dump yards *etc.*

Economic development is a must for the human welfare, but at the same time due care has to be taken of the environment. Maintains of its quality is equally essential otherwise the future generation of humanity will have to suffer in many ways.

Environment Education

Education at all levels should be devoted to knowledge about environmental problems. In the fragile eco-systems like Himalayas and deserts, environmental education has made great progress in India in recent years. Conservation of forests resources is beneficial too.

Education in environment protection is therefore a big source of investment. Combined with new eco-friendly technologies for industrial and agricultural production, education can open up new vistas for progress. The environment revolution world over holds great promise for mankind.

Studies of environmental education are getting lot of attention not only in the field of pollution control but also to sustain the life and nature. Environmental studies help us to understand the nature of environment and its components, nature of disturbing factors. The disturbing factors pressurize sustainability and natural living.

There are several environmental problems which have been solved by using environmental studies, such as global warming, population problems, depletion of the ozone layer, habitat destruction and species extinction, energy production, groundwater depletion and contamination, *etc.*

The Conflict

There has always been some conflict between the development bodies and the environmentalists. The environmentalists never want the ecosystem to be disturbed. Whereas, the development programmers like setting of large-scale industries, power plants construction of dams *etc.* cannot take place without causing some disturbance to the ecosystem. But these development programmes are must for betterment of national economy and larger well being of mankind. Simultaneously, it is also important that ecosystem showed not disturbed. So, it is very natural that a conflict may arise between the two bodies.

2013, Sustainable Approaches for Environmental Conservation *Pages* ***127–133***
Editors: **D.R. Khanna, A.K. Chopra, R. Bhutiani, Gagan Matta & Vikas Singh**
Published by: **BIOTECH BOOKS, NEW DELHI**

Chapter 18

Studies of Adsorption from the Adsorbent Prepared from Waste Material

M.P. Patil and P.P. Chahande

Department of Chemistry, Sevadal Mahila Mahavidyalaya, Nagpur, M.S.

Activated carbon is made from various agricultural waste by physical and chemical activation. The preparation of activated carbon from waste material could increase economic return and also provide an excellent method for solid waste disposal, thereby reducing pollution. The adsorption of potassium permanganate on prepared activated carbon was investigated. Freundlich and Langmuir adsorption were studied. Photocolorimetry technique was also applied. The result indicates that it can act as a good adsorbent.

Keywords: *Activated carbon, Peanut shells, Adsorption.*

Introduction

Activated carbon has long been recognized as one of the most versatile adsorbents to be used for effective removal of low concentrations of organic compounds from aqueous solutions. The activation of carbon provides many of its useful properties (Chermisinoff and Moressi, 1980). Nowadays interest is growing in the use of other low cost and abundantly available agricultural and other waste. In tropical countries a vast variety of fruit and agricultural wastes are likely precursors for the manufacturing of activated carbon from maize cob, almonds shells, ground nut husk, palm seed coat *etc.* It was found that colour removed by activated carbon is upto 94 per cent. It was found that the color removal by coir pith carbon was 94 per cent which is comparable with commercial activated carbon (90 per cent) at pH 8.6 (Mittal and Venkobachar, 1989). Dyes are harmful to all living [organisms] systems polluting water and soil. The uptake of dyes from wastewater using adsorption was reported in literature (Sivamani and Grace Leena, 2009).

Adsorption is one of the techniques that would be comparatively more useful and economical. Several adsorption methods have been developed and tested ranging from low cost waste material such as moss peat (Sharma and Forster, 1993), hazelnut shell (Kobya, 2004), rice husk carbon (Srinivasan *et al.*, 1988) to more sophisticated adsorbents such as modified clay (Tavani and Volzone, 1997), modified steel slag (Ocholo and Moo-Young, 2004), nanoscale magnetic material (Hu *et al.*, 2005). Many methods have been reported for the carbonization and activation of waste organic material and these are adequately covered in the text by Hassler (1974). India is producing peanut on large scale. During use the shells are removed as a solid waste. The abundant and easily available solid waste material peanut shells can be converted to an effective low cost adsorbent to remove colour from wastewater.

Experimental Method

Activated carbon was prepared from peanut shells. Peanut shells were collected, washed, dried. It was crushed and roasted. Then it was kept in 25 per cent calcium chloride solution overnight. Later, it was washed several times with distilled water and dried. Adsorption of oxalic acid on activated carbon was studied from which Freundlich and Langmuir adsorption was verified. Six stopper bottles were prepared as mentioned in observation table. After shaking for one hour it was filtered. 10 ml of filtrate was titrated against standard alkali (0.05 N).

Observations and Calculations

Table 18.1: Adsorption Capacity of Activated Carbon of Oxelic Acid

Bottle No.	*Wt.of Charcoal (gm)*	*Acid + Water*	*Vol. of NaOH Before Adsor-ption for 10ml (A)*	*Vol. of NaoH After Adsor-ption for 10ml (C)*	*Amount of Acid Adsorbed in Terms of NaoH (x) = A-C*	*x/m*	*Logx/m*	*log C*	*1/x/m*	*1/C*
1	1	50+0	20	9.3	10.7	10.7	1.0293	0.9684	0.0934	0.1075
2	1	40+10	16	12.5	3.5	3.5	0.5440	1.0969	0.22857	0.0800
3	1	30+20	12	8.3	3.7	3.7	0.5682	0.9190	0.2702	0.1204
4	1	20+30	8	7.8	0.2	0.2	0.698	0.8920	5.000	0.1282
5	1	10+40	4	1.6	2.4	2.4	0.3802	0.2041	0.4166	0.625
6	1	5+45	2	0.3	1.7	1.7	0.2304	0.5228	0.5882	3.333

Freundlich adsorption can be written as $x/\mathrm{m} = \mathrm{KC}^{1/n}$

where, x = the amount of adsorbate by m adsorbent,

C = Equilibrium conc. of adsorbate.

K and n are constants.

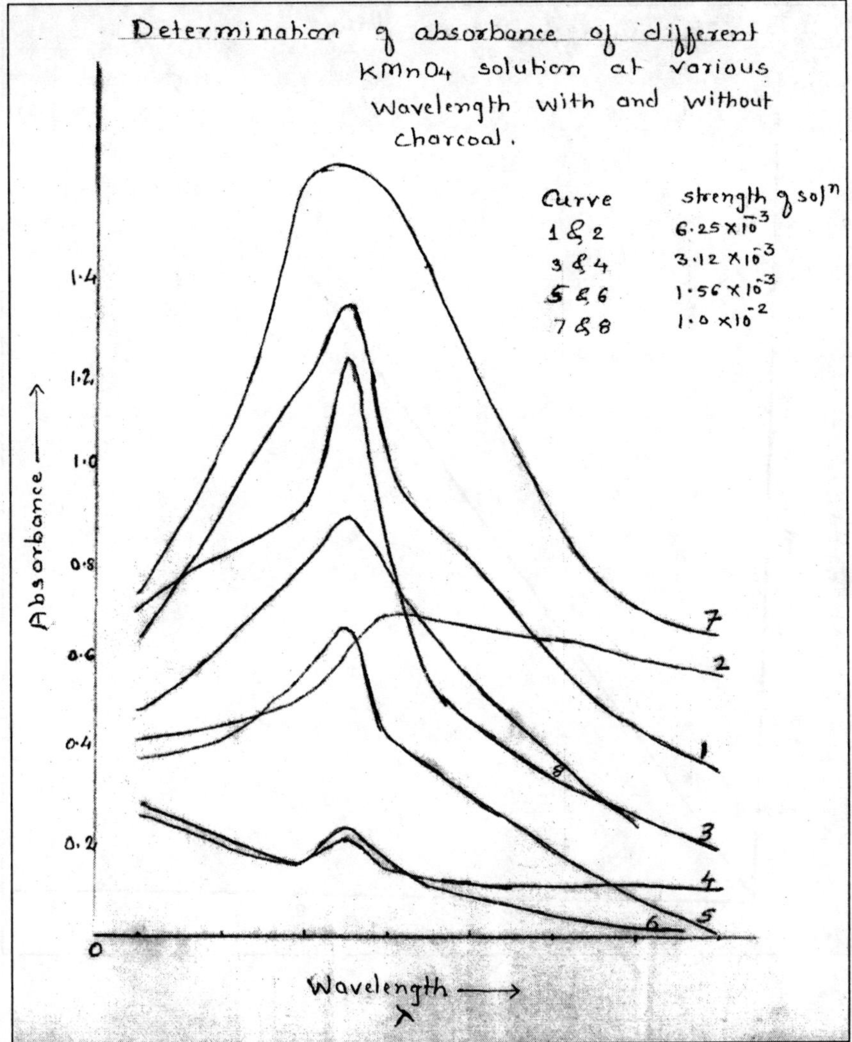

Figure 18.1

Log x/m =log K + 1/n log C

Plot log x/m versus log C. If it gives a straight line with slope equal to log K, Freundlich adsorption isotherm was verified.

Langmuir Adsorption Isotherm

Langmuir adsorption isotherm assumes a unimolecular layer adsorption on the surface of adsorbsent.

For solution the isotherm is $x/m = \dfrac{K_1 K_2}{1+KC}$

where, x is the mass of solute adsorbed on mass m of adsorbent and K are constants.

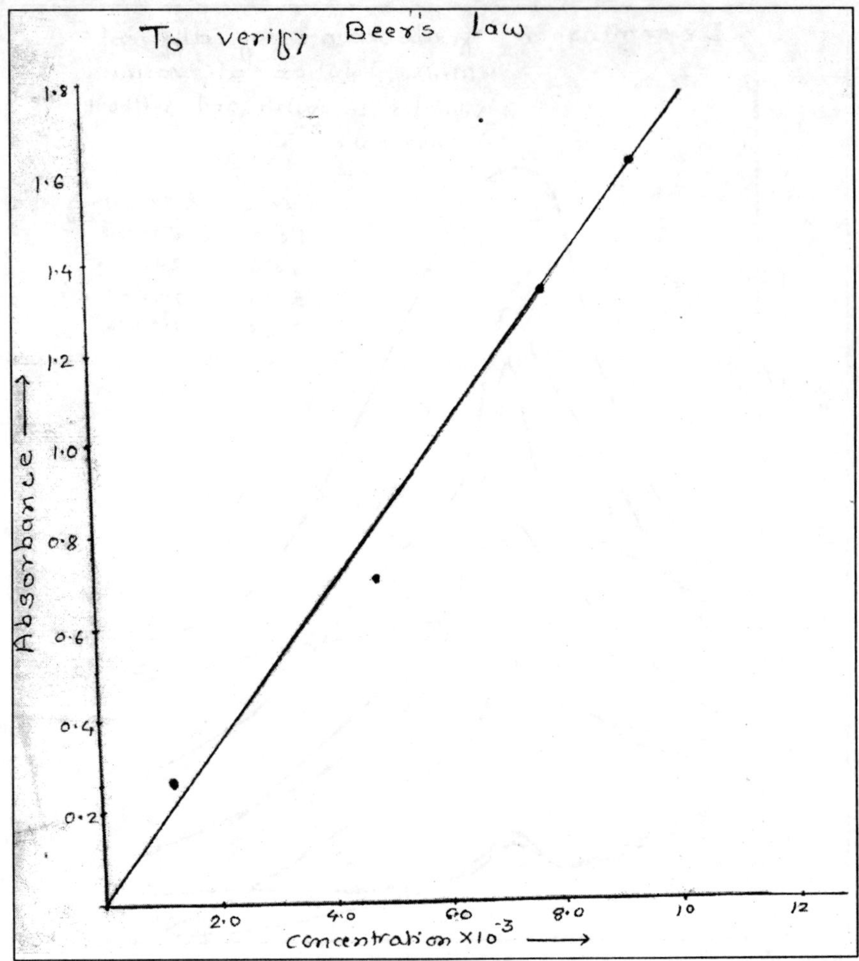

Figure 18.2

$$\frac{C}{x/m} = \frac{1}{K_1 K_2} + \frac{1}{K_2}$$

A straight line plot of $1/c$ versus $1/x/m$ confirms Langmuir adsorption isotherm.

Colorimetric Analysis

Colorimetry is the oldest known technique for determining any colour. The intensity of colour of a substance is in direct proportion to its concentration, which is in terms of transmittance (Optical density). Concentration of various solution can be determined by colorimetric technique. 0.01M Potassium permanganate solution was prepared. On diluting the solution 6.25×10^{-3}, 3.12×10^{-3}, 1.56×10^{-3} was prepared. The absorbance of different concentrated solution bto hwithout and with prepared activated carbon was recorded.

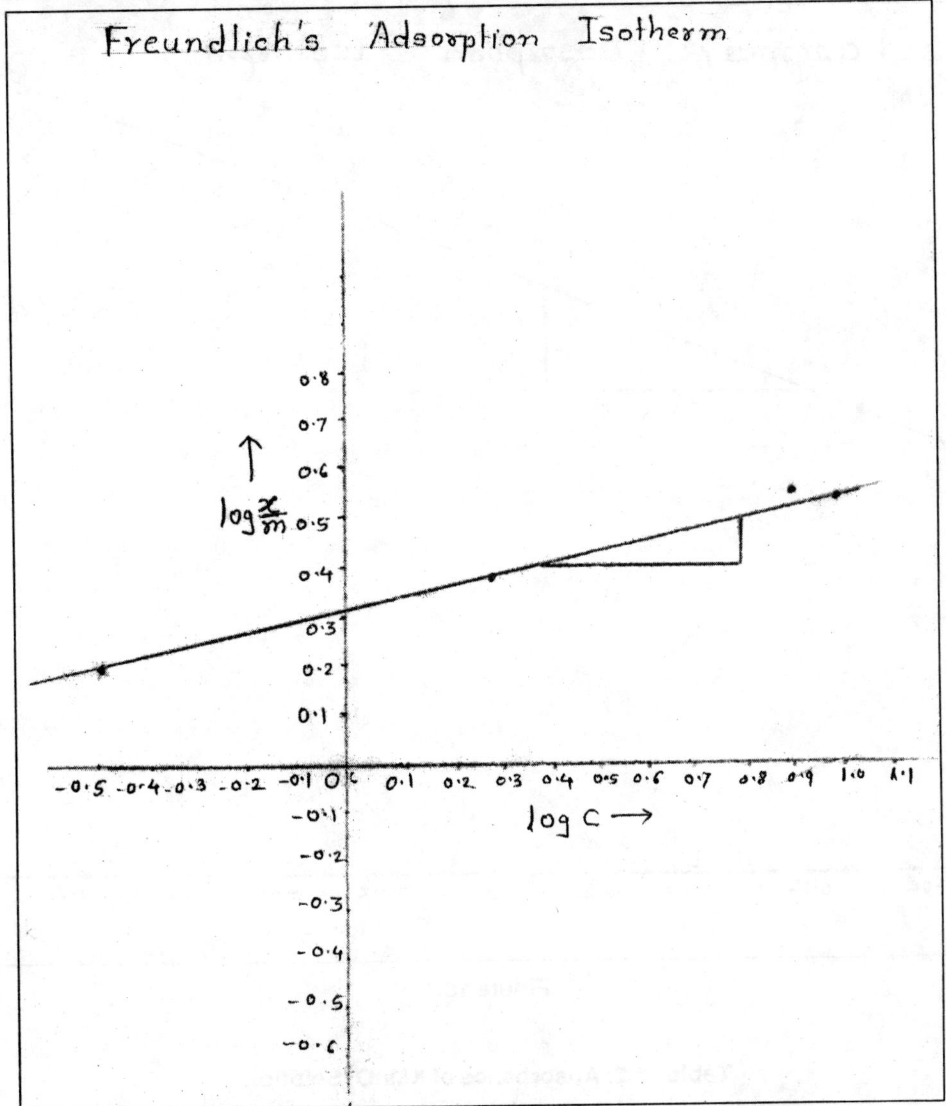

Figure 18.3

Concentration of unknown solution can be calculated as follows:

Concentration of solution: $\frac{1}{I_0} x$ Concentration of standard solution

For colour determination of different solutions five bottles of different concentrations were prepared. 50 ml of 1x10^{-2}m, 6.25 x10^{-3}m, 3.12x10^{-3}m, 1.56 x10^{-3}m solutions and 0.2 gm activated charcoal was added in each bottle. The absorbance was recorded before addition of charcoal and after shaking the bottles for 15 minutes at different wavelengths.

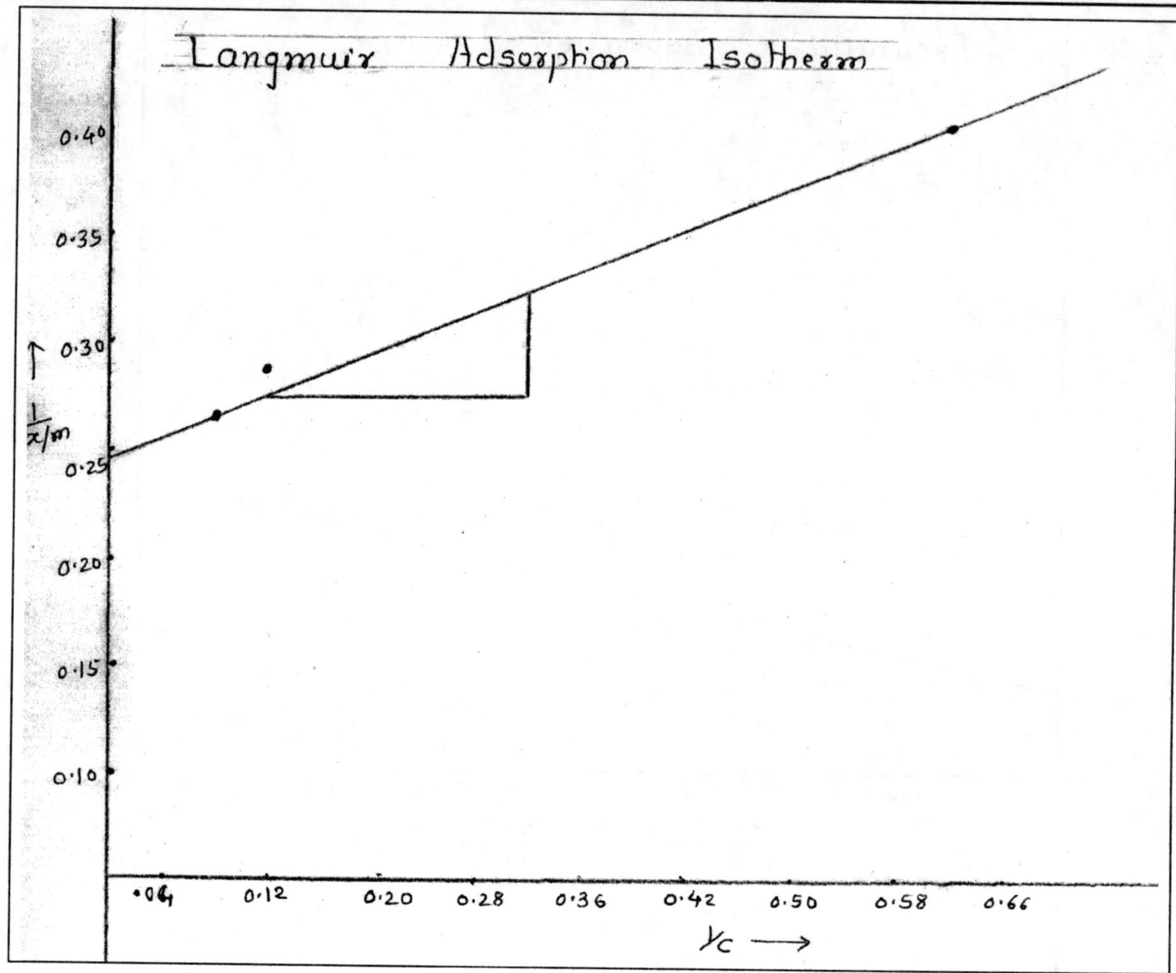

Figure 18.4

Table 18.2: Absorbance of $KMnO_4$ Solution

Wavelength λ	*1x10⁻²*		*6.25 x10⁻³*		*3.12x10⁻³*		*1.56 x10⁻³*	
	Before	*After*	*Before*	*After*	*Before*	*After*	*Before*	*After*
400	0.81	0.77	0.70	0.47	0.54	0.33	0.43	0.29
420	0.65	0.43	0.45	0.14	0.21	0.04	0.09	0.02
480	1.80	1.19	1.30	0.57	0.88	0.18	0.61	0.18
500	1.00	1.38	1.50	0.69	1.03	0.24	0.74	0.26
520	1.78	0.94	1.18	0.67	0.74	0.16	0.48	0.18
540	1.36	0.61	1.06	0.26	0.63	0.13	0.41	0.04
620	0.86	0.33	0.56	0.00	0.34	0.07	0.15	0.01
680	0.72	1.00	0.40	0.53	0.21	0.11	0.02	0.12

Table 18.3: Wavelength and Concentration Before and After Adsorption

Original Wavelength	Concentration			
	1.0×10^{-2}	6.25×10^{-3}	3.12×10^{-3}	1.56×10^{-3}
400	9.50×10^{-3}	4.19×10^{-3}	1.90×10^{-3}	1.05×10^{-3}
420	6.61×10^{-3}	1.94×10^{-3}	5.94×10^{-4}	3.46×10^{-4}
480	6.66×10^{-3}	2.74×10^{-3}	6.38×10^{-4}	4.60×10^{-4}
500	1.38×10^{-3}	2.87×10^{-3}	7.26×10^{-4}	5.48×10^{-4}
520	5.28×10^{-3}	3.54×10^{-3}	6.74×10^{-4}	5.85×10^{-4}
540	3.86×10^{-3}	1.53×10^{-3}	6.43×10^{-4}	1.52×10^{-4}
620	3.83×10^{-3}	-	6.42×10^{-4}	1.04×10^{-4}

Results and Discussion

The graph of log x/m versus log C was plotted. The graph was straight line with slope 0.20, intercept 0.32 and the value of K was found to be 2.08. Freundlich adsorption isotherm was verified. The graph of $1/x/m$ versus $1/c$ was plotted. The graph was straight line with slope 2.77, intercept 0.24 and the value k_1=11.33 and k_2=0.36. Thus Langmuir adsorption isotherm was verified. For determination of colour adsorption photocolorimetry technique was used. From observation table it is clear that the concentration of $KMnO_4$ decreases when the content was shaken for 15 minutes along with the activated charcoal prepared from waste peanut shells. Thus activated charcoal prepared from waste peanut shells can act as adsorbent.

References

Chermisinoff, N. P. and Moressi, A. C. (1980). Carbon adsorption and applications. Carbon Adsorption Handbook. Ann. Arbor Michigen, 1-54.

Hassler, J. W. (1974). Purification with activated carbon, 2nd Edn, Chemical Publishing Co. Inc., New York, 179-185.

Hu, J. Chen G. H. and Lo TMC (2005). Water Res., 39: 4528.

Kobya, M. (2004). Bioresour Technol, 91(3): 317.

Mittal, A. K. and Venkobachar, C. (1989). Studies on adsorption of dyes by sulphonated coal and Ganoderma lucidum. Indian J. Environ. Hlth, 31: 105-111.

Ocholo, C. E. and Moo-Young, H. K. (2004). Environ. Sci. Technol, 38: 616.

Sharma, D. C. and Forster, C. F. (1993). Water Res. 27: 1201.

Sivamani, S. and Grace Leena, B. (2009). Removal of dyes from wastewater using adsorption. Int. J. Bio Sci. Tech. 2, 47.

Srinivasan, K., Balsubramanian, N. and Ramkrishna, T. V. (1988). Indian Environ. Health, 30(4): 376.

Tavani, E. L. and Volzone, C. (1997). Leather Technol Chem, 81: 143.

2013, Sustainable Approaches for Environmental Conservation *Pages* ***135–143***
Editors: **D.R. Khanna, A.K. Chopra, R. Bhutiani, Gagan Matta & Vikas Singh**
Published by: **BIOTECH BOOKS, NEW DELHI**

Chapter 19

Correlates of Dietary Intake of Pregnant Women with the Birth Weight of Newborns

***Meghali Joharapurkar*[1] *and Rekha Sharma*[2]**

[1]*Department of Food and Nutrition, Sevadal Mahila Mahavidyalaya, Nagpur, M.S.*
[2]*UGC-Academic Staff College, Rashtrasant Tukadoji Maharaj, Nagpur University, Nagpur, M.S.*

Pregnancy is one of the most critical and unique period in a women's life cycle. The present study was carried out among 215 pregnant women in the last trimester. The samples were selected from Government Medical College and Hospital of Nagpur City. Information was collected on socio-demographic characteristics and 24 hours recall method of diet survey was applied for the collection of dietary information by personal interview method using questionnaire. Birth weight of newborns were noted from hospital records. The mean intake of different food stuff consumed was computed for a day and compared with the Recommended Dietary Intake. The nutrient intake was calculated and the mean intake of nutrient was compared with the Recommended Dietary Allowances. The data obtained was analyzed using percentages, means and standard deviations. Correlation were computed using Pearson's Product Moment Coefficients. The results of study indicated that the diet of pregnant women was deficient in all foods, except roots and tubers and other vegetables. The intake of cereals, pulses, green leafy vegetables and fats and oils ranged between 61-75 per cent of RDI. The intake of fruits and milk and milk products was found to be 31-33 per cent of RDI. It was found that the mean intake of energy, protein, iron, calcium, folic acid were lower than RDA values. Energy, protein, fat and carbohydrate intake of pregnant mothers were positively and significantly correlated with birth weight. With respect to micronutrient thiamin, niacin, calcium, magnesium and vitamin B_{12} intake of mother showed positive and significant correlation with birth weight of newborn.

Keywords: Pregnancy, Socio-demographic, Diet, Nutrition, Recommended Dietary Allowances.

Introduction

Pregnancy is one of the most critical and unique period in a women's life cycle. Pregnant women have been widely recognized as a vulnerable group from health point of view. They need more food than normal person for the proper nourishment of growing foetus. Maternal nutrition and health is considered as the most important regulator of human foetal growth. A healthy mother can produce a healthy child. If women are not well nourished, they are more likely to give birth to weak babies resulting in high infant mortality rate (Sahoo and Panda, 2006).

WHO signified that a women's normal nutritional requirements increases during pregnancy to meet the needs of the growing foetus and of the maternal tissues associated with pregnancy (Mridula *et al.*, 2002). Adequate nutrition before and during pregnancy is very important for a long term health of women and child. A women who has been well nourished before conception begins her pregnancy with reserves of several nutrients so that the needs of the growing foetus nourished in the womb, have enhanced chance of entering life in good health. Mother's diet should provide adequate nutrients so that maternal stores do not get depleted (Singh *et al.*, 2009).

A significant correlation between the quality of mother's diet and birth weight of infants was observed by many researchers (Mohapatra *et al.*, 1993; Roslo, 1978; Lechtig and Yarbrough, 1975). In the present study an attempt was made to study the correlates of dietary intake of pregnant women with the birth weight of newborns.

Materials and Method

The present study was conducted on 215 pregnant women, who were in their last trimester of the pregnancy period. They were selected from Government Medical College and Hospitals of Nagpur City. Purposive sampling method was adopted for the selection of the sample. A pretested structured interview schedule was used for collection of general information. Twenty four hours recall method of diet survey was used for the collection of dietary information. Birth weight of newborns was noted from hospital records. The mean intake of different foodstuff consumed was computed for a day and compared with Recommended Dietary Intake (RDI). The nutrient intake was calculated and the mean intake of nutrient was compared with the Recommended Dietary Allowances (RDA). The data obtained were analyzed using percentages, means and standard deviations. Correlations were computed using Pearson's Product moment coefficients. The data thus collected was analyzed statistically and results were interpreted accordingly.

Results and Discussion

Socio-demographic Profile

It was observed that 55.35 per cent respondents belong to 20-25 age group, followed by respondent of 25-30 years of age (34.42 per cent). The mean was ascertained to be 24.43ü3.24 years (Table 19.1). The husbands age in majority case (44.65 per cent) were between 30-35 years. 67.91 per cent respondents got married within 20-25 years of age. Majority of the respondent (40.47 per cent) belonged to nuclear family. On the basis of size of family, 48.84 per cent had family size 4-6 members, 47.44 per cent had family size between 1-3 and rest were from large family having member 7-9. 76.28 per cent respondent were Hindu religion. The educational status revealed that only 3.72 per cent of pregnant women were illiterate, 58.14 per cent pregnant women had education up to SSC. Out of the total 51.16 per cent respondents husbands were engaged as labourers and 27.44 per cent were in service. 91.16 per cent women were housewife. The mean per capita income of pregnant women was Rs.678.60±554.25.

Table 19.1: Distribution of Respondents on the Basis of Socio-Demographic Characteristics

Sl.No.	*Respondents*	*Categories*	*Pregnant Women N=215*	
			No	*per cent*
1.	**Age (Years)**	< 20	1	0.47
		20-25	119	55.35
		25-30	74	34.42
		< = 30	21	9.77
2.	**Age of Husband (Years)**	< 25	8	3.72
		25-30	86	40
		30-35	96	44.65
		35-40	18	8.37
		Above 40	7	3.26
3.	**Age of Marriage (Years)**	Below 20	51	23.72
		20-25	146	67.91
		25-30	17	7.91
		30-35	1	0.47
4.	**Type of Family**	Nuclear	87	40.47
		Nuclear + 1 Dependent	40	18.6
		Nuclear + 2 Dependent	44	20.47
		Joint	44	20.47
5.	**Family Size**	1-3	102	47.44
		4-6	105	48.84
		7-9	07	3.26
		Above 9	01	0.47
6.	**Type of Religion**	Hindu	164	76.28
		Muslim	07	3.26
		Buddha	43	20.00
		Christian	01	0.47
7.	**Education of Pregnant Women**	Illiterate	8	3.72
		Up to Primary	3	1.4
		Up to SSC	125	58.14
		Up to HSSC	37	17.21
		Graduate	35	16.28
		Post Graduate	7	3.26
8.	**Education of Husband**	Illiterate	6	2.79
		Up to Primary	4	1.86
		Up to SSC	125	58.14
		Up to HSSC	42	19.53
		Graduate	28	13.02
		Post Graduate	10	4.65

Contd...

Table 19.1–Contd...

Sl.No.	Respondents	Categories	Pregnant Women N=215	
			No	per cent
9.	**Occupation of Husband**	Labour	110	51.16
		Farmer	15	6.98
		Business	31	14.42
		Service	59	27.44
10.	**Occupation of Pregnant Women**	Housewife	196	91.16
		Working	19	8.84
11.	**Per Capita Income**	< 500	119	55.35
		501-1000	71	33.02
		1001-1500	10	4.65
		> 1500	15	6.98

Food Habits

A majority of (55.35 per cent) pregnant women were vegetarian whereas 36.28 per cent pregnant women were taking eggs respectively. It was noted that only 8.37 per cent were non vegetarian (Table 19.2). The consumption of non vegetarian foods was once or twice in a month. However the consumption of non-vegetarian foods was avoided during pregnancy due to spiciness of these foods.

Table 19.2: Distribution of Pregnant Women According to Food Habits, Dietary Pattern and Frequency of Eating Out

Sl.No.	Categories	Pregnant Women N=215	Sl.No.	Categories	Pregnant Women N=215
	Food Habits			**Dietary Pattern**	
1.	Vegetarian	119 (55.35)	1.	3 Meals	20 (9.30)
2.	Eggitarian	78 (36.28)	2.	4 Meals	78 (36.28)
3.	Non-vegetarian	18 (8.37)	3.	5 Meals	75 (34.88)
	Frequency of eating out		4.	6 Meals	42 (19.53)
1.	Sometimes	0 (0.00)			
2.	Very often	3 (1.40)			
3.	Not at All	212 (98.60)			

Mridula *et al.* (2003) also observed that majority of (67.5 per cent) the subjects were consuming vegetarian diet. Their diet was mainly wheat and rice based with limited amounts of pulses, vegetables and milk. Selected pregnant women (20.83 per cent) consumed egg once or twice a week. Although they were taking daily before pregnancy, egg intake was reduced during pregnancy as it was considered to cause miscarriage because egg is believed to be a hot food. Only 11.57 per cent mother's were consuming non-vegetarian items once or twice per week.

Majority of pregnant mother (36.28 per cent) were found to be consuming four meals pattern *viz.*, morning tea, lunch and dinner with either breakfast or snacks. 9.30 per cent pregnant women were consuming 3 meal patterns. Only 1.40 per cent pregnant women were found to be consuming foods outside and *Panipuri* was the most preferred food (Table 19.2).

Food Intake

The mean daily intake of foodstuff of pregnant women was compared with recommended dietary intake (ICMR Spl. Rep. Ser. No. 42) and has been presented in Table 19.3. Intake of all food stuffs by pregnant women except roots and tubers, other vegetables and sugar and jaggery, fell short of the required amount. The mean intake of cereals, leafy vegetables, roots and tubers, fruits, milk and milk products and fats and oils of pregnant women were 206.35 gm, 83.51 gm, 101.58 gm, 62.91 gm, 169.30 ml and 22.74 gm per day, respectively. The per cent adequacy of pulses for pregnant women were 61.43 per cent of RDA. The per cent adequacy of pregnant women for cereals, pulses, green leafy vegetables and fats and oils ranged between 55-75 per cent of RDI. The intake of fruit and milk and milk products were 31.45 and 33.86 per cent of RDI respectively. However the adequacy for roots and tubers, sugar and jaggery and other vegetables ranged between 101.58 to 151.76 of RDA.

Table 19.3: Mean Daily Food Intake of Pregnant Women

Sl.No.	*Food Stuff (gm/day)*	*RDI #*	*Actual Intake N=215 Mean ± SE*	*Per Cent Adequacy per cent*
1	Cereals (gm)	300	206.35 ± 2.05	68.78
2	Pulses and Legumes (gm)	60	36.86 ± 1.37	61.43
3	Green Leafy Vegetables (gm)	150	83.51 ± 5.46	55.67
4	Root and Tubers (gm)	100	101.58 ± 2.21	101.58
5	Other Vegetables (gm)	120	182.12 ± 3.73	151.76
6	Fruits (gm)	200	62.91 ± 3.42	31.45
7	Milk and Milk Products (gm)	500	169.30 ± 7.77	33.86
8	Fats and Oils (gm)	30	22.74 ± 0.29	75.81
9	Sugar and Jaggery (gm)	20	20.65 ± 0.32	103.26

*p < (0.05), ** p < (0.01); # ICMR Spl. Rep.Ser.No.42.

Jood *et al.* (2002) also reported the intake of cereals, pulses, roots and tubers, vegetables and sugar and jaggery by the rural pregnant women of Haryana State were significantly lower than the prescribed Indian Recommended Dietary Intakes (RDI). The consumption of milk and milk products and fats and oils was significantly higher than that of RDI whereas, green leafy vegetables and fruits were the most limited foods items. As the diets of rural pregnant women were inadequate with respect to some food groups, it resulted in lower intake of protein, beta-carotene and ascorbic acid.

The consumption of all the foods was found to be lower than the RDA of the ICMR, except for fats and oils and sugar and jaggery even before and after counseling (Poone and Lakashmi, 2007).

Nutrient Intake of Pregnant Women

Table 19.4 indicates the average intake of different nutrients in comparison to RDA. As per RDA the energy consumption should have been 2175 kcal/day. The mean daily intake of energy of pregnant women was 1631.74 ± 16.38 kcal. During pregnancy, protein rich diet promotes optimum foetal growth. RDA for protein for pregnant women is 65 gm per day. In this study the mean protein intake of pregnant mother's was 44.01 ± 0.60 gm per day. However, the intake of fat was found to be higher than that of RDA *i.e.,* 41.21 gm of pregnant mother's. The mean fat intake includes both visible and invisible source, hence the amount has exceeded RDA. The mean intake of vitamin A was 2401.87 ± 223.32 µg/day for pregnant mother's, whereas the mean intake of vitamin C of pregnant mother's was 116.89 ± 5.09 mg/day. Excess intake of vitamin was the reflection of adequate intake of citrus fruits and vegetables specially cabbage. RDA for folic acid for pregnant women was 400 µg per day the mean intake of folic acid of pregnant mother was 170.84 ± 4.94 µg. The average intake of niacin and vitamin B_{12} of pregnant mother were 11.22 ± 0.11 mg and 0.24 ± 0.01µg respectively. Low intake of vitamin B_{12} among the pregnant mother's might be due to lack of non-vegetarian foods in their diet. The mean daily iron intake obtained by the pregnant mother's was 14.88 ± 0.34 mg. The mean calcium intake of pregnant mother's were 603.66 ± 17.13 mg. This might be due to the inclusion of inadequate amounts of green vegetables and dairy foods.

Table 19.4: Mean Daily Nutrients Intake of Pregnant Women

Sl.No.	*Nutrients*	*RDA (1989)*	*Pregnant Women N = 215 Mean ± SE*	*Per Cent Adequacy (per cent)*
1.	Energy (kcal)	2175	1631.74 ± 16.38	75.02
2.	Protein(g)	65	44.01 ± 0.60	67.71
3.	Fat (g)	30	41.21 ± 0.74	137.36
4.	Carbohydrate (g)	-	260.15 ± 2.70	-
5.	Vitamin A (µg)	2400	2401.87± 223.32	100.08
6.	Vitamin C (mg)	40	116.89 ± 5.09	292.24
7.	Thiamin (mg)	1.1	1.21 ± 0.01	110.23
8.	Riboflavin (mg)	1.3	1.31 ± 0.10	100.45
9.	Niacin (mg)	14	11.22 ± 0.11	80.17
10.	Folic acid (µg)	400	170.84 ± 4.94	42.71
11.	Vitamin B_6 (mg)	2.5	0.20 ± 0.01	7.96
12.	Vitamin B_{12} (µg)	1	0.24 ± 0.01	23.70
13.	Iron (mg)	38	14.88 ± 0.34	39.17
14.	Calcium (mg)	1000	603.66 ± 17.13	60.37
15.	Zinc (mg)	14	8.22 ± 0.20	58.73
16.	Magnesium (mg)	300	382.44 ± 4.53	127.48

*P < (0.05), ** P < (0.01)

Pregnant mother's intake of energy and protein were 75.02 and 67.71 per cent of RDA, respectively. The per cent adequacy for vitamin B_{12}, iron and folic acid ranged between 23.70 to 42.71 per cent, whereas for calcium and niacin it was 60.37 per cent and 80.17 per cent, respectively. The intake of vitamin B_6 was (7.96) lowest. For rest of the nutrients *i.e.,* for vitamin A and vitamin C adequacy ranged between 100.08 per cent to 292.24 per cent of RDA.

Table 19.5: Distribution of Newborns According to Birth Weight

Sl.No.	*Birth Weight (kg)*	*N = 215*	*N=215 Mean ± SD*
1.	< 2	19 (8.84)	1.65 ± 0.21
2.	2-2.5	63 (29.30)	2.25 ± 0.15
3.	2.5-3	100 (46.51)	2.62 ± 0.13
4.	> = 3	33 (15.35)	3.18 ± 0.21
	Mean ± SD		2.51 ± 0.43

A study conducted by Mridula *et al.* (2003) on 120 expectant mother revealed that the average energy intake was 1954 Kcal and protein was 44 gm per day which was lower as compared to recommended dietary allowances. Another study conducted by Kharode and Antony (2002) on 30 expectant mothers showed that protein intake was 44.6 gm per day which was low as compared to RDA. According to Dahiya (2002) the diet of pregnant women was deficient in folic acid (203 mg), iron (20 mg), niacin (12.7 mg, 79.3 per cent of RDA), calories (2174 Kcal, 86.9 per cent of RDA) and riboflavin (1.28 mg, 85.3 per cent of RDA).

The average energy consumption of pregnant women of rural areas of Balasore District, Orissa was also considerably low (13.58 per cent) as compared to RDA. The mean protein intake was only 8.38 per cent as compared to RDA. The mean daily iron intake obtained by the mothers was 33.86 per cent deficits than RDA. Mean calcium intake of the mothers was lesser by 20.66 per cent than RDA. The percentage of iron deficit was high (33.86 per cent) when compared to the calcium deficit (20.66 per cent) among the pregnant women. The average daily intake of carotene was also deficit by 72.8 per cent. The mean intake of vitamin C was excess by 248 per cent than RDA. Excess intake of vitamin was the reflection of adequate intake of fruits like ripe tomato, orange, lemon *etc.* (Sahoo and Panda, 2006).

Birth Weight of Newborn

Birth weight is a reliable and sensitive predictor of a newborn's chances for survival, growth and long term physical and psychosocial development (Biswas *et al.*, 2008). In the present study the birth weight of newborns ranged between 1.2 to 3.7 kg. Majority of neonates (46.51 per cent) weighed between 2.5 to 3.0 kg whereas only 15.35 per cent weighed more than 3 kg. The mean birth weight of newborn was found to be 2.51 ± 0.43 kg.

Correlation Coefficient Between Food Intake and Birth Weight

Table 19.6 demonstrate a positive and significant coefficients between birth weight and mothers intake of cereals ($r = 0.175, p < 0.05$), leafy vegetable ($r = 0.171, p < 0.05$), milk and milk products ($r = 0.185, p < 0.01$), fats and oil ($r = 0.288, p < 0.01$) and sugar jaggery ($r = 0.160, p < 0.05$).

Correlation Coefficient Between Nutrient Intake and Birth Weight of Newborn

Table 19.6 reveals a positive and significant correlation between birth weight of neonates and energy ($r = 0.329, p < 0.01$), protein ($r = 0.296, p < 0.01$) and fat ($r = 0.282, p < 0.01$) intake of mothers. Amongst vitamins mothers intake of vitamin A ($r = 0.160, p < 0.05$), thiamin ($r = 0.233, p < 0.01$), niacin ($r = 0.230, p < 0.01$) and vitamin B_{12} ($r = 0.185, p < 0.01$) correlated positively and significantly with the birth weight. Amongst minerals the intake of calcium and magnesium correlated significantly and positively ($r = 0.269, p < 0.01; r = 0.159, p < 0.05$).

Table 19.6: Correlation Coefficient Between of Maternal Food and Nutrient Intake with Birth Weight of Newborn

Food Intake		*Nutrient Intake*	
Foodstuff	*Correlation with Birth Weight*	*Nutrients*	*Correlation with Birth Weight*
Cereals (g)	0.175*	Energy (kcal)	0.329**
Pulses and Legumes (g)	0.059	Protein (g)	0.296**
Leafy Vegetables (g)	0.171*	Fat (g)	0.282**
Roots and Tubers (g)	0.052	Carbohydrate (g)	0.303**
Other Vegetables (g)	0.045	Vitamin A (μg)	0.160*
Fruits (g)	0.114	Vitamin C (mg)	0.023
Milk and Milk Products (g)	0.185**	Thiamin (mg)	0.233**
Fats and Oil (g)	0.288**	Riboflavin (mg)	0.047
Sugar and Jaggery (g)	0.160*	Niacin (mg)	0.230**
		Folic Acid (μg)	0.126
		Vitamin B_6 (mg)	0.059
		Vitamin B_{12} (μg)	0.185**
		Iron (mg)	0.037
		Calcium (mg)	0.269**
		Zinc (mg)	0.007
		Magnesium (mg)	0.159*

* p< (0.05), ** p< (0.01)

Conclusion

The diet of pregnant mothers was found to be deficient in all foodstuffs except roots and tubers and other vegetables. The intake of cereals, pulses, green leafy vegetables and fats and oil ranged between 61 to 75 per cent of RDI. The intake of fruits and milk and milk products was found to be about 31 to 33 per cent of RDI. The nutrient intake of pregnant mothers revealed that the diet was deficient in energy, protein, iron, calcium and folic acid. Birth weight was found to be positively and significantly correlated with cereals, fat and sugar intake of pregnant mother. Energy, protein, fat and carbohydrate intake of pregnant mothers were positively and significantly correlated with birth weight. With respect to micronutrient intake, thiamine, niacin, calcium, magnesium and vitamin B_{12} intake of mothers showed positive and significant correlation with birth weight. There is a strong need to provide nutrition education to mothers to increase their awareness and knowledge to include appropriate dietary practices and to include cheap locally available nutrient dense foods in their daily diet.

Acknowledgement

Authors, are thankful to Dr. P.N. Charde, Principal, Sevadal Mahila Mahavidyalaya, Nagpur and Dr. (Mrs.) Sabiha Vali, Professor and Former Head, Department of Home Science, Rashtrasant Tukadoji Maharaj Nagpur University, Nagpur for guidance.

References

Biswas, R., Dasgupta, A., Sinha, R. N. and Chaudhuri, R. N. (2008). 'An epidemiological study of low birth weight newborns in the District of Puruliya, West Bengal.' *Indian J Public Health,* 52 (2): 65-71.

Dahiya, S. (2002). 'Nutritional status assessment of pregnant women from Hisar city of Haryana'. Nutrition and Health. 16 (3): 239-47. [Internet] Available from: <http://www.ncbi.nlm.nih.gov/> [Accessed 5 February 2009].

Gopalan, C., Rama Sastri, B.V. and Balasubramanian, S.C. (1989). *Nutritive value of Indian foods,* National Institute of Nutrition. India Council of Medical Research Hyderabad.

Gupta, S.G. and Kapoor N.K. (1982). 'Fundamentals of mathematical statistics'. 8th Edition S. Chand and Sons, 724-766.

ICMR (1989). *Nutrient Requirement And Recommended Dietary Allowances For Indians,* A Report of Expert committee of the Indian Council of Medical Research Hyderabad: NIN.

Jood, S., Bishnoi, S. and Khetarpaul, N. (2002). 'Nutritional status of rural pregnant women of Haryana state, Northern India'. *Nutrition and health,* 16 (2): 121-131. [Internet] Available from: < http://www.refdoc.fr/Detailnotice?idarticle=9580924/> [Accessed 2 March 2009].

Kharade, P. P. and Antony, U. (2002). 'Nutritional status and outcome of pregnancy in young and older mothers in Mumbai'. *Ind. J. Nutr. Dietet,* 39: 26-30.

Lechtig, A. and Yarbrough C. (1975). 'Influence of maternal nutrition on birth weight'. *Amer. J. Clin. Nutr., 28, 11.*

Mohapatra, P. Mohapatra, S.C., Agrawal, D.K. and Agrawal, K.N. (1993). 'The effect of nutritional supplementation on anthropometric parameters of antenatal women'. Ind. J. Comn. Med. 17, 28-22.

Mridula, D., Mishra, C.P. and Chakraverty, A. (2003) 'Dietary intake of expectant mothers'. *Ind. J. Nutr. Dietet,* 40 (1): 24-30.

Mridula, D., Mishra, C.P. and Chakraverty, A. (2002). 'Effect of mother's dietary intake on birth weight of newborn'. *Ind. J. Nutr. Dietet,* 39 (2): 327-332.

Poone, S. and Lakashmi, U. K. (2007). 'Effect of periconceptional counseling on serum folate levels of women planning pregnancy'. *Ind. J. Nutr. Dietet,* 44 (2): 124-130.

Roslo, P.Luke. (1978) 'Influence of maternal weight gain on the incidence of foetal growth retardation'. *Amer. Jr. Clin. Nutr. 3, 690.*

Sahoo, S. and Panda, B. (2006). 'A study of nutritional status of pregnant women of some villages in Balasore District, Orissa'. *J. Hum. Ecol,* 20 (3): 227-232.

Singh, M., Jain, S. and Choudhary, M. (2009). 'Dietary Adequacy of Pregnant Women of Four District of Rajasthan', *J Hum Ecol,* 25(3): 161-165.

Venkatachalam, P.S. and Rebello, L.M. (2004). *Nutrition for Mother and Child.* National Institute of Nutrition, Hyderabad. ICMR Special Report Series. No. 42: 9-17.

WHO, (1989). 'The prenatal and immediate pool pattern periods'. Bulletin OMS Supplementary volume 67; 9-18.

2013, Sustainable Approaches for Environmental Conservation *Pages* ***145–148***
Editors: **D.R. Khanna, A.K. Chopra, R. Bhutiani, Gagan Matta & Vikas Singh**
Published by: **BIOTECH BOOKS, NEW DELHI**

Chapter 20

Lyocell: An Eco-Friendly Fibre

Jyoti Joshi and Alka Goel

Department of Clothing and Textiles, College of Home Science,
G.B. Pant University of Agriculture and Technology, Pantnagar, Uttarakhand

The textile industry is considered as the most ecologically harmful industry in the world. The eco-problems in textile industry occur during some production processes and are carried forward right to the finished product.The concern for the degrading environment conditions due to irresponsible use of chemical products have led to worldwide efforts to develop eco-friendly fibres in the ever expanding horizon of textile fibres with a vision to bring about a drastic reduction in global consumption of harmful non-biodegradable products. As the ecological parameters becoming more stringent, it becomes the prime concern of the textile processors to be conscious about quality and ecology.

There has been a growing demand for absorbent fibres with the need hinging on comfort and fashion. Since cotton production cannot go beyond a particular level due to limited land availability, the other obvious options are viscose and the likes. But again, with the increasing awareness of eco-friendly concepts, viscose is not quite highly rated because its manufacturing plants have inherent problem of effluent generation. Carbon disulphide, which is used in significant quantity in viscose manufacturing process, is a source of major environmental problem. Thus a good water absorbent regenerated cellulosic fibre was developed known as "lyocell". Lyocell is obtained by spinning of dissolved wood pulp in organic solvent which is an eco- friendly process.

Keywords: *Non-biodegradable, Effluent, Eco-friendly fibres, Viscose, Lyocell.*

Introduction

Textile Industry occupies a vital place in the Indian economy and contributes substantially to its exports earnings. Textiles exports represent nearly 30 per cent of the country's total exports. Textiles production and processing use intensive chemical applications during the fibre production like in viscose and wet processing such as use of bleaching agents and other harmful chemicals along with dyes. Associated social and environmental concerns are growing towards increasing industrial Pollution, industrial waste disposal problems, economic and social health of the farmers and industrial worker. Petroleum-based products are harmful to the environment. In order to safeguard our environment from these effects, an integrated pollution control approach is needed. Luckily there is an availability of more substitutes. The increasing awareness of eco-friendly concepts, viscose is not quite highly rated because its manufacturing plants have inherent problem of effluent generation. Carbon disulphide, which is used in significant quantity in viscose manufacturing process, is a source of major environmental problem. Thus a good water absorbent regenerated cellulosic fibre was developed known as "lyocell". Lyocell is obtained by spinning of dissolved wood pulp in organic solvent which is an eco- friendly process.

Manufacturing Process of Lyocell Fibre

Raw Material

The main ingredient of lyocell is cellulose, a natural polymer found in the cells of all plants. The cellulose for lyocell manufacturing is derived from the pulp of hardwood trees. The pulp is typically from a mix of trees chosen for their cellulosic properties such as the colour and amount of contaminants. The solvent used in the manufacturing process is an amine oxide.

Preparing the Wood Pulp

The hardwood trees grown for lyocell production are harvested and trucked to the mill. At the mill, the trees are cut to 20 ft (6.1m) lengths and debarked by high-pressure jets of water. The logs are fed into a chopper that chops them into squares. The chips are loaded into a vat of chemical digesters that soften them into a wet pulp. This pulp is washed with water and may be bleached. Then, it is dried in a huge sheet and rolled onto spools. The roll of cellulose is enormous, weighing some 500 lb.

Dissolving the Cellulose

The roll of cellulose is unrolled from several spools and broken them into one inch squares. The workers then load these squares into a heated, pressurized vessel filled with N-methyl morpholine oxide (NMMO).

Filtering

After a short time soaking in the solvent, the cellulose dissolves into a clear solution. It is pumped out through a filter to insure that all the chips are dissolved.

Spinning

After filtering, solution is pumped through spinneret and long strands of fibre come out. The fibres are then immersed in another solution of amine oxide, diluted this time. This sets the fibre strands.

Drying and Finishing

The lyocell fibre next passes to a drying area, where the water is evaporated from it. The strands at this point pass to a finishing area, where a lubricant is applied. This may be soap or silicone or other agent, depending on the future use of the fiber.

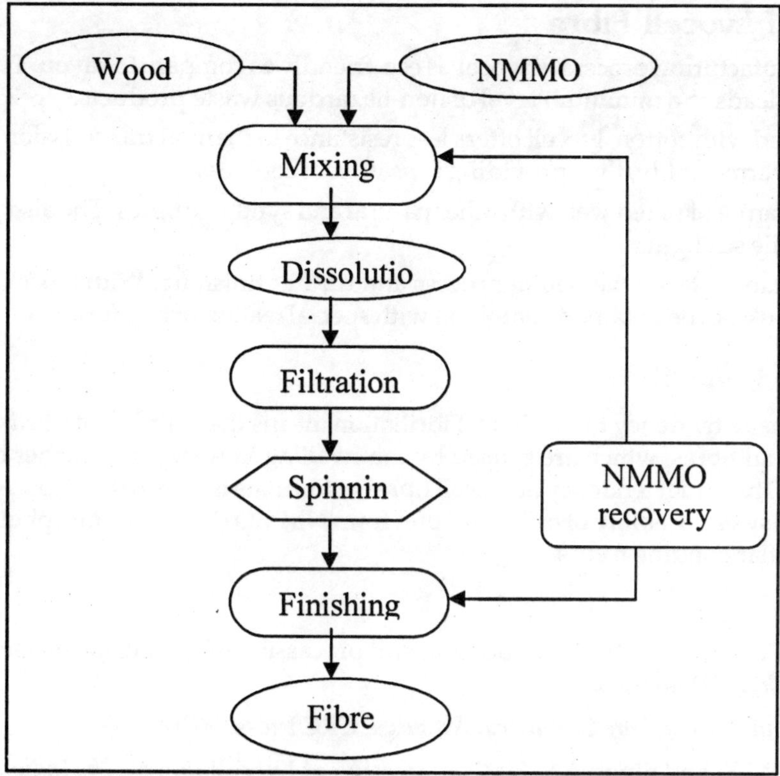

Figure 20.1: Manufacturing of Lyocell Fibre

Recovery of the Solvent

The amine oxide used to dissolve the cellulose and set the fiber after spinning is recovered and re-used in the manufacturing process. The dilute solution is evaporated, removing the water and the amine oxide is routed for re-use in the pressurized vessel. Ninety-nine percent of the amine oxide is recoverable in the typical lyocell manufacturing process.

Properties of Lyocell Fibre

- ☆ Lyocell is breathable, absorbent and generally comfortable to wear. In fact, lyocell is more absorbent than cotton and silk, but less so than wool, linen and rayon.
- ☆ It can take high ironing temperatures, but like other cellulosic fibres will scorch, not melt, if burned and is susceptible to mildew and damage by silverfish.
- ☆ Cellulosic fibres are not resilient, which means they wrinkle. Lyocell has moderate resiliency. It does not wrinkle as badly as rayon, cotton or linen and some wrinkles will fall out if the garment is hung in a warm moist area, such as a bathroom after a hot shower. A light pressing will renew the appearance, if needed.
- ☆ Lyocell has strength and durability. It is the strongest cellulosic fiber when dry, even stronger than cotton, rayon or linen and is stronger than cotton when wet.
- ☆ Lyocell has its luster and soft drape which makes it an aesthetically pleasing fiber.
- ☆ Fibrillate during wet processing to produce special textures.

Advantages of Lyocell Fibre

- ☆ The manufacturing process of lyocell is eco-friendly as compare to rayon. Total recycling of solvents leads to a minimum level of non-hazardous waste products.
- ☆ Compared with cotton, lyocell offers less resistance to thermal transmission and has higher vapour permeability, thus providing sensation of coolness.
- ☆ Lyocell can be blended well with other natural and synthetic fibres. The also yields high tear and tensile strength.
- ☆ Lyocell can undergo the denim process and further finishing. With lyocell the traditional denim look can be obtained combined with special feeling of comfort.

Fibrillation in Lyocell

Lyocell fibre has a tendency to fibrillate. Fibrillation means the splitting of fibrils along the fibre surface of individual fibres, which are caused by the swelling in water and further by the action of mechanical stress. The actual tendency of lyocell fibres to fibrillate is due to the unusually high degree of crystallinity and swelling ability of cellulose spun from NMMO (N-methyl morpholine oxide), plus the weak interfibrillar bonding forces.

References

Chavan, R.B. and Patra, A.K. 2004. Development and processing of lyocell. *Indian Journal of Fibre and Textile Research*. 29(4):483-492.

Lewin, M. 2007. *Handbook of Fibre Chemistry*. America. CRC Press. 667p.

Sayed, U., Pratap, M.R. and Singh A.S. 2002. Processing of Lyocell Fibres. *The Indian Textile Journal*. 112(7):13-20.

http://ohioline.osu.edu/hyg-fact

http://www.weyerhaeuser.com/Businesses/CelluloseFibers/Textiles

http://en.wikipedia.org/wiki/Lyocell

2013, Sustainable Approaches for Environmental Conservation *Pages* **149–160**
Editors: **D.R. Khanna, A.K. Chopra, R. Bhutiani, Gagan Matta & Vikas Singh**
Published by: **BIOTECH BOOKS, NEW DELHI**

Chapter 21

Carrying Capacity of a Temperate Grazing Land as Affected by Nomadic Grazing in Kumaun Himalaya

Prem Prakash

Department of Botany, Govt. P.G. College, Dwarahat, Almora, Uttarakhand

Carrying capacity or grazing capacity of a grazingland is essential for proper management of cattle and the sustainability of the grazingland. It is expressed in terms of livestock per acre on a given kind of cover. Grazing influences the structure and function of grasslands in different ways. Overgrazing activity has caused the replacement of herbaceous plants by shrubs and palatable species by unpalatable ones. The actual time spent for grazing by animal depends on their diet, bite size and as well as food availability in the grazingland. The bite size was directly proportional to the amount of available forage biomass. Dry matter consumption was found to be highest for buffaloes and minimum for goats. The present carrying capacity of the study area is 4.610 cows ha^{-1} yr^{-1} for animal upto 2 year, and 5.13 cows ha^{-1} yr^{-1} for animals of more than 2 years of age on the basis of 50 per cent utilization. Present capacity of the grazingland for the animals of up to two years was highest in September (18.187 cow ha^{-1}) followed by June (14.135 cow ha^{-1}), July (11.853 cow ha^{-1}) and August (11.04 cow ha^{-1}). The minimum capacity was in April (0.0006 cow ha^{-1}), while during February, October, November and December had negative or zero carrying capacity.

Keywords: *Carrying capacity, Grazing land, Ground cover, Bite size, Grazing period, Consumption.*

* Corresponding Author: E-mail: drpp_bot@yahoo.co.in

Introduction

The alpine, montane and temperate grazinglands of the Kumaun Himalaya have been influenced in various ways but extensive grazing is the biggest factor too. The estimation of grazing capacity is a step towards the range management of deteriorating grazinglands. Grazing capacity of a range is regarded as the maximum animal number that can graze each area of the range for a specific number of days, without inducing a downward trend in forage production, forage quality or soil. Grazing capacity, although sometimes used synonymously with carrying capacity is defined as the total number of animals, which may be sustained on a given area, based on total forage resources available, including harvested roughages and concentrates. It may be expressed as animal units (AU) per acre, AU per section or AU ha^{-1}.

Bor (1942) has rightly pointed out that excessive grazing by goats, sheep, camels and cattle may not only destroy the forest crop and ground cover but converts the area into barren ground. Grazing ecological studies also derive strength from the concept of co-evolution of pasture plants and herbivore (Mc Naughton, 1985; Heady and Childs, 1994).

Study Area

The study was conducted during the years 2010-2011 near Suraiketh located in district Almora (Latitudes 30° 22'-30° 14' and Longitudes 78° 56'-78° 47'), 20 km north of Dwarahat, Uttarakhand, India extending from 1200-1500 m above mean sea level. The region comprises temperate zone of Himalaya with *Pinus* as dominant tree species with graminoides as predominant under canopy vegetation and therefore, used as grazing land by local inhabitants. The study area consists of well-covered grazingland being exploited by the villagers of Suraiketh, Bitholi, Talli Bitholi, Kande, Walna, Naarh, Narsingh, Bawan, Banoli, Phaldwari and Bayela. The climate of the area is warm-temperate with moderate summers and severe winters with an annual precipitation of 250 cm, most of which commences during the rainy season.

The study was conducted at two different sites in their degree of disturbance: Ungrazed site extending over an area of 200 m x 100 m, has been permanently fenced by stone wall to avoid any type of domestic animals since 2008 and a grazed site with an area of 5 ha was exposed to extensive grazing.

Methodology

With the view of maintaining cattle population, an attempt to estimate the grazing capacity was undertaken in 2010-2011 near Suraiketh and its adjacent villages in Dwarahat forest division. After completing the survey for total animal population, they were classified into two categories *i.e.*, up-to two years and animals above two years. This categorization was done on the basis of veterinary science by which we can estimate the age of the animals (Banerjee, 1988). During the sampling all the grazing activities and other information were recorded for the two animal groups by the method proposed by Singh (1981).

Time Budget for Animal Activities

Hours Spent on Grazing/Day = Grazing hours/day – Time spent on walking + Time spent on resting + Time spent on other activities.

Bite Count of Different Animal Groups

While the animals bite off the plants, a sound occurred with each bite, therefore, number of sounds produced/15 minutes was recorded from a distance of 1-3 m to the animals at 9.30 am, 12.30 pm and 3.30 pm on the sampling date. Later on the average bite frequency is taken for the calculations.

Bite Size of Different Animal Groups

$$\text{Bite Size} = \frac{\text{Weight of intact patch} - \text{Weight of grazed patch}}{\text{Number of bites on the grazed patch}}$$

Computation

Forage Dry Matter Consumption

1. Consumption/animal/day = Bite size x Total number of bites during hours spent on grazing/day

2. $\text{No. of visit/day} \times \text{Time spent on grazing/visit/2500 m}^2 = \dfrac{\text{Total grazing period}}{2500\ \text{m}^2\text{/day}}$

3. $\dfrac{\text{Consumption/animal/day}}{\text{Time spent for grazing/day}} \times \dfrac{\text{Total grazing period}}{2500\ \text{m}^2\text{/day}} = \dfrac{\text{Consumption}}{2500\ \text{m}^2\text{/day}}$

4. $\dfrac{\text{C}}{2500\ \text{m}^2} = \text{Consumption/m}^2\text{/day}$

5. $\text{d} \times 30\ \text{(days of month)} = \text{Consumption/m}^2\text{/month}$

Standardization of Different Animal Groups against Cow

$$\frac{\text{Food consumption of cow/day}}{\text{Food consumption of a particular animal group/day}} = \text{No. of cows}$$

Grazing Capacity of the Site (Brown, 1954)

Grazing capacity of the site (animal ha^{-1}) =

$$\frac{\text{Total forage production/ha} \times \text{Proper forage use factor}}{\text{Animal requirement}}$$

(*a*) Dry matter consumption/animal/day is computed maximum during August for each animal group and that has been considered as the standard requirement of the animal.

(*b*) Proper forage use factor has been considered as 48 per cent of the total above ground production.

Results

Animals spent maximum time in the grazingland during the rainy season followed by winter and summer. The actual time spent by animal depends on their diet, bite size and as well as food availability in the grazingland. Cows spent maximum time in walking and resting while goats always grazed for the maximum time. For up to two years goats spent maximum 7.2±0.8 hour day^{-1} and cows spent a minimum 5.8±0.7 hour day^{-1} in August when the forage production was highest. In the month of June cows actually grazed for 4.0±0.9 hours day^{-1} only while goats for 5.3±0.5 hour day^{-1} only (up to two years). While for above two years age group these data were 4.6±0.5 hours day^{-1} for cows and 5.6±0.5 hours day^{-1} for goats in the month of June.

Table 21.1a: Animal Bite Frequency (hour^{-1}) in Different Months (For Up to 2 Years)

Month	*Cow*	*Bullock*	*Buffalo*	*Goat*
January	690 ± 25	695 ± 16	673 ± 20	808 ± 16
February	670 ± 30	688 ± 12	660 ± 14	794 ± 18
March	685 ± 15	697 ± 13	658 ± 19	823 ± 26
April	660 ± 12	672 ± 12	638 ± 20	790 ± 32
May	630 ± 14	642 ± 16	627 ± 16	781 ± 25
June	595 ± 22	623 ± 15	617 ± 12	753 ± 28
July	675 ± 24	687 ± 12	670 ± 17	844 ± 32
August	730 ± 12	745 ± 16	705 ± 18	956 ± 30
September	728 ± 18	736 ± 22	694 ± 22	915 ± 28
October	698 ± 23	712 ± 18	682 ± 20	887 ± 32
November	695 ± 20	701 ± 17	672 ± 18	823 ± 31
December	696 ± 21	690 ± 11	665 ± 17	812 ± 24

Among the entire animal group bite frequency was found maximum for goat and minimum for buffalo (Table 21.1b). The bite size was directly proportional to the amount of available forage biomass. The bite size or dry matter intake (g bite^{-1}) varied considerably for all the animal groups. For up to two years the maximum bite size was recorded for buffalo (1.00 g bite^{-1}) in August and minimum for goat (0.10 g bite^{-1}) in June (Figure 21.1a). For the animals of more than two year bite size was maximum in August for buffalo (1.20 g bite^{-1}) and minimum for goat (0.14 g bite^{-1}) in June (Figure 21.1b).

Table 21.1b: Animal Bite Frequency (hour^{-1}) in Different Months (For More Than 2 Years)

Month	*Cow*	*Bullock*	*Buffalo*	*Goat*
January	606 ± 16	617 ± 23	602 ± 21	785 ± 17
February	601 ± 9	597 ± 19	597 ± 26	763 ± 18
March	615 ± 12	607 ± 12	622 ± 12	777 ± 20
April	592 ± 10	585 ± 18	596 ± 20	764 ± 20
May	511 ± 11	578 ± 20	563 ± 22	748 ± 27
June	615 ± 10	570 ± 24	557 ± 12	720 ± 25
July	592 ± 11	628 ± 23	598 ± 18	798 ± 26
August	690 ± 14	678 ± 28	676 ± 25	877 ± 28
September	645 ± 12	671 ± 17	667 ± 23	860 ± 33
October	638 ± 11	647 ± 16	648 ± 22	842 ± 30
November	605 ± 16	626 ± 14	630 ± 20	818 ± 28
December	610 ± 13	623 ±20	618 ± 24	797 ± 18

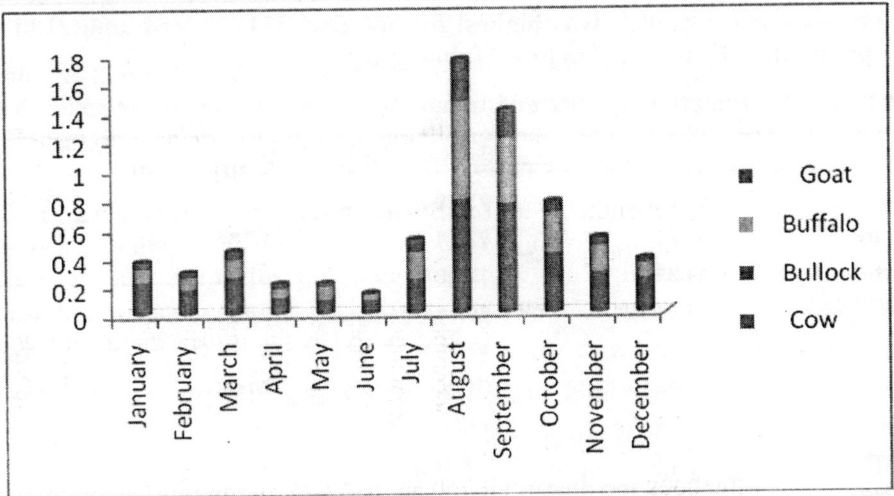

Figure 21.1a: Dry Matter Intake (g bite^{-1}) of Different Animal Groups (Bite Size) (For Up to 2 Years)

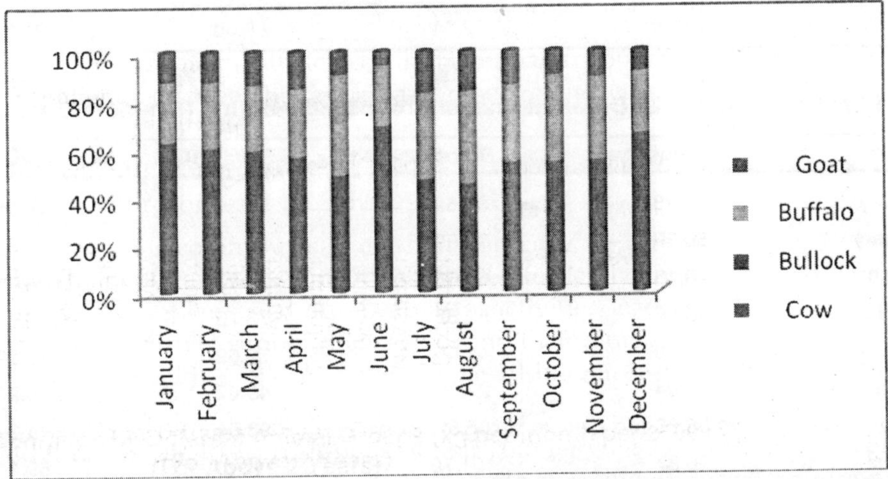

Figure 21.1b: Dry Matter Intake (g bite^{-1}) of Different Animal Groups (Bite Size) (For More Than 2 Years)

Visit per day of all the animal groups into the marked 50 x 50 m plot and time spent on grazing per visit varied for each animal group. For up to two years maximum numbers of visits were observed for goats (5 visits day^{-1}) during August followed by cows, bullocks and buffaloes (4 visits day^{-1}) each, in the same month. While minimum numbers of visits were noted during the months of May and June *i.e.,* 1 visit day^{-1} for bullocks and buffaloes and 2 visits day^{-1} by cows and goats in both the months. For animals of more than two year visits day^{-1} recorded were maximum for goats (7 visits day^{-1}) in August and minimum for buffaloes (1 visit day^{-1}) in June.

Dry matter consumption is directly correlated with forage availability or the amount of green biomass. For up to two years, dry matter consumption was found to be highest for buffaloes (137.45 kg month^{-1}) in August and minimum for goats (13.58 kg month^{-1}) in May (Table 21.2a) and for animals of more than two years, consumption was highest for buffaloes (146.01 kg month^{-1}) in August and minimum for goats (16.93 kg month^{-1}) in June (Table 21.2b).

Table 21.2a: Consumption by Different Animal Groups (kg month^{-1}) (For Up to 2 Years)

Month	*Cow*	*Bullock*	*Buffalo*	*Goat*
January	67.27	75.89	83.28	31.50
February	54.27	72.24	77.61	23.22
March	49.32	72.03	81.14	39.50
April	41.87	60.48	71.58	28.44
May	34.02	47.18	51.72	13.58
June	21.42	31.39	39.98	23.94
July	80.79	79.34	85.70	44.81
August	107.96	124.71	137.47	58.50
September	108.32	121.99	130.54	45.45
October	87.94	91.20	109.09	39.22
November	80.27	82.94	90.72	34.22
December	63.89	72.84	86.58	31.86

Table 21.2b: Consumption by Different Animal Groups (kg month^{-1}) (For More than 2 Years)

Month	*Cow*	*Bullock*	*Buffalo*	*Goat*
January	57.26	86.62	81.36	38.26
February	50.39	74.59	83.72	32.22
March	48.15	69.65	78.37	34.12
April	40.06	52.82	73.75	27.91
May	36.16	41.00	52.69	23.02
June	27.04	27.36	46.87	16.93
July	53.86	67.82	93.64	44.24
August	96.87	101.70	146.01	69.14
September	98.68	101.35	140.37	58.61
October	76.25	82.49	99.72	45.97
November	65.88	75.45	95.25	42.40
December	61.76	73.82	88.67	38.85

Consumption by the animals of up to two year age group was recorded maximum for buffaloes in August (35.10 kg month^{-1}) and minimum values were obtained for goats (0.70 kg month^{-1}) in the month of May (Figure 21.2a). For the animals of above two-year age group the data once again showed highest consumption by buffaloes (50.61 kg month^{-1}) in August and minimum consumption obviously by goats (0.69 kg month^{-1}) in the month of June when the green biomass production is less (Figure 21.2b).

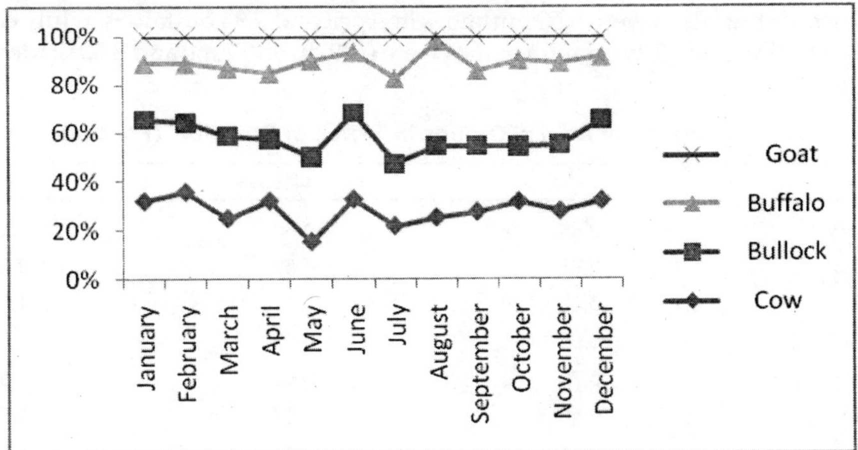

Figure 21.2a: Consumption by Different Animal Groups (kg per month per 2500 m²) (For Up to 2 Years)

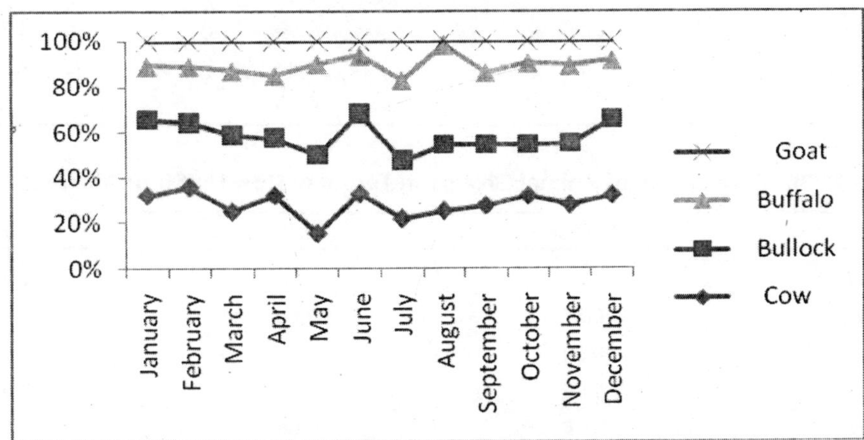

Figure 21.2b: Consumption by Different Animal Groups (kg per month per 2500 m²) (For Above 2 Years)

The dry matter consumption g m^{-2} day^{-1} during the most productive period (August) for the animals of up to two years was highest for buffaloes (0.468 g m^{-2} day^{-1}) and minimum for goats (0.208 g m^{-2} day^{-1}). During the least productive month (June) these values were highest for cows (0.036 g m^{-2} day^{-1}) and least for goats (0.012 g m^{-2} day^{-1}) (Figure 21.3a). At last for animals of more than two years consumption values were highest in August for buffaloes (0.672 g m^{-2} day^{-1}) and minimum for goats (0.292 g m^{-2} day^{-1}), and in the month of June buffaloes (0.088 g m^{-2} day^{-1}) attained peak consumption and goats (0.020 g m^{-2} day^{-1}) had the least amount of consumption during this period (Figure 21.3b).

Animals other than cow *i.e.*, bullock, buffalo and goat have been standardized into cow units on the basis of their amount of dry matter consumption. After making these conversions it became clear that, for the animals of up to two years this standardization showed minimum and maximum values for bullocks, as 0.68 bullocks in March and 1.01 bullocks in July equal to that of a cow. Minimum and maximum values for buffaloes were 0.52 buffaloes in April and 0.94 buffaloes in July equal to that of a cow. For goats minimum and maximum values were 0.89 during June and 2.50 in May equal to that of a cow (Table 21.3a).

Table 21.3a: Conversion of Animal Groups in Terms of Cow Unit (For Up to 2 Years)

Month	*Bullock*	*Buffalo*	*Goat*
January	0.88	0.80	2.13
February	0.75	0.69	2.33
March	0.68	0.61	1.24
April	0.69	0.52	1.47
May	0.72	0.65	2.50
June	0.68	0.53	0.89
July	1.01	0.94	1.80
August	0.86	0.78	1.84
September	0.88	0.82	2.38
October	0.96	0.80	2.24
November	0.96	0.88	2.34
December	0.97	0.73	2.07

Table 21.3b: Conversion of Animal Groups in Terms of Cow Unit (For Above 2 Years)

Month	*Bullock*	*Buffalo*	*Goat*
January	0.66	0.70	1.49
February	0.67	0.60	1.56
March	0.69	0.61	1.41
April	0.75	0.54	1.43
May	0.88	0.68	1.57
June	0.98	0.57	1.59
July	0.71	0.57	1.21
August	0.95	0.66	1.40
September	0.97	0.70	1.68
October	0.92	0.76	1.65
November	0.87	0.69	1.55
December	0.83	0.69	1.58

For the animals of above two years, the minimum and maximum values for bullocks were 0.66 in January and 0.97 in September as equal to that of a cow after conversion. For buffaloes, 0.57 and 0.76 individuals were equal to that of a cow during June-July and October respectively. For goats, this conversion value was 1.21 in July and 1.68 in September respectively (Table 21.3b).

Present capacity of the grazing land for the animals of up to two years was highest in September (18.187 cows ha^{-1}) followed by June (14.135 cows ha^{-1}), July (11.853 cows ha^{-1}) and August (11.04

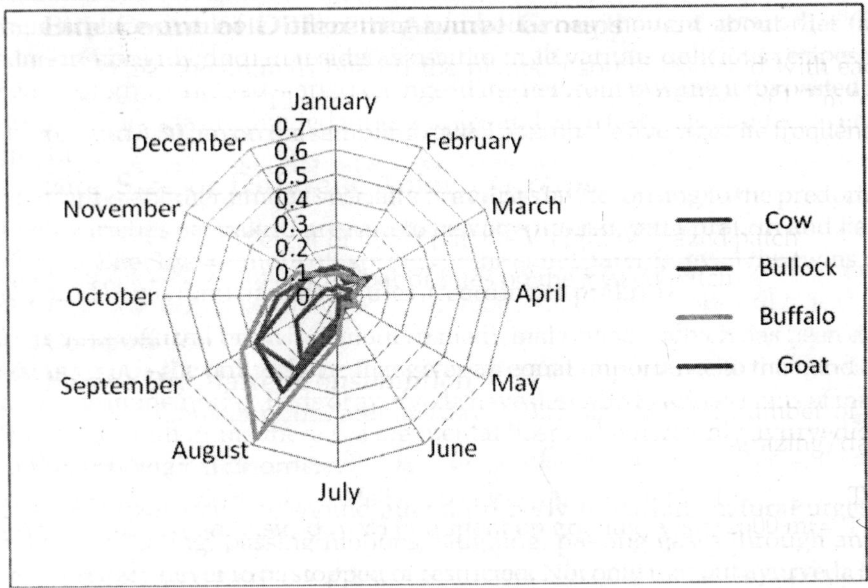

Figure 21.3a: Consumption by Different Animal Groups (g m^{-2} day^{-1}) (For Up to 2 Years)

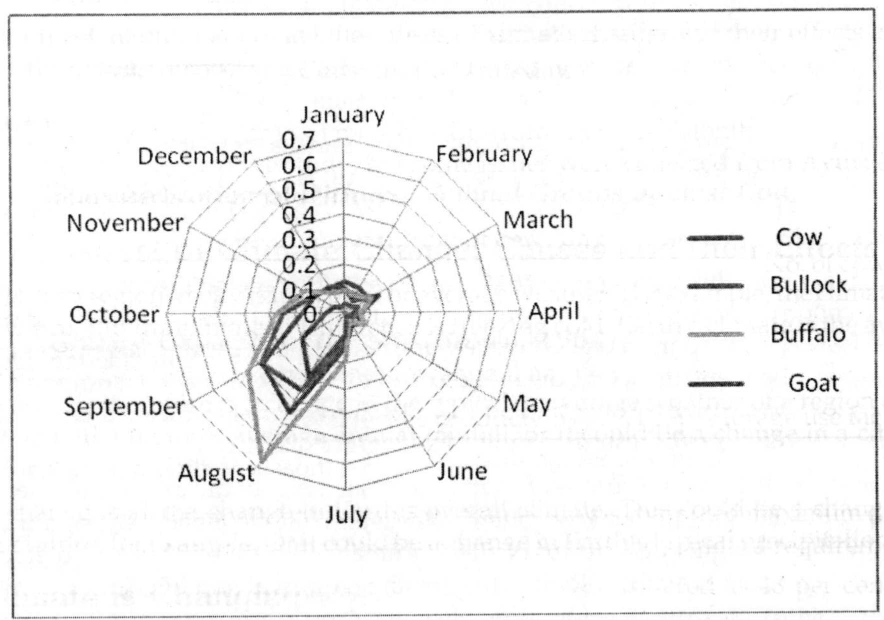

Figure 21.3b: Consumption by Different Animal Groups (g m^{-2} day^{-1}) (For Above 2 Years)

cows ha^{-1}). The minimum capacity was in April (0.0006 cow ha^{-1}), while during February, October, November and December had negative or zero carrying capacity.

For animals of more than two years of age, the present carrying capacity was highest in September (19.964 cows ha^{-1}), followed by June (15.516 cows ha^{-1}), July (13.021 cows ha^{-1}) and August (13.011

cows ha^{-1}), while the least capacity was found in April (0.0007 cows ha^{-1}) and January (0.002 cows ha^{-1}). Similar to other age groups of animals in February, October, November and December the present capacity remained zero or negative.

The potential carrying capacity for the animals of up to two years of age was highest in the month of July (33.066 cows ha^{-1}), when the forage yield was highest. It was followed by September (16.997 cows ha^{-1}) and June (5.480 cows ha^{-1}) while in March, April, May, October, November, December and January the potential capacity was negative or zero due to negative forage yield.

For the animals of over two years of age group the potential carrying capacity remained once again highest in July (36.256 cow ha^{-1}) followed by September (18.658 cow ha^{-1}), June (6.024 cow ha^{-1}) and August (5.564 cow ha^{-1}) and least in February (1.167 cow ha^{-1}). Remaining months of the year had negative production and therefore the potential capacity was also in negative values.

Discussion

Animals have a capacity to choose habitat, vegetation types within habitats and plant species or some parts of the vegetation as per food quality and availability. Intensity of grazing may influence the interspecific association among plant species through changes in habitat conditions (Singh, 1969).

The present study site is extensively used for grazing every year. The total numbers of animals present in all the 11 villages adjacent to the study site was 524. The grazing period allowed by villagers to their animals varies month to month according to the climatic conditions. The grazing time was found to be longest in the rainy season as compared with winter and summer season. Although the time allowed by villagers to their cattle was 7-9 hours per day in the grazingland. The time allowed for grazing by migratory grazers to different animals is comparable to tropical region (Pandey, 1981) and temperate zone (Agrawal and Dhasmana, 1989).

The green biomass was significantly correlated with time budget for all the age groups of cow ($r = 0.80$-0.89 at $p < 0.01$ and 0.001). In case of bullock the relationship was significant for upto 2 years of age ($r = 0.87 – 0.88$ at $p < 0.001$). The correlation ship between green biomass and time budget for all age groups of buffalo was usually significant at $p < 0.05$.

Cattle seem to prefer dicotyledons probably due to better forage quality, higher digestibility and good nutrient concentration particularly protein and other minerals in dicots (Sundriyal, 1994). Sundriyal and Joshi (1992) reported higher nutrient concentration in dicots than grasses and probably due to the reason cattle were attracted towards dicots. Lower dry matter consumption during later part of the study may be due to frost burn and poor availability of herbage during this period (Sundriyal, 1994; Shankar and Singh, 1996; Oesterheld and McNaughton, 2001).

Bite frequency for the two age groups of animals remained higher during the rainy season (August and September). The relationship between green biomass and bite frequency was significant at higher level of probabilities for higher age groups of all the animals. Variation in bite frequency in different months depends upon the available biomass for grazing (Agrawal and Dhasmana, 1989). The data recorded for dry matter intake per bite for different animal groups are in conformity with the results of Sundriyal (1994). The consumption is directly related to the amount of animal diet and the diet also varies according to the changing climatic conditions. The relationship between bite size and consumption were significant for all the animals at $p < 0.001$.

The potential carrying capacity of the present grazingland for animal of upto 2 years of age was 5.130 cows ha^{-1} yr^{-1} and for animals of more than 2 years this capacity was 5.639 cows ha^{-1} yr^{-1} (Table 21.4). But the present grazing capacity of the study area was 4.610 cows ha^{-1} yr^{-1} for animals of upto 2 year and 5.130 cows ha^{-1} yr^{-1} for more than 2 years of animal (Table 21.4). In the present study site the carrying capacity was found highest during the rainy season and least during winter and summer. The carrying capacity is directly proportional to the amount of production. The temperate meadows available to the livestock are not yet sufficient to provide adequate fodder for the number of animals present for grazing (Agrawal and Dhasmana, 1989; Mehta, 1990; McNaughton, 2001 b) and hence, they are under great grazing pressure. For better production of fodder, further management practices are needed.

Table 21.4: Carrying Capacity for the Ungrazed and Grazed Plots in the Year 2010-2011

Sl.No.	*Months*	*Ungrazed*		*Grazed*	
		Upto 2 Years	*Above 2 Years*	*Upto 2 Years*	*Above 2 Years*
1	February	1.060	1.167	-	-
2	March	-	-	0.0012	0.0014
3	April	-	-	0.0006	0.0007
4	May	-	-	0.0443	0.0486
5	June	5.480	6.024	14.135	15.516
6	July	33.066	36.256	11.853	13.021
7	August	5.069	5.564	11.104	13.011
8	September	16.997	18.658	18.187	19.964
9	October	-	-	-	-
10	November	-	-	-	-
11	December	-	-	-	-
12	January	-	-	0.0018	0.0020
	Average	5.130	5.639	4.610	5.130

References

Banerjee, G.C. 1988. *A Textbook of Animal Husbandry*. Eighth Edition (Reprinted) Oxfhord and IBH Publishing Co. Pvt. Ltd. New Delhi., Pp- 45-47.

Bor, N.L. 1942. The relict vegetation of Shillong plateau, Assam. *Indian For. Rec.* 3: 152-195.

Brown, D. 1954. *Methods of Surveying and Measuring Vegetation*. Commonwealth Agricultural Bureaux, England.

Heady, H.F. and Childs, R.D. 1994. *Rangeland Ecology and Management*. W.Junk Publishers, The Hague.

Joshi, S.P.; Anurag Raizada and M.M.Srivastava 1991a. Biomass and net primary productivity under a differed grazing pattern in a Himalayan high altitude grassland. *Range Management and Agroforestry* 12(1): 1-13.

McNaughton, S.J. 1985. Ecology of grazing ecosystem: The Serenageti. *Ecol. Monogr.* 55: 259-294.

McNaughton, S.J. 2001b. Herbivory and Trophic Interactions. Pp. 101-122. In: J. Roy, B. Saugier and H.A. Mooney (Eds.). *Terrestrial Global Productivity: Past, Present, Future.* Academic Press, San Diego.

Mehta, J.P. 1990. *Vegetation and bovine population interaction in burnt and unburnt forest grazingland at Pauri, Garhwal Himalaya.* D.Phil. Thesis, H.N.B. Garhwal University, Srinagar Garhwal, U.P. India.

Nebel, B.J., and R.T. Wright. Environmental Science: The Way the World Works. Seventh Edition. Prentice Hall, New Jersey, 2000.

Oesterheld, M. and S.J. McNaughton 2001. Herbivory in terrestrial ecosystems. In: O.E. Sala; I.A. Mooney and R.W. Howarth (eds.). *Methods in Ecosystem Science*. Springer Verleg, New York.

Pandey, A.N. 1981. Vegetation and bovine population interactions in the savanna grazinglands of Chandra Prabha sanctuary Varanasi. II. Seasonal behaviour of grazing animals and an assessment of carrying capacity of the grazinglands. *Trop. Ecol.* 22: 170-186.

Singh, J.S. 1969. Influence of biotic disturbance on the preponderence and interspecific association of two common forbs in the grasslands at Varanasi. India. *Canadian Journal of Botany* 47: 425-427.

Smith, G., A. Franks, *et al.* (2000). Impacts of domestic grazing within remnant vegetation. Native Vegetation Management in Queensland. S. L. Boulter, B. A. Wilson, J. Westrupet al. Brisbane, Department of Natural Resources.

Sundriyal, R.C. 1994. Vegetation dynamics and animal behaviour in an alpine pasture of the Garhwal Himalaya. Pp. 179-192. In: Y.P.S. Pangtey and R.S. Rawal (eds.). *High Altitude of the Himalaya,* Nanital.

Sundriyal, R.C. and A.P. Joshi 1990. Effect of grazing on standing crop, productivity and efficiency of energy capture in an alpine grassland ecosystem at Tungnath (Garhwal Himalaya), India. *Trop. Ecol.* 31: 84-97.

Sundriyal, R.C. and A.P. Joshi 1992. Interspecific relationships among plant species in an alpine grassland of the Garhwal Himalaya, India. *Bangladesh J. Bot.* 21: 81-92.

2013, Sustainable Approaches for Environmental Conservation *Pages* ***161–167***
Editors: **D.R. Khanna, A.K. Chopra, R. Bhutiani, Gagan Matta & Vikas Singh**
Published by: **BIOTECH BOOKS, NEW DELHI**

Chapter 22

A Survey of Nitrate Concentrations in Retila Fresh Leafy Vegetables from Daily Markets of Different Locations

J.C. Jana and P. Moktan

Department of Vegetable and Spice Crops, Faculiy of Horticulture, Uttar Banga Krishi Viswavidyalaya, Pundibari, Cooch Behar, West Bengal

In this study, nitrate content in three leafy vegetables namely radish, palak and amaranth were studied after a survey from four different markets at different elevations namely Gangtok of Sikkim State, Kalimpong, Siliguri and Cooch Behar of West Bengal. Average nitrate content of radish leaves (1521.13 $mgkg^{-1}$) was found to be maximum at Kalimpong market followed by Gangtok, Siliguri and Cooch Behar markets, whereas average nitrate contents of palak (2971.26 $mgkg^{-1}$) and amaranth leaves (1854.96 $mgkg^{-1}$) were maximum at Gangtok market followed by Kalimpong, Siliguri and Cooch Behar markets. Among the samples of the three leafy vegetables, nitrate content was higher than the ADI for an average 60 kg person if consumed 100 g per day in some palak samples collected from all survey markets and this incidence was maximum at Gangtok. Therefore, more care is needed for nitrogen fertilization in palak crop so that not to exceed leaf nitrate content over the ADI limit.

Keywords: Nitrate concentration, Leafy vegetables, Crop, Fertilizers.

Introduction

Nitrate is a naturally occurring form of nitrogen and is an integral part of nitrogen cycle in the environment. Nitrate is usually formed from fertilizers, decaying plants, manure and other organic residues. It has been found that due to the increased use of synthetic nitrogen fertilizers and livestock manure in the intensive agriculture, vegetable and drinking water may contain higher concentrations of nitrate than was found in the past. Exposure of human to nitrate is mainly exogenous which occurs mainly through the consumption of vegetables and to a lesser extent water and other foods. More than three quarter of our average nitrate intake comes from vegetables, which provide about 80 per cent of the average daily dietary intake.

In evaluating nitrate and nitrite, the Joint FAO/WHO Expert Committee on Food Additives (JECFA) set as 0 - 3.65 $mgkg^{-1}$ body weight the acceptable daily intake (ADI) for nitrate. Subsequently, the European Communities' Scientific Committee for Food (SCF) also set an ADI for nitrate of 0-3.65 $mgkg^{-1}$ body weight. Compared with this ADI, the ingestion of only 100 g of raw vegetables with a nitrate concentration of 2190 $mgkg^{-1}$ fresh matter (FM) corresponds to the whole nitrate ADI for a person of 60 kg though it must be observed that some components of vegetables (*e.g.* ascorbic acid, phenols *etc.*) have been reported to inhibit the toxic effects of nitrites (Santamaria, 2006).

Leafy vegetables grown under different agro-ecological conditions accumulate nitrate to potentially harmful concentrations. Generally, nitrate accumulating vegetables belong to the families Brassicaceae (rocket, radish and mustard), Chenopodiaceae (beetroot, Swiss chard and spinach), Amaranthaceae (Amaranthus), Asteraceae (lettuce) and Apiaceae (celery and parsley). Nitrate content can vary also within species, cultivars and even genotypes with different ploidy. Nitrate content differs in the various parts of a plant. Indeed, the vegetable organs can be listed by decreasing nitrate content as follows: petiole > leaf > stem > root > inflorescence > tuber > bulb > fruit > seed (Maynard *et al.*, 1976 and Santamaria *et al.*, 1999). The nitrate content of vegetables can be affected by processing of the food, the use of fertilizers, and growing conditions, especially the soil temperature and (day) light intensity (Gangolli *et al.*, 1994). Vegetables such as beetroot, lettuce, radish and spinach often contain nitrate concentrations above 2500 mg/kg, especially when they are cultivated in greenhouses. The differing capacities to accumulate nitrate can be correlated with differing location of the nitrate reductase activity, as well as to differing degree of nitrate absorption and transfer in the plant. An important part of the vegetable nutrition research has been directed towards efficient nitrogen use for high yields and quality together with minimal nitrate concentrations in the harvested commodity.

In view of the importance of vegetables to human health and fact that many people have now resulted to eating vegetables for the well being of their health, the aim of this study is therefore to determine nitrate concentrations in some common leafy vegetables with the following objectives:

1. To demonstrate ranges of nitrate content of some common leafy vegetables namely radish, palak and amaranth sold in different markets of different locations *i.e.*, North Eastern Hills of Sikkim and West Bengal and Terai region of West Bengal.
2. To assess the relative safety of some common leafy vegetables based on European Standard (EU) nitrate limits.

Materials and Method

Sampling

A sampling method which factors in type of vegetable (Radish, Palak and Amaranth) and markets with different locations (Gangtok of Sikkim State located between East 88° 03' 40" to 88° 51' 9" longitude and North 27° 03' 47" to 28° 07' 34" latitude, Kalimpong Darjeeling district of West Bengal located at latitude of 27.06° North to longitude of 88.47° East, Siliguri of Darjeeling district of West Bengal located at latitude of 26° 42' North and longitude of 88° 25' East and Cooch Behar of West Bengal located at latitude of 26° 20' North and at longitude of 89° 29' East) was established (Table 22.1). Sampling was replicated thrice within each market at intervals of two weeks and three different brands (sellers).

Table 22.1: Sampling Method for Market Survey

Factor	*Level*
Type of vegetables (3)	Radish, Palak, Amaranth
Market (4)	Three different markets of different elevations
Interval (2)	Intervals of two weeks
Brand (3)	Three different sellers per vegetables
Sample (3)	Three sub-samples per brand
Total number	3 x 4 x 2 x 3 x 3 = 216 samples

Pretreatment and Nitrate Analysis

All sub-samples were put into cooler boxes immediately after purchase and brought to the Post Graduate Laboratory, Department of Vegetable and Spice Crops, Faculty of Horticulture, Uttar Banga Krishi Viswavidyalaya, Pundibari, Cooch Behar. As needed, sub-samples were washed to remove soil and blotted on paper. Dead leaves and non-edible parts were removed and weighed.

The amount of nitrate in different samples was estimated according to rapid colorimetric determination of nitrate in plant tissue by nitration of salicylic acid (Cataldo *et al.*, 1975).This method is based on forming a nitro derivative of salicylic acid with nitrate. High (sub-milimolar) concentrations of nitrite could also form a nitro derivative with salicylic acid. Thus, nitrite should be removed from the sample. In the first step of the analysis a saturated solution of sulfamic acid is added to the sample to prevent interference from nitrite. Sulfamic acid converts nitrite to nitrogen gas. Second, salicylic acid reacts with nitrate under acidic conditions to form nitrosalicylic acid. Then the pH is brought to above 12 to form a chromophore with a maximum absorbance at 410-420 nm.

Standards

- ✰ Stock solution 0.25 g/l NO_3^-N (=250mg/l, 250 μg/ml)
- ✰ In a 1.0l standard Flask containing approximately 600 mL Type-I-water, 1.805 g potassium nitrate was dissolved.
- ✰ It was ensured that all potassium nitrate (KNO_3) is dissolved, made up to the mark with Type-I-water, mixed and stored in a suitably labelled plastic container.

Table 22.2: Prepare Standards Containing ~0 to 60 μg NO_3^-N in a 0.25 ml Aliquot

Amount of NO_3^-N	*Vol stock (ml)*	*Vol extractant (or H_2O)*
62.5	0.25	0.00
50.0	0.20	0.05
37.5	0.15	0.10
25.0	0.10	0.15
12.5	0.05	0.20
0.00	0.00	0.25

Blanks

A blank of 0.25 ml extractant (or H_2O) with the normal reagents was normally sufficient. For pigmented samples a separate blank was required and prepared for each sample. This blank consisted of the extract, 0.8 ml of concentrated sulphuric acid (minus salicylic acid) and 19 ml of 2 N NaOH.

Chemicals

NO_3^- Reagent A: Saturated solution of amidosulfonic (sulfamic) acid

NO_3^- Reagent B: 5 per cent (w/v) salicylic (2-hydroxybenzoic) acid in pure H_2SO_4

NO_3^- Reagent C: 4 M NaOH

Procedure

Before measuring the nitrate concentration in the sample, we made sure to dilute it to the above mentioned range.

1. 10 µl reagent A was added to cuvette (4 ml macro-cuvette)
2. 40 µl of sample was added on reagent A (pipette to mix reagent A and the sample)
3. 200 µl reagent B was added
4. Waited for 10 minutes
5. 2 ml reagent C was added
6. Waited for 20 min
7. Mixed with a clean pipette tip to remove the formed bubbles
8. Absorbance was read at 420 nm
9. The liquid was discharged to sink

The procedure was done with water instead of the sample for the "blank" measurement.

Calibration and Notes

Stock solutions were made of 1-5 mM, solutions B and C were cold and solution B was kept in the dark. Nitrate content was expressed as milligrams nitrate per kilograms on a fresh weight basis (mg NO_3/kg FW).

Statistical Analysis

The data collected were analyzed using Fisher's analysis of variance technique and differences among the various treatments were determined by using least significant difference test at 5 per cent probability level (Steel and Torrie, 1984).

Results and Discussion

Nitrate concentrations in leafy vegetables procured from different markets are summarized in Table 22.3 for radish leaves, Table 22.4 for palak leaves and Table 22.5 for amaranth leaves. The SCF has set ADI for nitrate ion on the basis of body weight of the consumer *i.e.* 3.65 $mgkg^{-1}$body wt day^{-1}.

Nitrate Concentration in Radish Leaves

Nitrate concentrations in fresh radish leaves procured from different markets are summarized in Table 22.3. The nitrate content ranged from 951.25 $mgkg^{-1}$ to 1847.58 $mgkg^{-1}$ fresh weight of the sample. In this study the average nitrate content of radish leaves was found to be highest at Kalimpong market with a value of 1521. 13 $mgkg^{-1}$ which was followed by Gangtok market at 1196.01 $mgkg^{-1}$ and Siliguri market at 1249.14 $mgkg^{-1}$. The average nitrate content of 882.17 $mgkg^{-1}$ was found to be the minimum at Cooch Behar market.

Table 22.3: Nitrate Content of Radish Leaves Collected from Different Markets.

Markets	*Range ($mgkg^{-1}$)*	*Average ($mgkg^{-1}$)*
Gangtok	951.25–1567.21	1196.01
Kalimpong	975.25–1749.23	1521.13
Siliguri	894.72–1847.58	1249.14
Cooch Behar	676.55–1121.23	882.17

The nitrate content of the sample procured from Kalimpong showed an increase of 72.43 per cent over the sample of Cooch Behar market. However, in all the four markets, the nitrate content in radish leaves did not cross the ADI limit of 3.65 $mgkg^{-1}$ body wt day^{-1} set by the Scientific Committee for Food (SCF).

Nitrate Concentration of Palak Leaves

The nitrate concentration of palak leaves procured from different markets are summarized in Table 22.4 and from the data in table, we find the average nitrate concentration in the range from 783.56 $mgkg^{-1}$ to 3731.53 $mgkg^{-1}$. The maximum nitrate concentration of 2971.26 $mgkg^{-1}$ was found to be in the sample procured from the Gangtok market followed by the sample procured from Kalimpong market at 2523.99 $mgkg^{-1}$ and Siliguri market at 2405.20 $mgkg^{-1}$. The minimum nitrate concentration of 1888.70 $mgkg^{-1}$ was found to be in the sample procured from the Cooch Behar market. The nitrate content of the sample procured from Gangtok market showed an increase of 57.32 per cent over the sample of Cooch Behar market.

Table 22.4: Nitrate content of palak leaves collected from different markets.

Markets	*Range ($mgkg^{-1}$)*	*Average ($mgkg^{-1}$)*
Gangtok	1984.95–3625.22	2971.26
Kalimpong	1184.57–3545.67	2523.99
Siliguri	1757.72–3414.79	2405.20
Cooch Behar	783.56–3731.53	1888.70

Vegetable nitrate content was higher than the ADI for an average 60 kg person if consumed 100 g per day in some palak samples mostly collected from Gangtok. It may be due to a complex interaction between growing season, production method and location in relation to the amount of sunlight. In low light intensity and low temperature condition the accumulated nitrate is not rapidly reduced as in other places with high temperature and bright sunshine.

It has also been reported that accumulation of large amounts of nitrate in palak may be due to their low Nitrate Reductase Activity (NRA), as evident by the negative relationship. Other causal factors may be related to variations in the uptake and distribution of nitrate needed for NRA, differences in generation of electron donors needed in the assimilative pathway (Cantliffe, 1973) in photosynthetic capacity (Behr and Wiebe, 1992) or ability to generate and translocate respiratory substrate and reducing equivalents.

Nitrate Concentration of Amaranth Leaves

The data presented in Table 22.5 show average nitrate concentrations in the range of 690.93 mgkg^{-1} to 2625.52 mgkg^{-1}. Among the four markets, the maximum recorded average nitrate concentration of 1854.96 mgkg^{-1} originated from the sample procured from the Gangtok market. This was then followed by the Kalimpong market at 1330.84 mgkg^{-1} and by the Siliguri market at 1233.95 mgkg^{-1}. The minimum average nitrate concentration of 1176.89 mgkg^{-1} was found to be in the sample procured from Cooch Behar market.

Table 22.5: Nitrate Content of Amaranth Leaves Collected from Different Markets

Markets	*Range (mgkg^{-1})*	*Average (mgkg^{-1})*
Gangtok	1243.67–2625.52	1854.96
Kalimpong	1021.42–1598.21	1330.84
Siliguri	884.63–1578.84	1233.95
Cooch Behar	690.93–1784.95	1176.89

The nitrate content of the sample procured from Gangtok market showed an increase of 57.62 per cent over the sample of Cooch Behar market. However, in all the four markets, the nitrate content in radish leaves did not cross the ADI limit of 3.65 mgkg^{-1} body wt day^{-1} set by the Scientific Committee for Food (SCF).

Variation in nitrate content between plant species and even between cultivars of the same species has been reported earlier (Blom-Zandstra and Eenink, 1986; Reinink *et al.*, 1994). Over expression of NR genes may be a useful approach to reduce the nitrate content of plants that have a propensity to accumulate the same and to improve their quality for human consumption, although genetic manipulation of activities of nitrate assimilatory enzymes may not increase yield and/or nitrogen use efficiency of plants (Quillere *et al.*, 1994).

Nitrate consumption through vegetables can be kept low by harvesting them at the proper time. Some earlier reports have advised the harvest of spinach crop in the afternoon of a sunny day when nitrate concentration in the leaves is low (Steingrover *et al.*, 1982 and Reinink, 1991).

Thus a careful selection of vegetable genotypes based on the relationship between nitrate and Nitrate Reductase Activity (NRA), coupled with due management of the nutrition and harvest regime, help avoid nitrate accumulation and the associated health hazards. Moreover, by cooking vegetables in water (with low nitrate concentration), at least 50 per cent of accumulated nitrate can be removed (Meah *et al.*, 1994).

Summary and Conclusion

Average nitrate content of radish leaves (1521.13 mgkg^{-1}) was found to be maximum at Kalimpong market followed by Gangtok, Siliguri and Cooch Behar markets, whereas average nitrate contents of palak leaves (2971.26 mgkg^{-1}) and amaranth leaves (1854.96 mgkg^{-1}) were maximum at Gangtok market followed by Kalimpong, Siliguri and Cooch Behar markets. Among the samples of the three leafy vegetables, nitrate content was higher than the ADI for an average 60 kg person if consumed 100 g per day in some palak samples collected from all survey markets and this incidence was maximum at Gangtok. Therefore, more care should be taken for nitrogen fertilization in palak crop not to exceed the leaf nitrate content over the ADI limit.

References

Behr U and Wiebe H J. 1992. Relation between photosynthesis and nitrate content of lettuce cultivars. *Sci. Hortic* 49: 175-179.

Blom Zandstra M and Eenink A H. 1986. Nitrate concentration and reduction in different genotypes of lettuce. *J Am Soc Hortic Sci.* 111: 908 – 911.

Cantliffe D J. 1973. Nitrate accumulation in table beets and spinach as affected by nitrogen, phosphorus, and potassium nutrition and light intensity. *Agron. J* 65: 563-565.

Cataldo D A, Haroom M, Schrader L E, Young VL. 1975. Rapid colorimetric determination of nitrate in plant tissues by nitration of salicylic acid. Comm. Soil Science and Plant Analysis 6(1): 71-80.

Gangolli S D, Van Den Brandt P, Feron V, Janzowsky C, Koeman J, Speijers G, Spiegelhalder B, Walker R and Winshnok J. 1994. Assesment of nitrate, nitrite and N-nitroso compounds. *Eur. Jr Pharmacol Environ Toxicol Pharmacol Sect* 292: 1-38.

Maynard D N, Barker A V, Minotti P L and Peck N H. 1976. "Nitrate accumulation in vegetables." *Advances in Agronomy* 28: 71-118.

Meah M N, Harrison N and Davies A. 1994. Nitrate and nitrite in foods and the diet. *Food Addit. Contam* 11: 519 – 532.

Quillere L, Dufoss C, Roux Y, Foyer C H, Caboche N and Marot-Gaudry J F. 1994. The effect of deregulation of NR gene expression on growth and nitrogen metabolism of Nicotiana plumbaginifolia plants. *J Exp Bot* 45: 1205 – 1211.

Reinink K, Vannes M and Groenwold R. 1994. Genetic variation for nitrate content between cultivars of endive. *Euphytica* 75: 41 – 48.

Reinink K. 1991. Genotype X Environment interaction for nitrate concentration in lettuce. *Plant Breed* 107: 39–49.

Santamaria P, Elia A, Serio F and E Todaro. 1999. A survey of nitrate and oxalate content in retail fresh vegetables. *Journal of Science and Food Agriculture* 79: 1882–1888.

Santamaria P. 2006. Nitrate in Vegetables: Toxicity Content, Intake and European Commission Regulation. *Journal of Science and Food Agronomy* 86: 10 – 17.

SCF (Scientific Committee on Food). 1997. Assessment of dietary intake of nitrates by the population in the European Union, as a consequence of the consumption of vegetables, in Reports on tasks for scientific cooperation: report of experts participating in Task 3.2.3, ed by European Commission, Brussels 34.

Steel R G and Torrie J H. 1984. *Principles and Procedures of Statistics: A Biometrical Approach.* 173 – 191. Mc Graw Hill Co. New York.

Steingrover E, Oosterhius R and Wieringa F. 1982. Effect of light treatment and nutrition on nitrate accumulation in spinach (*Spinacia oleracea* L.). *Z. Pflanzenphysiol* 107: 97–102.

2013, Sustainable Approaches for Environmental Conservation *Pages* ***169–179***
Editors: **D.R. Khanna, A.K. Chopra, R. Bhutiani, Gagan Matta & Vikas Singh**
Published by: **BIOTECH BOOKS, NEW DELHI**

Chapter 23

Environmental Biotechnology: Opportunities and Challenges

Poonam Khurana and Shagun Kaushik

J.P. Institute of Engineering and Technology, Meerut, Uttar Pradesh

This paper review possibilities of Environmental Biotechnology with their related issues and implications. Considering the number of problems that define and concretize the field of Environmental Biotechnology, the role of some Bioprocessers and Biosystems for environmental protection control and health, based on utilization of living organisms are analyzed. Environmental remediation, pollution prevention, detection and monitoring are evaluated considering the achievement, as well as perspectives in the development of biotechnology. The distinct role of environmental biotechnology in the future is emphasized considering the opportunities to contribute with new solution and directions in remediation of contaminated environment, minimizing future waste release and creating pollution prevention alternatives. To take advantage of these opportunities innovative new strategies which advance the use of molecular biological method and genetic engineering technology are examined. These methods would improve the understanding of existing biological processes in order to increase their efficiency, productivity and flexibility. Also, the contribution of environmental biotechnology to the progress of a more sustainable society is reveled.

Keywords: *Bioremediation, Phytoremediation, Heavy metal, Organic compounds, Biotechnology.*

Introduction

Biotechnology is versatile and has been assessed a key area which has greatly impacted various technologies based on the application of biological processes in manufacturing, agriculture, food processing, medicine, environmental protection, resource conservation (Chisti and Moo-Young, 1999;

EC, 2002; Evans and Furlong, 2003; Gavrilescu, 2004; Gavrilescu and Chisti, 2005). Environmental concern help derive the use of biotechnology not only for pollution control, but prevent pollution and minimize waste in first place, as well as for environmentally friendly production of chemicals, biomonitoring *etc.*

Environmental Biotechnology: Issues and Implications

Environmental biotechnology is concerned with the application of biotechnology as an emerging technology in the context of environmental protection since rapid industrialization, urbanization and other developments have resulted in a threatened clean environment and depleted natural resources. Advanced techniques or technologies are now possible to treat waste and degrade pollutants assisted by living organism or to develop product and processes that generate less waste as a result of improved treatment of solid waste and wastewater bioremediation (Olguin, 1999; EIBE 2000).The production processes themselves can assist in the reduction of waste and minimization of pollution within so called clean technologies based on biotechnological issues involved in reuse or recycle waste, generate energy sources or produce new viable products (Evans and Furlong, 2003; Gavrilescu and Chisti, 2005).

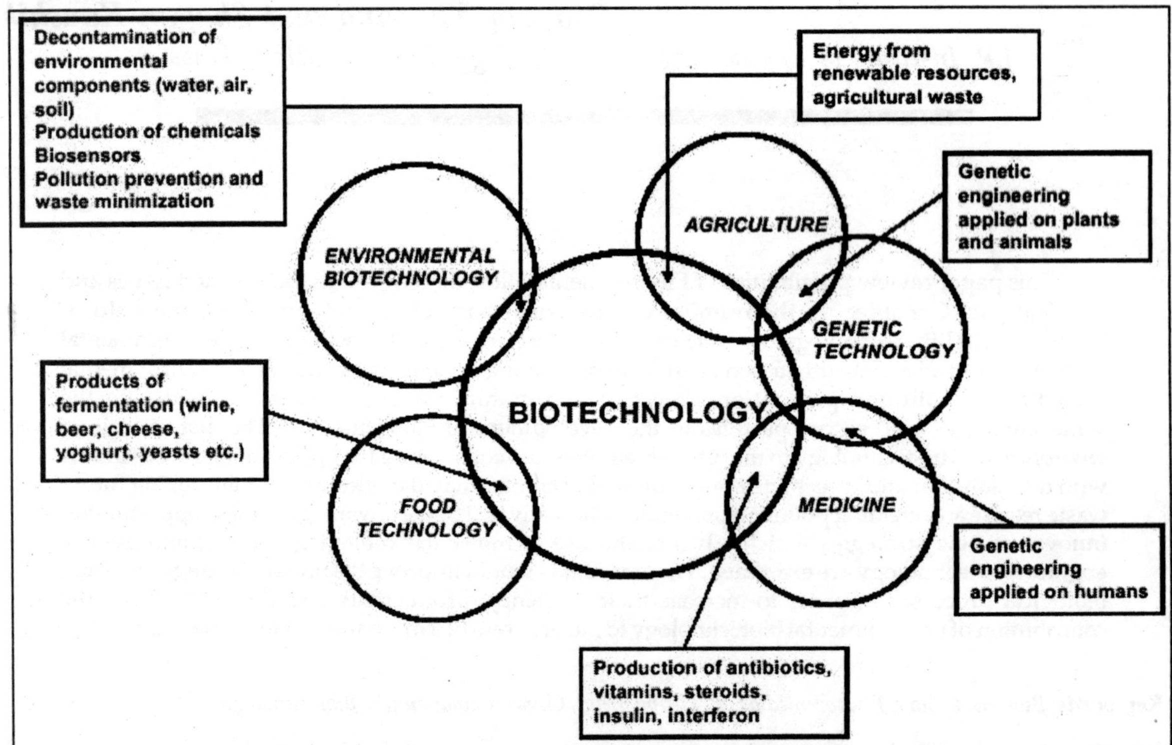

Figure 23.1: Applications of Biotechnology in Anthropogenic Activities (adapted from Sukumaran Nair, 2006)

Environmental Remedation by Biotreatment/Bioremediation

Bioremediation is defined by US Environmental Protection Agency as a managed or spontaneous practice in which microbiological processes are used to degrade or transform contaminants to less toxic or non toxic form, thereby remediating or eliminating environmental contamination (USEPA,

1994; Talley 2005). This method is applied to remove, degrade or detoxify pollution in environmental media, including water, air soil and solid waste. Removal of any pollutant from the environment can take place on following two routes: degradation and immobilization by a process which causes it to be biologically unavailable for degradation and so is effectively removed (Evans and Furlong 2003)

Microbes and Plants in Environmental Remediation

All forms of life can be considered as having a potential function in environmental biotechnology. However microbes and certain plants are of interest even as normally present in their natural environment or by deliberate introduction (Evans and Furlong, 2003). The generic term microbes including prokaryotes (Bacteria and Archaea) and Eukaryotes (Yeasts, fungi, protozoa and unicellular plants). The role of plants in environmental cleanup is excreted during the oxygenation of a microbe rich environment, filtration and solid to gas conversion. The use of organisms for the removal of contamination is based on the concept that all organisms could remove substances from the environment for their own growth and metabolism (Hamer, 1997; Saval, 1999, Dob *et al.*, 2004). Bacteria and fungi are very good at degrading complex molecules and the resultant waste are generally safe, algae and plants proved to be suitable to absorb nitrogen, phosphorus, sulpher and many minerals and metals from the environment. Microorganism used in bioremediation include aerobic and anaerobic type. Some have been isolated, selected, mutated and genetically engineered for effective bioremediation capabilities.

Two groups of factors can be identified that determine the success of bioremediation processes are (1) Nature and character of contaminant and (2) Environmental conditions (Saval, 1999; Sasikumar

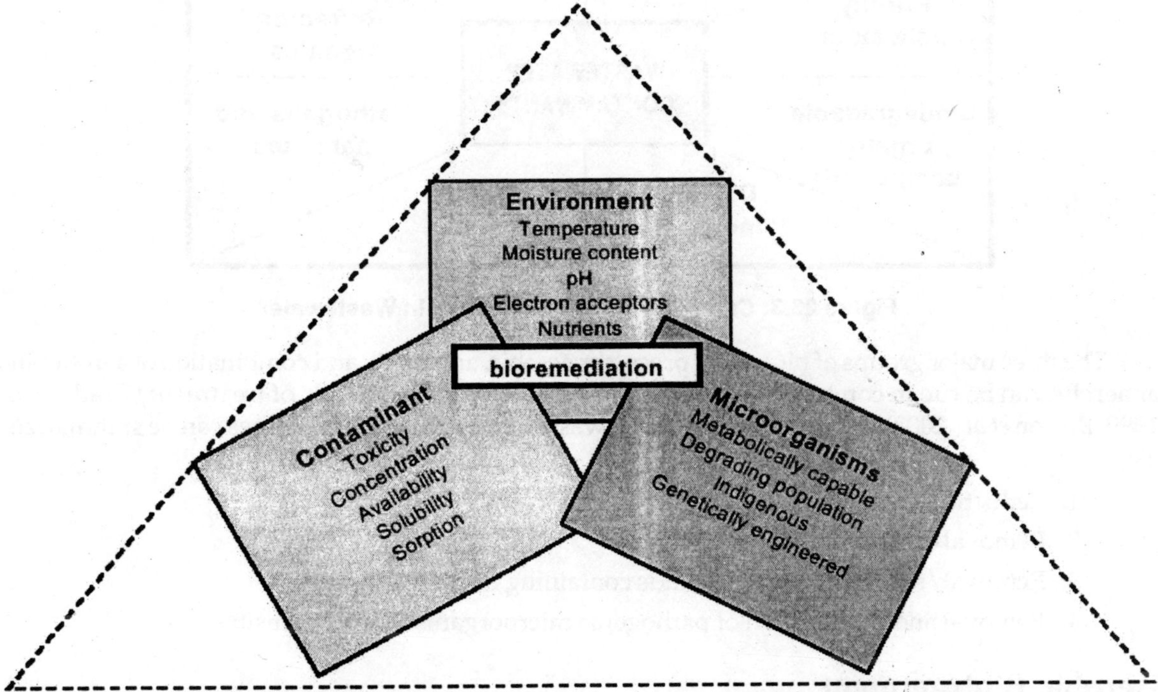

Figure 23.2: Main Factors of Influence in Bioremediation Processes

and Papinazath, 2003; Bitton, 2005) also bioremediation tend to rely on the natural abilities of microorganisms to develop their metabolism and to optimize enzyme activity.

Wastewater Biotreatment

The use of microorganisms to remove contaminants from wastewater is rarely dependent on wastewater source and characteristics. Wastewater is typically categorized in to one of the following groups (Wiesmann *et al.*, 2007)

1. Municipal wastewater (domestic wastewater mixed with effluents from commercial and industrial works)
2. Commercial and industrial wastewater
3. Agricultural wastewater

The effluent components may be of chemical, physical or biological nature and they can induce an environmental impact, which includes changes in aquatic habitat and species structure as well as biodiversity and water quality. Since many of the compounds present in wastewater are toxic to microorganisms, pretreatment may be required (Burton *et al.*, 2002). Biological treatment requires that effluent be rich in unstable organic matter so the microbes break up these unstable organic pollutants into stable products like CO_2, CO, NH_3, CH_4, H_2S *etc.* (Dunn *et al.*, 2003).

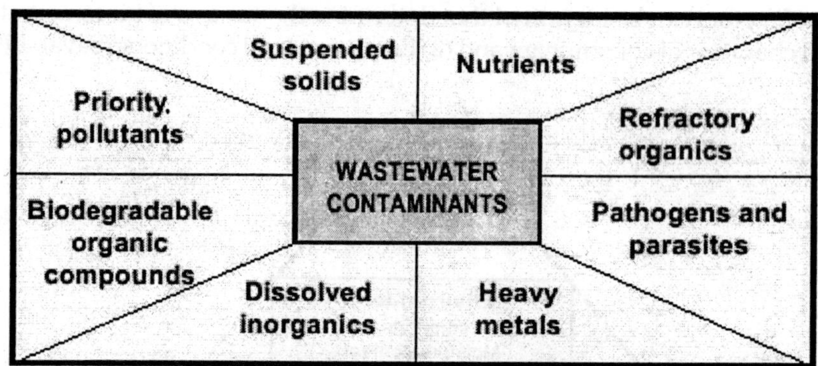

Figure 23.3: Categories of Contaminants in Wastewater

The three major groups of biological processes aerobic, anaerobic and combination of aerobic and anaerobic can be run in combination or in sequence to offer greater levels of treatment (Grady *et al.*, 1999, Burton *et al.*, 2002). The main objectives of wastewater treatment processes can be summarized as:

1. Reduction of biodegradable organic content (BOD)
2. Removal of heavy/toxic metals.
3. Removal/reduction of compounds containing P and N nutrient.
4. Removal and inactivation of pathogenic microorganisms and parasites.

Aerobic Biotreatment

Aerobic processes are often used for municipal and industrial wastewater treatment. Easily biodegradable organic matter can be treated by this system (Dob and Kumar, 2005; Russell, 2006). The basic reaction in aerobic treatment plant is represented by the reaction (1,2)

$$\text{Organic material} + O_2 \longrightarrow CO_2 + H_2O + \text{new cells} \quad (1)$$

Microbial cells undergo progressive auto-oxidation of cell mass:

$$\text{Cells} + O_2 \longrightarrow CO_2 + H_2O + NH_3 \quad (2)$$

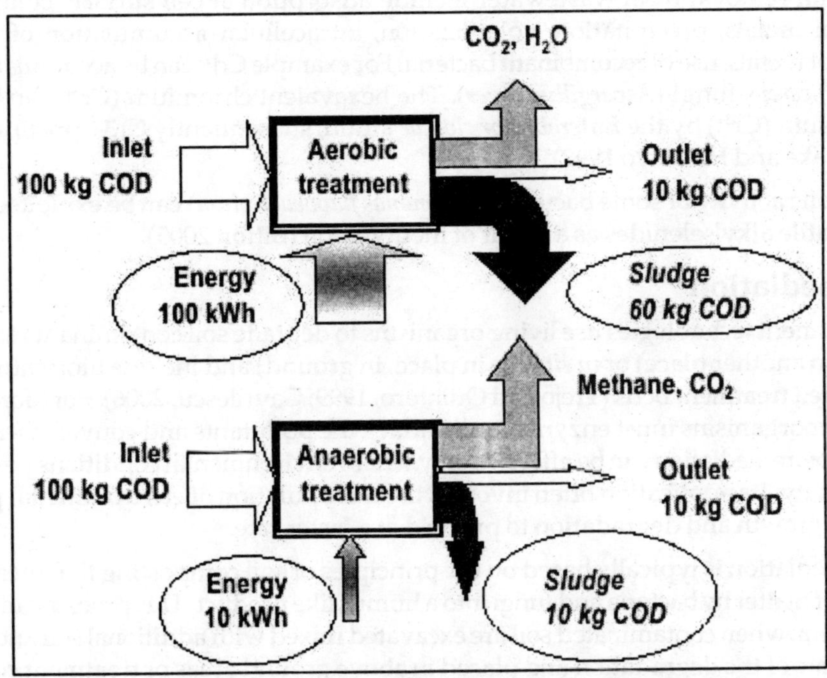

Figure 23.4: Comparison of Aerobic and Anaerobic Biological Treatment

Anaerobic Biotreatment

Anaerobic treatment of wastewater does not generally lead to low pollution standards and it is often considered a pretreatment process, devoted to minimization of oxygen demand and excessive formulation of sludge. Highly concentrated wastewater should be treated anaerobically due to the possibility to recover energy as biogas and low quantity of sludge (Gallert and Winter 1999)

High loads of wastewater treated by anaerobic technologies generates low quantities of biological excess sludge with a high treatment efficiency, low capital costs, no oxygen requirement, methane production, low nutrient requirement. High rate anaerobic wastewater treatment technologies can be applied to treat dilute concentrated liquid organic wastewater which are discharged from distilleries, breweries, paper mills, petrochemical plants *etc.* even municipal wastewater can be treated using high rate anaerobic technologies.

Metals Removal by Microorganism from Wastewater

Heavy metal come in wastewater treatment plants from industrial discharges, storm water *etc.* Toxic metals may damage the biological treatment process. However there are microorganisms with metabolic activities resulting in solubilization, precipitation, chelation, biomethylation, volatilization of heavy metals (Bremer and Geesey, 2001; Bitton, 2005).

Metals from wastewater such as iron, copper, cadmium, nickel, uranium can be mostly complexed by extracellular polymer produced by several types of bacteria. Subsequently metals can be accumulated and then released from biomass by acidic treatment. Nonliving immobilized bacteria, fungi, algae are able to remove heavy metals from wastewater (Eccles and Hunt 1986, Bitton 2005). The mechanism involved in metal removal from wastewater include adsorption at cell surface, complexation and solubilization of metals, precipitation, volatilization, intracellular accumulation of metals, redox transformation of metals, use of recombinant bacteria. For example Cd^{2+} can be accumulated by bacteria such as E.coli, *B.cereus* fungi (*Aspergillus niger*). The hexavalent chromium ($Cr^{6+)}$ can be reduced to trivalent chromium (Cr^{3+}) by the *Enterobactor cloacae* strain; subsequently Cr3+ precipitates as metal hydroxide (Ohtake and Hardoyo 1992)

The metabolic activity of some bacteria (*Aeromonas flavobacterium*) can be exploited to transform selenium to volatile alkylselenides as a result of methylation (Bitton 2005).

Soil Bioremediation

Soil biotreatment technologies use living organisms to degrade soil contaminants either *ex situ* (*i.e* above ground, in another place) or *in situ* (*i.e* in place, in ground) and include biotreatment cells, soil piles and prepared treatment beds (Trejo and Quintero, 1999; Gavrilescu, 2006). For bioremediation to be effective microorganisms must enzymatic ally attack the pollutants and convert them to harmless products. Since bioremediation can be effective only where environmental conditions permit microbial growth and activity. Its application often involves the manipulation of environmental parameters to allow microbial growth and degradation to proceed at a faster rate.

Soil bioremediation is typically based on the principles of soil composting that means controlled decomposition of matter by bacteria and fungi into a humus like product. This process can be performed in an *ex situ* system, when contaminated soil are excavated mixed with additional soil and/or bacteriaa to enhance the rate of the degradation and placed in above ground areas or treatment compartments. Another type of soil biotreatment consist of an *in situ* process when a carbon source such as manure is added in an active or passive procedure depending upon whether the carbon source is applied directly to the undisturbed soil surface (*i.e* passive) or physically mixed in to the soil surface layer (*i.e* active).

Bioremediation of land is often cheaper than physical method and its products are harmless if complete mineralization takes place. Bioremediation using plants, identified as phytoremediation is presently used to remove metals from contaminated soil and groundwater and is being further explored for the remediation of other pollutants. Certain plants have also been found to absorb toxic metals such as mercury, lead and arsenic from polluted soil and water and scientist are hopeful that they can be used to treat industrial waste.

Vidali (2001) described five types of phytoremediation technique: classified based on the contaminant fate-phytoextraction, phytotransformation, phytostabilization, phytodegredation, rhizofilteration and summarizes some phytoremediation mechanism and applications (Table 23.1).

Solid Waste Biotreatment

Biowaste is generated from various anthropogenic activities and can be categorized as manures, raw plant matter, and process waste. The approach involves carefully selecting organisms known as biocatalyst which are enzymes that degrade specific compounds and define the conditions that accelerate degradation process which contribute to:

☆ Reducing the potential for adverse effects to the environment.

☆ Reclaiming valuable minerals for reuse.

☆ Generating useful end products.

Advantages of the biological treatment include: stabilization of waste, reduced volume in the waste material, destruction of pathogen in the waste material and production of biogas for energy use.

Table 23.1: Overview of Phytoremediation Applications

Technique	*Plant Mechanism*	*Surface Medium*
Phytoextraction	Uptake and concentration of metal *via* direct uptake into the plant tissue with subsequent removal of the plants	Soils
Phytotransformation	Plant uptake the degradation of organic compounds	Surace water, groundwater
Phytostabilization	Root exudates cause metal to precipitate and become less available	Soils, groundwater, mine tailing
Phytodegradation	Enhances microbial degradtion in rhizosphere	Soils, groundwater within rhizo-sphere
Rhizofiltration	Uptake of metals into plant roots	Surface water and water pumped
Phytovolatilization	Plants evapotranspirate selenium, mercury and volatile hydrocarbons	Soils and groundwater
Vegetative cap	Rainwater is evapotranspirated by plants to prevent leaching contaminants from disposal sites	Soils

Biotreatment of Gaseous Stream

In the waste gas treatment, biotechnology has been applied to find green and low cost environmental processes. Odorous emission represents a serious problem related to biowaste treatment facilities as they may be a trouble to local residents. Biofilters are one of the main biological system used which work at normal operating conditions of temperature and pressure. Therefore they are relatively cheap with high efficiencies. Microbial population in biofilters are very diverse, these type of reactors can simultaneously remove complex mixture of pollutants (Cox and Deshusses 2001; Shareefdeen *et al.*, 2005).

Biodegradation of Hydrocarbons

Various types of microorganisms can degrade hydrocarbons: bacteria, yeasts, filamentous fungi but none of them degrade all the possible hydrocarbon molecules at the same rate. Due to different hydrophobicity and low solubility in water of the hydrocarbons, the process should be intensified by enhancing physical contact between microorganism and oil by adding adjuvant to improve the contact area by injecting of mixtures of microorganisms during the so called bioaugmentation (Malina and Zawierucha, 2007).

Biosorption

Biosorption is fast and reversible process for the removal of toxic metal ions from wastewater by live or dried biomass, which resemble adsorption and in some cases ion exchange. Biosorption of heavy metals by algal biomass is an advantageous alternative, an appropriate and economically feasible method used for wastewater and waste cleanup (Vilar *et al.*, 2007).

Bioindicators/Biomarkers

Some organisms or communities may react to an environmental effect by changing a measurable biological function and their chemical composition. This way it is possible to infer significant environmental change and their responses are referred as *bioindicators.* Biomarkers are thus used in biomonitoring programmes to give biological information.

Environmental Biotechnology for Pollution Prevention and Cleaner Production

Since biotechnology can contribute to the elimination of hazardous pollutants at their source before they enter the environment, industrial and environmental biotechnology use biological processes to make industrially useful products in a more efficient environmentally friendly way by cutting waste byproducts, air emission, energy consumption and toxic chemicals in several industries (Gavrilescu and Chisti, 2005). Because biotechnological processes once set up are considered cheaper than traditional methods, changes in production processes will not only contribute to environmental protection but also help save money and continuously improve their public image.

In the context of pollution prevention, biotechnology can contribute to substitute multistep chemical processes using genetically modified organisms (GMO) as well. This action should have other beneficial results because land disposal of hazardous waste, wastewater loading air emission and production coast are greatly reduced.

Process Modification and Production Innovation

The technique of modern molecular biology is applied in the industry and environment to improve efficiency and diminish the environmental impacts. Process innovation, the development of new biological processes and the modification of existing processes by the introduction of biological steps based on microbial or enzyme action are increasingly being used in industrial operations as an important potential area of primary pollution prevention (Olguin, 1999; Gavrilescu and Nieu, 2005).

Environmental Biotechnology and Eco Efficiency

Eco efficiency analysis can offer comprehensible information for a large number of applications concerning multifactorial problems with in relatively short time and at relatively low cost. World business council for sustainable development (WBCSD) developed eco-efficiency as a way for an operational sustainable development driving force from a business perspective (WBSCSD, 2000). Eco efficiency is more and more becoming heart of success in the economic world as a way to maximize efficiency, while minimizing the impact on the environment. It is achieved in practice by means of three key objectives that regard increasing product or service value, optimizing the use of resource, reducing environmental impacts (Gabriel and Braune, 2005; Bidoki *et al.*, 2006). Biotechnology in general and environmental biotechnology in particular can be considered one of the useful means to attain eco efficiency and for decision making because offers a number of practical benefits (Saling, 2005).

Challenges and Perspectives

New environmental challenges continue to evolve and new technologies for environment protection and control are currently under development. Also, new approaches continue to gain more and more ground in practice harnessing the potential of microorganism and plants as eco efficient and robust cleanup agents in variety of practical situations such as:

- ☆ Enzyme engineering for improved biodegradation
- ☆ Designing strains for enhanced biodegradation
- ☆ Design wastewater treatment based on decentralized sanitation and reuse.
- ☆ Implementation of anaerobic digestion to treat biowaste.
- ☆ Emerging and growing up technological applications of soil remediation and clean up of contaminated sites.

Since environmental biotechnology proves to have a large potential to contribute to the prevention, detection and remediation of environmental pollution and degradation, it is a sustainable way to develop clean processes and products, less harmful, with reduced environmental impacts than their forerunners. Since some new techniques make use of genetically modified organisms, regulations to guarantee safe application of new or modified organisms in the environment is important.

A wide range of biological methods are already in use to detect pollution incidents and for the continuous monitoring of pollutants, but new developments are expected. Environmental and economic benefits that biotechnology can offer in manufacturing, monitoring and waste management are in balance with technical and economic problem which still need to be solved. All this is being achieved with reduced environmental impact and enhanced sustainability.

An evaluation of the consequences, opportunities and chalienges of modern biotechnology is important both for policy makers and the industry.

References

Bidoki, S.M., Wittlinger, R., Alamdar, A.A. and Burger, J., 2006. Eco–efficiency analysis of textile coating materials. *Journal of the Iranian Chemical Society*, 3: 351–359.

Bitton, P.R.O., 2005. *Wastewater Microbiology*. Wiley-Liss, John Wiley and Sons, New Jersey, USA, 766 pp.

Bremer, J.P. and Geesey, G.G., 2001. Laboratory based models of microbiologically induced corrosion of copper. *Applied and Environmental Microbiology*, 57: 1956–1962.

Burton, F.L., Stensel, H.D. and Tchobanoglous, G., 2002. *Wastewater Engineering Treatment and Reuse*. Metcalf and Eddy Inc., Mc-Graw Hill Professional, 1848 pp.

Chisti, Y. and Moo-Young, M., 1991. Fermentation technology, bioprocessing, scale up and manufacture. In: *Biotechnology: The Science and the Business*, 2nd Edn. (Eds.) V. Moses, R.E. Cape and D.G. Springham. Harwood Academic Publishers, New York, pp. 177–222.

Cox, H.H. and Deshusses, M.A., 2001. Biotrickling filters. In: *Bioreactors for Waste Gas Treatment*, (Eds.) C. Kennes and M.C. Veiga. Kulwer Academic Publishers, The Netherlands, pp. 99–131.

Doble, M., Kruthiventi, A.K. and Galker, V.G., 2004. *Biotransformations and Bioprocesses*. Marcel Dekker, New York Basel, 371 pp.

Dunn, I.J., Heinzle, E., Ingham, J. and Prenosil, J.E., 2003. *Biological Reactions Engineering: Dynamic Modelling Fundamentals with Simulation Examples*. Wiley-VCH, Weinheim, 532 pp.

EC, 2002. Life sciences and biotechnology: A strategy for Europe communication from the commission to the Europe Parliament the council, the economic and social committee and the committee of the regions, online at http://ec.europa.eu/biotechnology/pdf/com2002-27_en.pdf.

EIBE, 2000. *Biotechnology and Environment*. European initiative for biotechnology education. Available online at: http://www.ipn.uni-kiel.de/eibe/unit16EN.pdf.

Evans, G.M. and Furlong, J.C., 2003. *Environmental Biotechnology: Theory and Application*. John Wiley and Sons, Chinchester, 300 pp.

Gabriel, R. and Braune, A., 2005. Eco-efficiency analysis: Applications and user contacts. *Journal of Industrial Ecology*, 9: 19–21.

Gallert, C., and Winter, J., 1999. Bacterial metabolism in wastewater treatment system. In: *Biotechnology*, 2nd Edn. (Eds.) H–J. Rehm G. Reed. Wiley–VCH verlag GmbH Weihheim, pp. 17–94.

Gavrilescu, M., 2004. Cleaner production as a tool for sustainable development. *Environmental Engineering and Management Journal*, 3: 346–362.

Gavrilescu, M., 2006. Overview of *in situ* remediation technology for sites and groundwater. *Environmental Engineering and Management Journal*, 5: 79–114.

Gavrilescu, M. and Chisti, Y., 2005. Biotechnology: A sustainable alternative for chemical industry. *Biotechnology Advances*, 23: 471–499.

Grady, L.J.R., Daigger, G.T. and Lim, H.C., 1999. *Biological Wastewater Treatment*. Marcel Dekker, New York, 1076 pp.

Hamer, G., 1997. Microbial consortia for multiple pollutant biodegradation. *Pure and Applied Chemistry*, 69: 2343–2356.

Ohtake, H. and Hardoyo, J.K., 1992. New biological methods for detoxification and removal of hexavalent chromium. *Water Science and Technology*, 25: 395–405.

Olguin, E.J., 1999. Cleaner bioprocesses and sustainable development. In: *Environmental Biotechnology and Cleaner Bioprocesses*, (Eds.) E.J. Olguin, G. Sanchez and E. Hernandez. Taylor and Francis, Boca Raton, pp. 318.

Russell, D.L., 2006. *Practical Wastewater Treatment*. John Wiley and Sons, Hoboken, New Jersy, 288 pp.

Saling, P., 2005. Eco-efficiency analysis of biotechnological processes. *Applied Microbiology and Biotechnology*, 68: 1–8.

Sasi kumar, C.S. and Papinazath, T., 2003. Environmental management: bioremediation of polluted environment. In: *Proceedings of the Third International Conference on Environment and Health*, (Eds.) M.J. Bunch, V.M. Suresh and T.V. Kumaran. Chennai, India, p. 15–17.

Saval, S., 1999. Bioremediation: clean up biotechnologies for soil and aquifers. In: *Environmental Biotechnology and Cleaner Bioprocesses*, (Eds.) E.J. Olguin, G. Sanchez and E. Hernandez. Taylor and Francis Boca Raton, pp. 155–166.

Shareefdeen, Z., Herner, B. and Singh, A., 2005. Biotechnology for air pollution control: An overview. In: *Biotechnology for Odour and Air Pollution Control*, (Eds.) Z. Shareefdeen and A. Singh. Springer, Berlin, Heidelberg, pp. 3–16.

Talley, J., 2005. Introduction to recalcitrant compounds. In: *Bioremediation of Recalcitrant Compounds*, (Eds.) W. Jaffrey and J. Talley. CRC Press, Boca Raton, pp. 1–9.

Trejo, M. and Quintero, R., 1999. Bioremediation of contaminated soil In: *Environmental Biotechnology and Cleaner Bioprocesses*, (Eds.) E.J. Olguin, G. Sanchez and E. Hernandez. Taylor and Francis Boca Raton, pp. 179–190.

USEPA, 1994. *Assessment and Remediation of Contaminate Sediments (ARCS) Program*. Final summary report, EPA–905–S–94–001, USEPA, Chicago.

Vidali, M., 2001. Bioremediation: An overview. *Pure and Applied Chemistry*, 73: 1163–1172.

Vilar, V.J.P., Botelho, C.M.S. and Boaventura, R.A.R., 2007. Kinetics and equilibrium modeling of lead uptake by algae Gelidium and algal waste from agar extraction industry. *Journal of Hazardous Materials*, 143: 396–408.

WBCSD, 2000. *Eco-efficiency: Creating More Value with Less Impact*. World Business Council for Sustainable Development, Geneva, Switzerland.

Wiesmann, U., Choi, I.S. and Dombrowski, E–M., 2007. *Fundamentals of Biological Wastewater Treatment*. Wiley-VCH, Weincheim, 391pp.

2013, Sustainable Approaches for Environmental Conservation *Pages* **181–185**
Editors: **D.R. Khanna, A.K. Chopra, R. Bhutiani, Gagan Matta & Vikas Singh**
Published by: **BIOTECH BOOKS, NEW DELHI**

Chapter 24

Tiny Civic Bombs and Environmental Pollution: A Survey Report of Ujjain City, M.P., India

Shobha Shouche and Minal Lodha

Department of Zoology, Microbiology, Bioinformatics
Govt. P.G. Madhav Vigyan Mahavidyalaya, Ujjain, Madhya Pradesh

Environmental protection is a practice of protecting the environment by an individual and at organizational level for the benefit of the natural environment and humans. Environmental protection cannot be achieved until and unless it is started among the people or masses. Sometimes small but important issues are left behind. One such issue which has its importance in the environmental protection is the use of Crackers, which can also be consider as the TINY CIVIC BOMB.

The use of crackers occurs on the large scale and it has its various ill-effects on the environment, individual and also on the government at economic level.

Keywords: *Crackers (Tiny civic bomb), Pollution, Economy, Child labour.*

Introduction

Crackers are given the name TINY CIVIC BOMB as they are small in size but are more common among the society and bomb because they contain explosives and cause harm at a large scale.

A firecracker is a small explosive device primarily designed to produce a large amount of noise, especially in the form of a loud bang. They are fused and are wrapped in a heavy paper casing to contain the explosive compound.

Crackers contain gunpowder; fireworks are packed with heavy metals and toxins that produce their sparkling showers of colors. Various metals that used are strontium which produces red color, Aluminum that produces white color, Copper that produces Blue color, Barium produces Green color *etc.*

Firecrackers are commonly used in the celebrations of festivals like Diwali, Christmas and other occasion like New Year, Marriages and Elections *etc.*

These cause sounds and explosion from all directions, lightening of colors fascinate many people but in turns fire crackers are known to cause many health hazards, air pollution, noise pollution *etc.*

Materials and Method

For survey, the criteria was made in which total expenditure on crackers was calculated and based on three major areas of considerations like:

Festivals, Marriages, Elections

Festivals

Following festivals that are well associated with the crackers are:

- ☆ Dushehra
- ☆ Diwali
- ☆ Christmas
- ☆ Others

Diwali

It is the main festival associated with the crackers and the amount spent will be considered as per the population of the Ujjain city.

Calculation

Total population=7,00,000 lac

If 60 per cent of population celebrates and burns crackers excluding the children below 6 yrs and old age persons.

Then the total population=700000*60/100=420000.0

Amount is calculated by categorizing the total population into three category-Lower class, Middle class and Higher class.

According to the survey Avg. amount calculated=Rs.5000

Table 24.1: Observation

Sl.No.	*Festivals*	*Amount*	*Total Amount (Annually)*
1.	Diwali	420000*Rs5000	2100000000cr.
	Total		**2100000000cr.**

Table 24.2: Calculated as the Total Amount per annum for Total Population

Sl.No.	*Festivals*	*Amount*	*Total Amount (Annually)*
1.	Dushehra	5 areas*avg. Rs. 50000	250000 lac
2.	Christmas	200000	200000 lac
3.	Others	300000 lac	300000 lac
	TOTAL		**750000 lac**

Marriages

Table 24.3: Observation

Sl.No.	*Marriages*	*Amount*	*Total Amount (Annually)*
1.	Season from Nov. to June	1000*avg Rs.5000	500000 lac
	TOTAL		**500000 lac**

Elections

Table 24.4: Observation

Sl.No.	*Name of Election*	*Amount*	*Total Amount (Annually)*
1.	Loka Sabha- 01 seat	1*Rs.100000	100000 lac
2.	Vidhan Sabha-02 seat	2*Rs.100000	200000 lac
3.	Panchayat	3* Rs.50000	150000 lac
4.	Nagar Nigam	54* Rs.10000	540000 lac
5.	College	6* Rs.2000	12000
	TOTAL		**10,02,000 lac**

Net Total=Festivals+Marriages+Election

2100750000+500000+1002000=**210,22,52,000cr.**

Health Hazards Caused Due to the Usage of Crackers

Survey was conducted in various hospitals like

- ☆ G.D Birla
- ☆ Civil Hospital
- ☆ S.S Hospital

According to the survey.

1. *Fire Crackers Caused Burns*: Carelessness during the burning of crackers may lead to burns and sometimes also due to the manufacturing defect it may burst and causes burns. Approx 5-6 cases of burns due to fire crackers come in each hospital during Diwali.

2. *It also causes hearing loss*: It may be permanent or temporary hear loss. Average of 4-5 cases comes in each hospital.
3. Fire cracker bang causes high blood pressure in some aged group persons. Average of 6 out of 10 people. High Blood Pressure during Diwali.
4. *Accidents*: Smog caused by fire crackers can cause reduced visibility which may lead to accidents. Approx 2-3 cases come every year for the accidents.
5. *Air Pollution*: The major effect that occurs due to fire crackers is air pollution. The level of suspended particles in the air increases alarmingly during the bursting of crackers affecting the health and environment simultaneously. About 60 per cent of air pollution increases during the Diwali and about 10 per cent during the Elections. It also causes breathing problems in some people and leads to asthama and lung irritation. Every year approx 10 cases for asthma comes and 2-3 cases for breathing problems comes during Diwali.
6. *Noise Pollution*: Bursting of Crackers cause very high Noise pollution. High decibels levels result in restlessness, anger, impulsive behavior and over reaction to simulation. Most Crackers used to have more than 80Db noise that causes temporary hearing loss. Children, pregnant woman and those suffering from respiratory problems suffer most due to excessive noise. High decibel levels of crackers annoy even the foetus in the womb.
7. *Child Labour*: Child labour is the disastrous side of making of crackers. Under the survey it was found that about 6/10 are child labour in various firecrackers industries. *i.e.* about 70 per cent of chidren are working as child labour in fire cracker industries. These children are deprived of their studies and are at the risk of their health daily.

Results

The survey report shows that total of Rs. 2102252000 cr. is spended per year on crackers which includes 57 per cent during festivals, 14 per cent during marriages, and 7 per cent during elections. That means if Rs. 10000 is the per capita income of the total population, 30 per cent of the total income of a population is wasted on crackers per year which is a great loss to the economy of a city as a whole and also it is causing various environmental as well as health problems in the society and the most disastrous effect is child labour which spoils health and life of 70 per cent of the children.

Conclusion

Ujjain is a holy city where there is no industrial set up and very less job opportunities. In a city like this if total of 30 per cent of income of a population is wasted on crackers, it is a mere loss to the society. Instead this money could be utilized in many ways like:

- ✰ Setting up of small scale industries, which cost Rs.1000000 then about 2000 industries could be established and which would provide employment to 10000 people yearly.
- ✰ If one plate of food cost Rs.40 then about 68500 people can have one time of food for a year.
- ✰ If Rs.20000 is the amount for schooling of a child then about 20000 children can get full education.

Likewise many other fruitful work could be done like providing clothing, medicine *etc.* And most importantly if we do not burn a cracker in a year we can reduce the environmental pollution for about 70 per cent approx. and with the money we can improve the environment by planting trees *etc.*

Acknowledgement

Authors are thankful to the Principal Govt. M.V.M Ujjain, M.P, India for providing necessary facilities.

I also give my thanks to Dr. Tavar Surgeon at G.D Birla Hospital and to Dr. Sushil Gupta at S.S Hospital and to Dr.Vyas at Civil Hospital Ujjain, who has given their time and support in the survey.

I also give my thanks to Mr. Nawab Nazim, Babji Fireworks, Mr.Hussain, Badshah Fireworks and other who have helped in industrial survey.

2013, Sustainable Approaches for Environmental Conservation *Pages 187–200*
Editors: **D.R. Khanna, A.K. Chopra, R. Bhutiani, Gagan Matta & Vikas Singh**
Published by: **BIOTECH BOOKS, NEW DELHI**

Chapter 25

Climate Change: Knowledge and Attitude of Teachers and Students

Animesh Kumar Mohapatra

Department of Life Sciences,
Regional Institute of Education (NCERT), Bhubaneswar, Odisha

Fifteen years ago it was prudent to discuss climate change in tentative terms. More recently, what was considered a debatable has become regarded by most scientists to not only a real phenomenon, but also one that is an increasing threat to the world's environmental, social and economic stability. Students are thought to have an insecure knowledge about the "science" of climate change. Overcoming student's misconception may be a challenge when teaching about phenomenon such as climate change. Teachers have the responsibility for transmitting knowledge to the students for sound understanding of environmental problem. This study aims to examine the attitude and knowledge level of teachers' and students' about climate change. The research instruments employed explored both teachers and students tend to cite short term weather effects as evidence to support or refuge long term climate transformation, which display a fundamental misunderstanding about weather and climate distinction. The results found in the study indicated that not only students but teachers do not fully comprehend the under lying concepts related to climate change.

Keywords: *Anthropogenic global warming, Continental drift, Tectonic plates, Volcanoes, Greenhouse gases, Solar radiation, Pollen, Geo-engineering.*

Introduction

In recent years, there has been much national and international concern voiced about the environment. Environment can be identified as an external shelf where all the creatures exists (Basal,

2003). There have been more than 30 years of intense discussion and research on environmental crisis. Nevertheless, the problem continue worsen daily because humans are the cause of environmental problems occurring almost everywhere (Sudarmadi *et al.*, 2001). As quoted famously by an astronaut and replace in environmental literature "if you observe the earth from the space, there no borders that separates the countries". With this perspective we can understand that environmental problem also have no boundaries and they affect the whole world in the same way. The environmental problem which affects the whole humanity at same time and at the same level are called "global environmental problem". All global environmental problem share a common structure in which key roles are played by concepts such as human values, carrying capability, chains of causes, effects, norms, social dilemmas and policy instruments (De Groot, 1993).

While there seems to be a great deal of expressed concern about environmental degradation, nationally and globally (Dunlap *et al.*, 1992; Krause, 1993; Kempton *et al.*, 1995), most individual see the responsibility for changing environmentally destructive behaviors belonging to technological development and industrial practices rather than to changing their personal behavior (Dunlap *et al.*, 1992). The international agreements developed to reduce human interference with the earth's atmospheric system, the *Montreal protocol* on substances that deplete the ozone layer and the United Nations Framework convention on climate change, recognize the environmental consequences of human activities and look to both societal and personal behavior changes to reduce environmental destruction. Both agreements call upon educational programmes to assist in moving human behavior in a sustainable direction (Carter, 1998).

Global climate change has been the focus of appreciable scientific research efforts for more than three decades. Beginning in the early 1990s, global climate change was one of many topics highlighting in the *Benchmark for Science Literacy*, with a single benchmark addressing the idea of detrimental impacts on earth's environment as a result of human activities. Media attention to climate change has increased dramatically over the same period, with spikes in coverage occurring in 1997 and 2006, associated with the negotiation of the Kyto protocol and release of documentary film *An Inconvenient Truth* (Boykoff, 2007), respectively.

As students encounter global climate change in school and the media, they often approach the phenomenon with misconception. Research studies involving both students and the public have revealed several of these misconceptions, including:

1. Confusing weather and climate (Read *et al.*, 1994; Gowda *et al.*, 1997; Papadimitriou, 2004).
2. Identifying stratospheric ozone depletion as the primary contributor to increase global temperature (Bostrom *et al.*, 1994; Read *et al.*, 1994; Gowda *et al.*, 1997; Rye *et al.*, 1997; Papadimitriou, 2004; Osterlind, 2005; Keller, 2006), and
3. Linking unrelated pollution effects (*e.g.* litter, photochemical smog, and radioactive waste disposal) to global climate change (Read *et al.*, 1994; Gowda *et al.*, 1997; Papadimitriou, 2004; Keller, 2006).

Schools should be not only able to play a vital role in teaching about the environment and environmental issues but also in encouraging children to be advocates for their environment and actively participate in positive action towards the environment. However according to the literature (Jensen and Schnack, 1997) a clear understanding of environmental issues is a pre-requisite for taking positive action. Results of Chuckran *et al.* (1993) and Mohapatra (2009) indicated that many students appear to confuse on certain major environmental problems (the green house effect, global warming, ozone depletion, loss of biodiversity and nuclear power) in terms of causality, consequence, and in ways to alleviate these problems.

How do such misconceptions arise? Hills (1989) suggested that misconception can come from "untutored beliefs", *i.e.*, student's lack sufficient understanding because of lack of exposure to environmental issues and lack sufficient information with which to develop correct understanding. This challenges the idea that students receive an adequate education from either schools or from popular media presentations. Do these misconceptions arise from incorrect instruction given by teachers who do not have correct understanding of these phenomena for themselves? Since virtually no information is available on perception, knowledge, awareness and attributes with regard to climate change of teachers. It was very important to explore differences in perception, knowledge and attitude between teachers and students on climate change. Based on the above, this study was undertaken to explore cause of misconception prevailing among students.

Procedure

Background

The present study aimed to explore the level of knowledge, understand attitude of the teachers' and higher secondary students' about the issue of climate change. The rationale for choosing teachers and students is due to India's rapid economic development and quickly deteriorating environment. India is also emerging as one of the main contributors of green house gases. However, the question of reducing poverty and raising economic standards while at the same time conserving the environment and natural resources is a dilemma not only for India but for all countries in the world.

Instrument

The research tool employed for this study was a closed-form questionnaire contained thirty two questions in the form of statements arranged in three sections. The section A contained twelve statements related to possible reasons while section B contained ten statements about possible consequences of an exacerbation of the climate change. Within each of these sections equal numbers of scientifically acceptable and scientifically unacceptable statements were interspersed at random. The section C of the questionnaire contained ten knowledge based statements.

Randomly schools were identified in different cities of Odisha. The questionnaire was completed by 309 teachers (all streams) and 763 higher secondary students of class XII (all streams). Questionnaire was completed under examination conditions, although no time limit was imposed. The respondents were asked to respond by ticking boxes labeled 'I fully agree', 'I think so', 'I don't know', 'I don't think so' and 'I fully disagree'.

Comparisons of the responses by teachers and students to different questions were made by Chi-square analysis. In order to this, two positive responses ('I fully agree' and 'I think so') were combined to provide a measure of the proportion who affirmed an idea. Similarly, to indicate those who did not affirm an idea, the other three responses ('I don't know', 'I don't think so' and 'I fully disagree') were combined.

Results and Discussion

Responses of Teachers and Students to the Possible Reasons for Climate Change

The responses of teachers and students to the possible reasons for climate change expressed in percentage are plotted graphically in Figures 25.1 and 25.2. Majority of teachers and students are aware of the fact that energy sector is responsible for emission of 3/4 of the CO_2, 1/5 CH_4 and large quantity of NO_2 (T= 86.5 per cent and S = 82 per cent); 1/4 of the total CH_4 emission are said to come from domesticated animals (T= 85 per cent and S = 83.5 per cent) and burning of fossil fuels emits large amount of greenhouse gases (T = 88 per cent and S = 81 per cent, $p< 0.05$). The result showed significant difference between teachers and students an approving that global warming is caused because of the heat rays coming from the earth can not escape through the atmosphere out into space (T = 90 per cent and S = 69.6 per cent, $p<0.01$) and that a large amount of NO_2 emission has been attributed to fertilizer application (T = 73.9 per cent and S = 67.3 per cent, $p< 0.05$). The result showed that three-fourth of the teachers (75.6 per cent) and more than four-fifth of students (84 per cent) were ignorant of the fact that CH_4 is released from paddy fields that are flooded during the showing and maturing periods.

The findings of the present study showed that some ideas held by teachers and students on possible reasons of climate change are well known. For example, majority of teachers and students knew that large quantities of GHGs are emitted by energy sector, burning of fossil fuel and domesticated animals that cause warming of the climate. Similar observations have also been reported by Jeffries and Stanisstreet (2001) while studying knowledge level of British students on green house effect. While majority of teachers were aware of the fact that energy from sun (largely in the visible part of spectrum, but also some in the ultraviolet and infrared portions) is absorbed by the land, seas, mountains *etc.* If all this energy were to be absorbed completely, the earth would become gradually hotter and hotter. But actually, the earth both absorbs and simultaneously releases it in the form of infrared waves. All this rising heat is not lost to space, but is partially absorbed by some gases present in small quantities in the atmosphere called greenhouse gases. Greenhouse gases re-emit some of this heat to the earth's surface. Increase in amount of greenhouse gases would absorb heat and that would lead to global warming and it was not known to about one- third of students.

Surprisingly, very few teachers (20.6 per cent) and students (13.2 per cent) had clear idea that changes in land use pattern led to a rise in the emission of CO_2. Similarly, very less percentage of teachers (14 per cent) and students (16.5 per cent) has clear knowledge about the emission of CO_2 and deforestation. The result showed that both teachers and students have poor understanding of the effect of natural phenomena like volcanism, variation in solar radiation, tectonic plate movement on climate change (T = 80 per cent and S = 87 per cent) and that changes in the tilt of the earth would effect the severity of the seasons (T = 72.9 per cent and S = 78 per cent). However, many misconceptions are prevailing in the mind of teachers and students that gases and dust particles released into atmosphere during volcanic eruption partially block the incoming rays of the sun, which led to warming of the climate (T = 61 per cent and S = 64 per cent) and that warming of the climate is due to holes in the ozone layer (T = 55 per cent and S = 74 per cent). The results of the present study revealed that both teachers and students believe that climate change is entirely because of anthropogenic activities. Surprisingly, teachers were ignorant about the effect of natural phenomena like volcanism, continental drift, deviation in earth orbit, tilting of the earth and variation in solar radiation on climate change. Because of this inadequacy found in teachers, students have difficulties in understanding the concepts and to develop idea of natural phenomenan causing climate change (Mohapatra, 2009).

Responses of Teachers and Students to Possible Consequences of Climate Change

The data pertaining to teachers' and students' perceptions about possible consequences of climate change are shown graphically in Figure 25.3 and 254. The results of the study reflected that teachers and students were having sound knowledge about some of the general possible consequences of

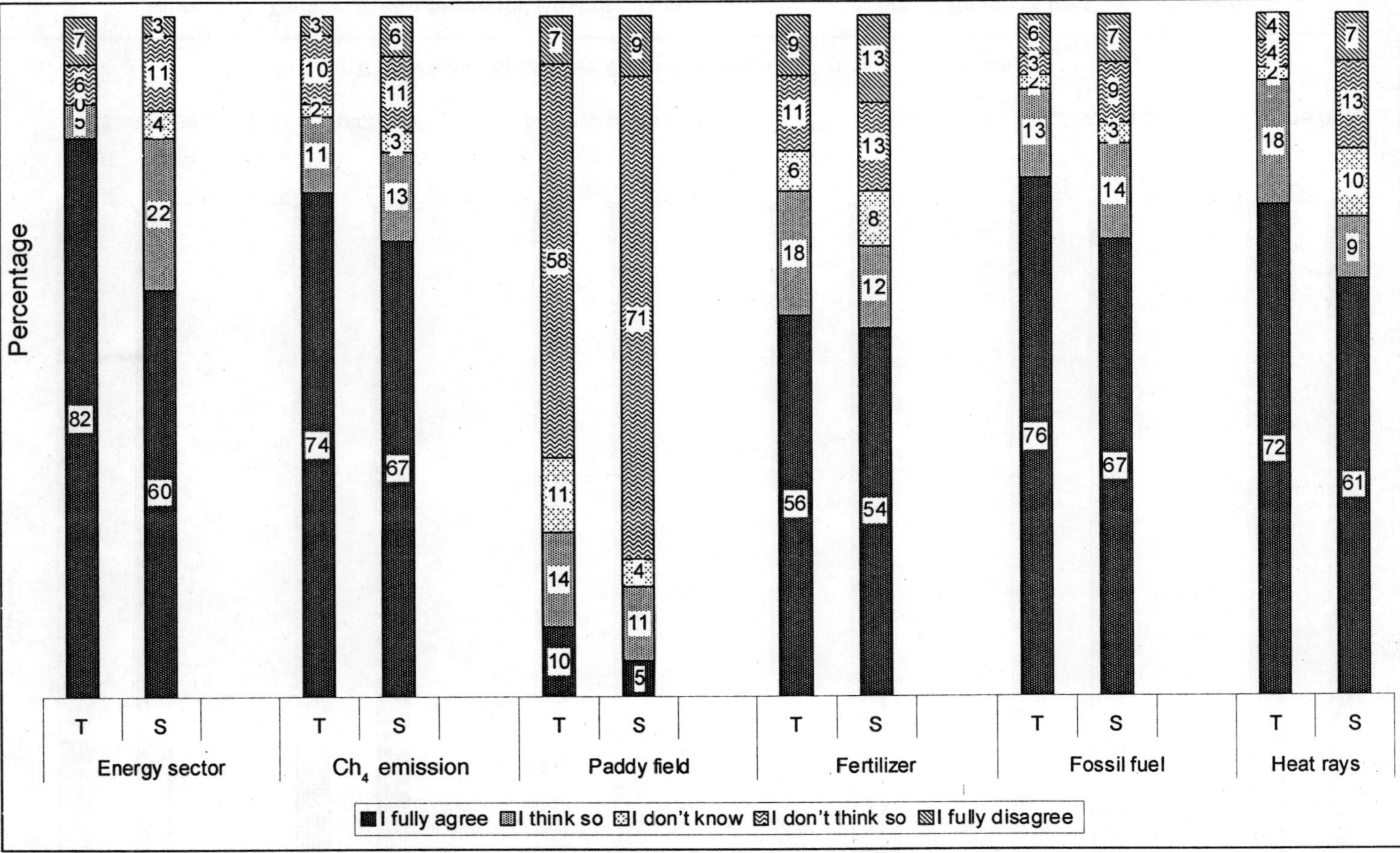

Figure 25.1: Teachers' and Students' Responses on Scientifically Possible Reasons for Climate Change

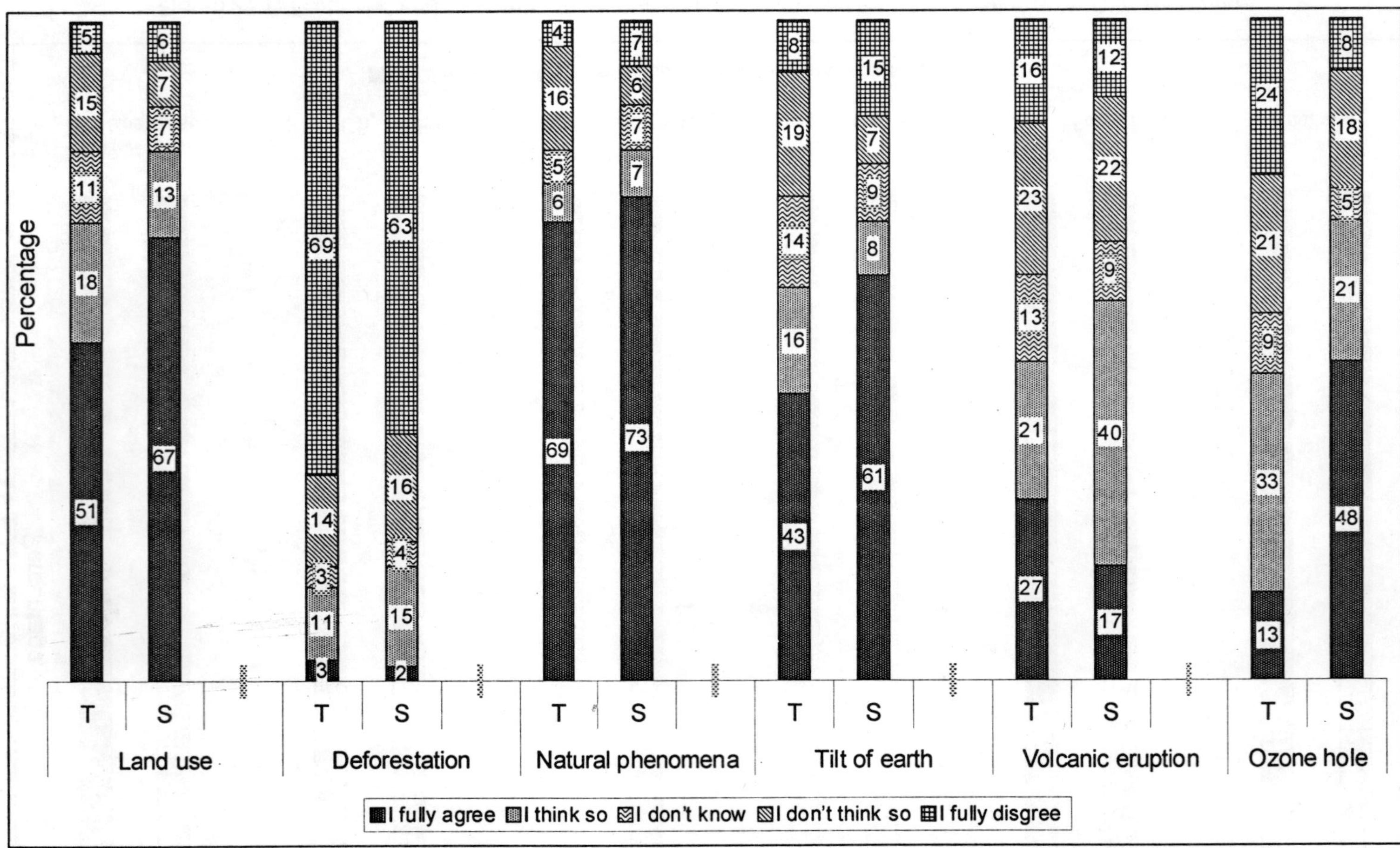

Figure 25.2: Teachers' and Students' Responses on Scientifically Incorrect Reasons for Climate Change

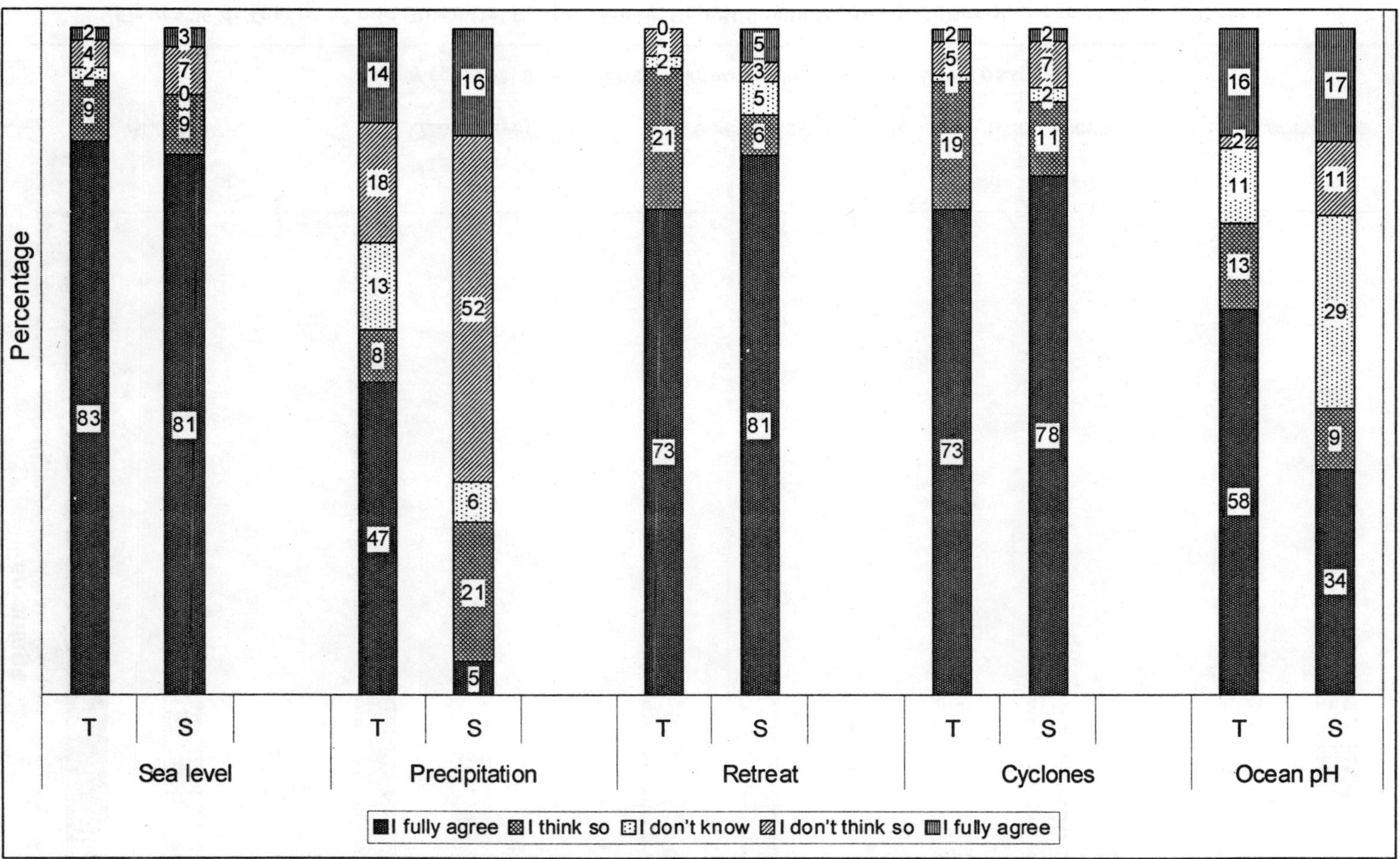

Figure 25.3: Teachers' and Students' Responses on Scientifically Possible Consequences of Climate Change

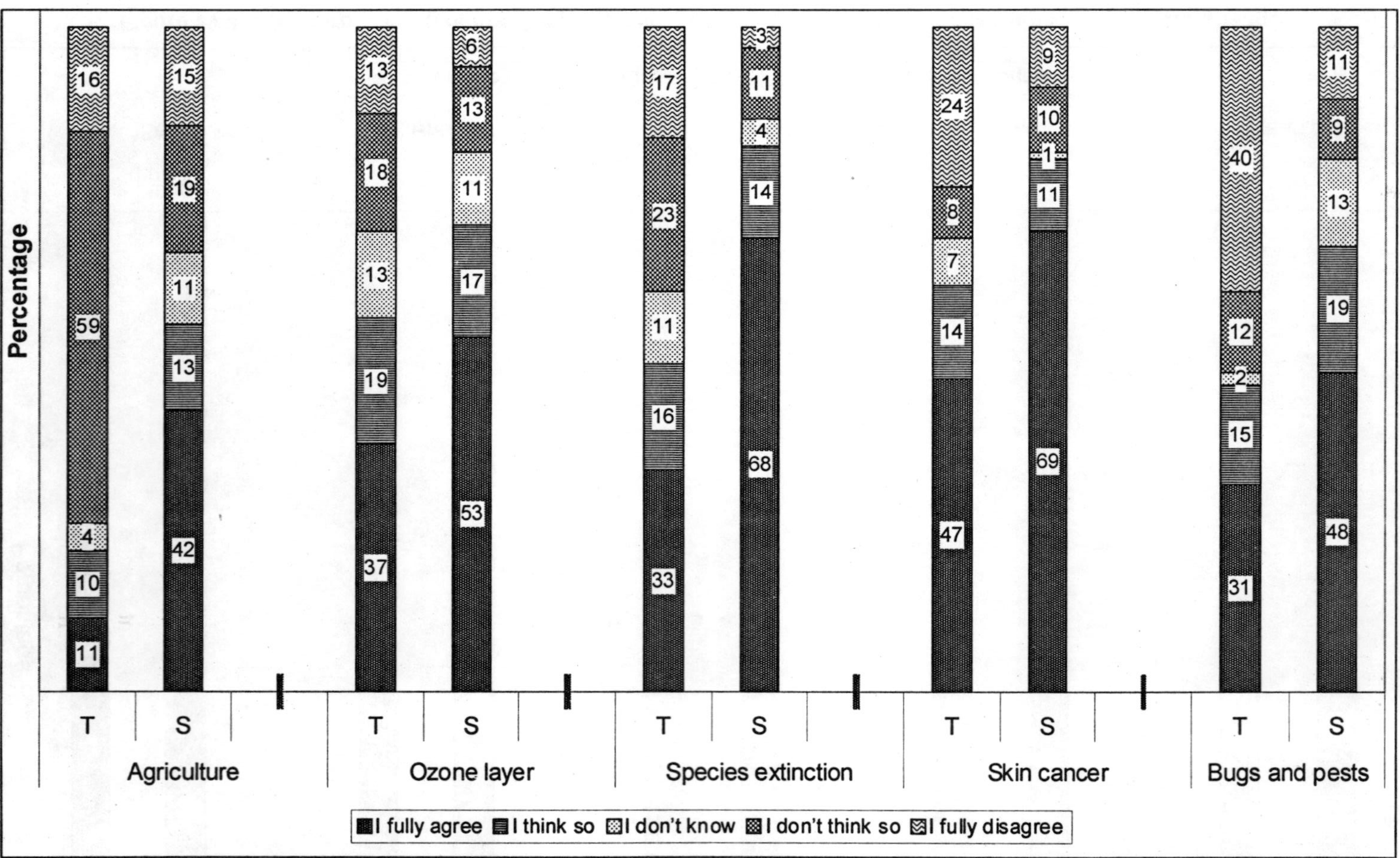

Figure 25.4: Teachers' and Students' Responses on Scientifically Incorrect Consequences of Climate Change

climate change like an increase in global temperature would cause sea level to rise (T = 91.6 per cent and S = 87 per cent), warming will cause retreat of glaciers, permafrost and sea ice (T = 94 per cent and S = 85 per cent, $p < 0.05$) and variation in climate pattern would increase the frequency and intensity of extreme events like cyclones (T = 90 per cent and S = 89 per cent). However, a remarkable difference was observed between teachers and students about amount and pattern of precipitation in relation to change in global temperature ($p < 0.01$). Three- fifth of the teachers (61 per cent) affirmed that amount and pattern of rainfall would change because of global warming while three- fourth of the students (74 per cent) rejected it. Significant difference was observed between teachers and students in affirming ocean surface pH have decreased as ocean absorbed more CO_2 (T = 71 per cent and S = 43 per cent, $p < 0.01$). The simple fact that increased atmospheric CO_2 increases the amount of CO_2 dissolved in the oceans. CO_2 dissolved in the oceans reacts with water to form carbonic acid, resulting in ocean acidification was not known to them.

About four–fifth of the teachers (79 per cent) rightly rejected that climate change will favor agricultural yields all over the world while more than half of the students (55.2 per cent) agreed to it and showed poor understanding about pattern of precipitation. Both teachers and students erroneously linked warming of climate with ozone layer damage (T = 69 per cent and S = 80.2 per cent); with skin cancer (T = 68 per cent and S = 91 per cent) and with reduction of the population of bugs and pests on crops (T = 48 per cent and S = 79 per cent). They also thought that there would be no extinction of species because of climate change as majority of plants and animals would tolerate it and showed poor knowledge about biology of living organisms (T = 59.7 per cent and S = 86 per cent). A number of misconceptions pertaining to consequences of climate change are prevailing in the mind of teachers as well as students. They erroneously linked global warming with depletion of ozone layer and with skin cancer. Even they were not aware of the fact that warming would lead to an increase in population of insects and pests. Similar findings have been reported by Boyes and Stannisstreet (1993, 1997) and Mohapatra (2009).

Responses of Teachers and Students to the Knowledge Based Statements

The responses of teachers and students to the knowledge based questions expressed in percentages are plotted graphically in Figures 25.5 and 25.6. The results of the present study revealed that more than half of the teachers (54.5 per cent) and majority of the students (91.5 per cent) could not differentiate between weather change and climate change. Nearly, same numbers of teachers and students wrongly affirmed that if there would be any change in present weather in comparison to last year is climate change. The data of the present study showed that majority of our teachers and students have inadequate knowledge about the science of climate change. They are ignorant of the processes by which climate change can be studied like air trapped in bubbles in Antarctic ice sheet which would reveal the CO_2 variation in the atmosphere over several millennia (T = 70 per cent and S = 68 per cent); growth ring patterns in trees (T = 65 per cent and S = 64 per cent) and changes in the type of pollen found in different sedimentation levels in rivers and lakes (T = 68 per cent and S = 78 per cent). Majority of teachers and students were aware of the common remedial actions fro checking climate change such as using nuclear power energy (T = 80 per cent and S = 70.7 per cent) and by wind energy (T = 86 per cent and S = 91 per cent) instead of coal energy. Studies of Jeffries and Stanisstreet (2001) revealed similar results while studying the understanding level of British students on green house effect. However, they are ignorant of many recently developed techniques/ideas which would help us to reduce the causative agent of climate change such as carbon sequestration (T = 61 per cent and S = 71 per cent) and solar radiation management (T = 62 per cent and S = 67 per cent). About three - fourth

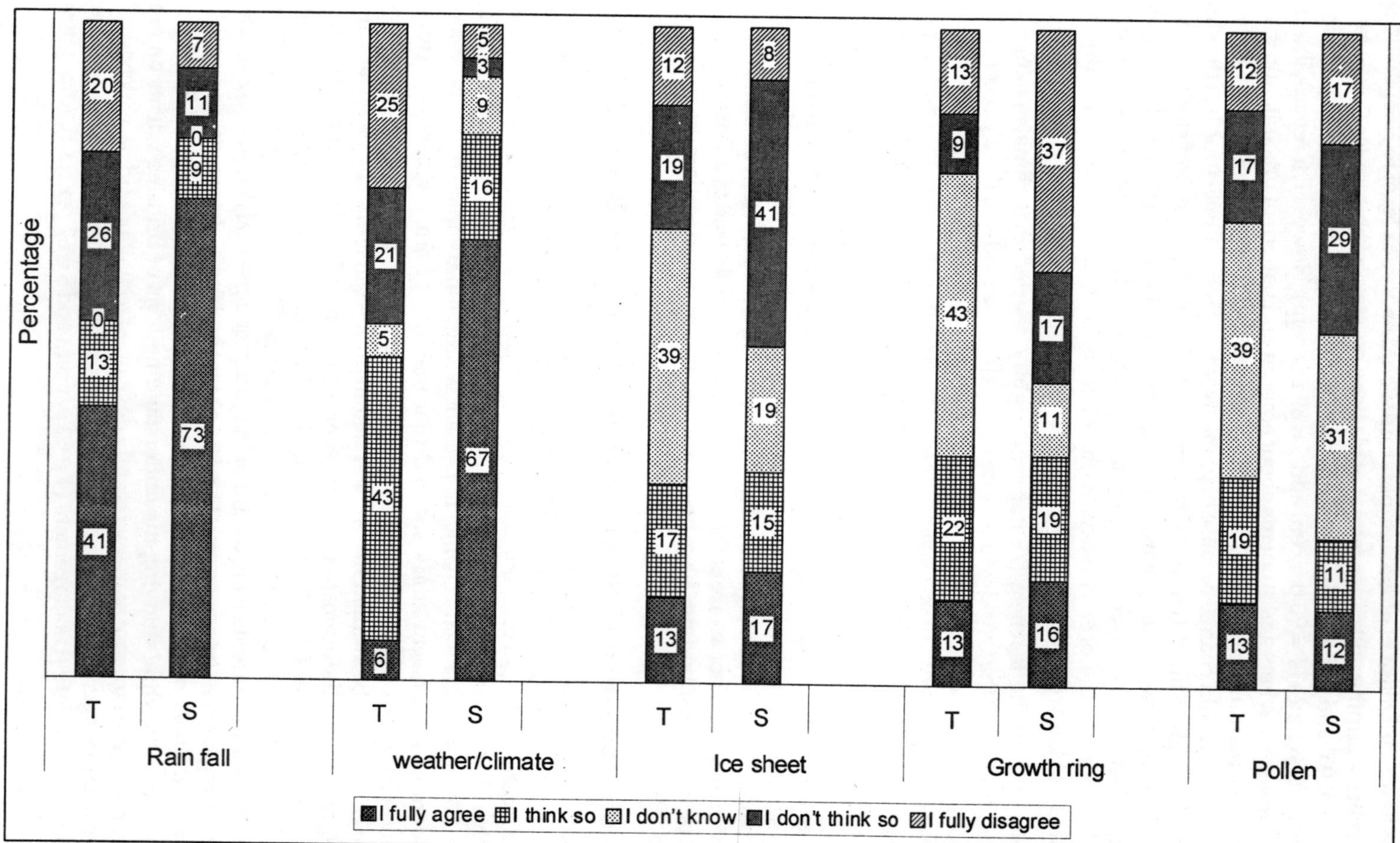

Figure 25.5: Teachers' and Students' Responses on Possible Remedial Actions and Detection of Climate Change

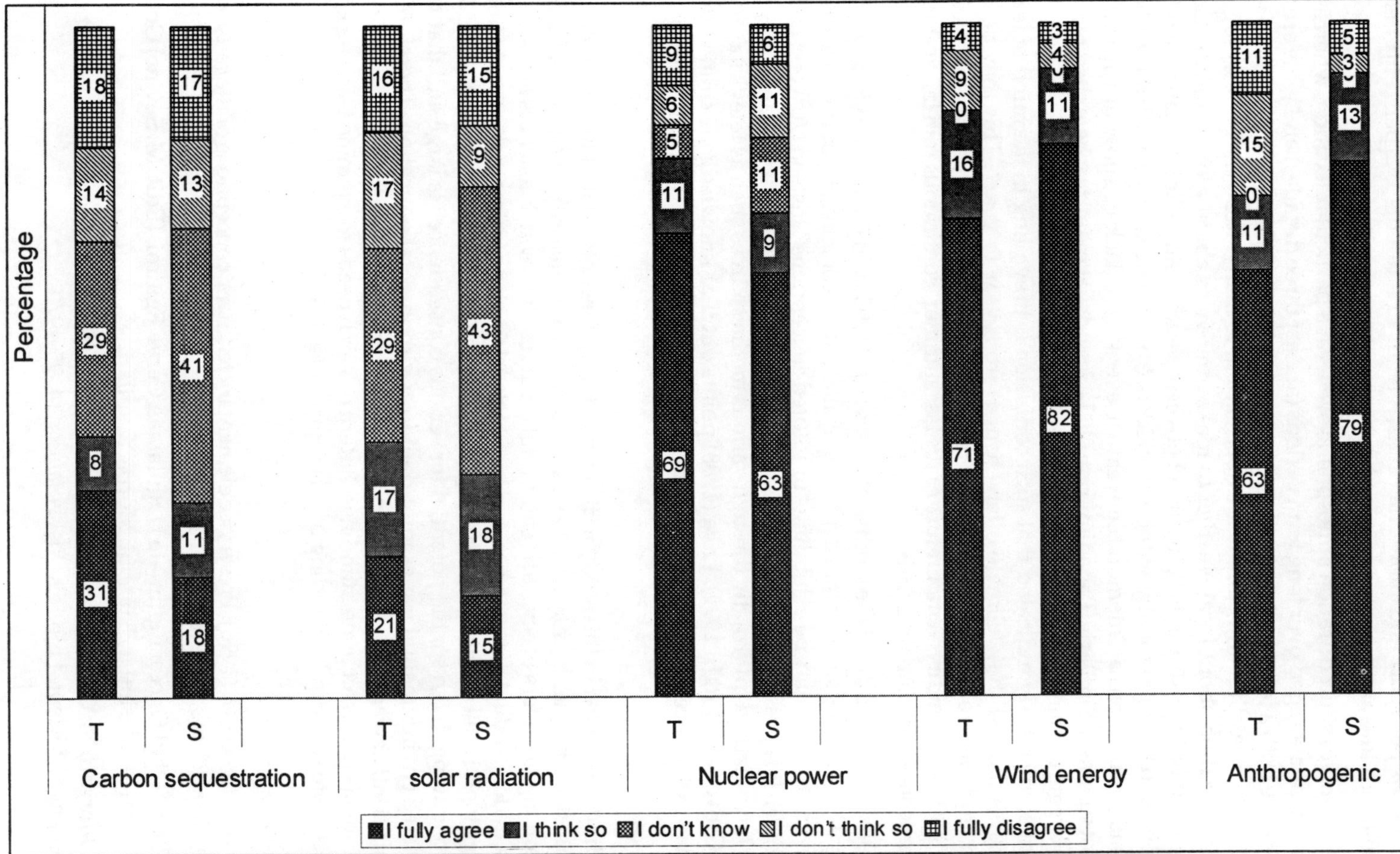

Figure 25.6: Teachers' and Students' Responses on Possible Remedial Actions and Detection of Climate Change

of the teachers (73 per cent) and more than nine – tenth of students (92 per cent) believed that anthropogenic activities are the sole cause of climate change.

Weather involves short duration atmospheric events at a particular location, whereas climate is weather conditions averaged over long term periods (at least three decades) and wide areas. Climate and weather have both a temporal and geographic aspect. Several research studies have shown that distinction between weather and climate are not clear-cut to many. Read *et al.* (1994) suggested that people use information about local weather to make inferences about global climate. Gowda *et al.* (1997) reported about the students confusion between weather and climate. They also found that students claimed to have personally witnessed evidence of climate change, with the authors stating that such claims arise from a "memorable weather event" (*e.g.* lack of snow at Christmas, severe flooding, a hot summer). Similarly, Papadimitriou (2004) found that many teachers cited recent weather events (*e.g.* extreme temperatures for a particular summer) as evidence of global warming. The results of the present study clearly indicated that such confusion is prevailing in the mind of teachers and students. They indicated that "climate often changes from year to year". This may be because of teachers and students are using weather data from short-term periods, specifically "local weather excursions" like high humidity, temperature of the last summer, continuous rain fall and severe flood to make judgments about climate trends.

The results of this study and the studies of Boyes *et al.* (1993) and Groves and Pugh (1999) are not encouraging. These results suggest that a number of alternative conceptions or misconceptions are at work. However, students may be exhibiting "untutored conceptions" instead (Atwood and Atwood, 1996; Hills, 1989). It is possible that the experiences of the teachers on climate change may be limited to what they have learned informally through various media sources such as television and popular magazines (Groves and Pugh, 1999). Lave (1988) points out that knowledge is context-bound, and that "everyday experience is the major means by which culture impinges on individuals".

Conclusion

The findings presented in this study indicate that many concepts related with climate change are not well understood neither by many of the teachers nor by the majority of students. Inadequacy is found even among members of scientific community as a result of which teachers and students have difficulties in understanding the concepts on which to develop idea of climate change. At this time, the climate change topic is still debatable, with disagreements over many facts, its magnitude and potential effects still continuing. Even so, the potential for serious consequences is large and that makes this and other related environmental issues important to environmental education. Therefore, teachers need to be sufficiently aware of these issues and their interrelationships so that they can more properly guide their students to correct understandings. Future research needs to examine teaching approaches which best promote solid understanding of this complex issue.

References

Atwood, R., and Atwood, V. (1996). Preservice elementary teachers' conceptions of the cause of seasons. *Journal of Research in science teaching* 33: 553-563.

Basal, H.A (2003). Okul Oncesi Egitimate Uygulamali Cerve Egitimi, (Edit. M. Sevinc) Gelisimde ve Egitimde Yeni Yaklasimlar, Istanbul: Morpa Yaymlari.

Bostrom, A., Morgan, M. G., Fischhoff, B., and Read, D. (1994). What do people know about global climate change? 1 Mental models. *Risk Analysis*, 14: 959-970.

Boyes, E. and Stanisstreet, M. (1993). The 'greenhouse effect': children's perceptions of causes, consequences and cures, *International Journal of Science Education*, 15: 531-552.

Boyes, E. and Stanisstreet, M. (1997). Children's models of understanding of two major global environmental issues (ozone layer and greenhouse effect), *Research in Science and Technological Education*, 15: 19-28.

Boykoff, M. (2007). From convergence to contention: United States mass media representations of anthropogenic climate change science. *Transactions of the Institute of British Geographers*, 32: 477-489.

Carter, M. L. (1998). Global Environmental Change: Modifying Human Contributions Through Education. *Journal of Science Education and Technology*, 7(4): 297-309.

Chuckran, D., Boyes, E., and Stanisstreet, M. (1993). How do high school students perceive global climatic change: what are its manifestations? What are its origins? What corrective action can be taken? *Journal of Science Education and Technology*, 2: 541- 557.

De Groot, W. T. (1993). *Environmental decision Making*, Belgium, Vrije Universiteit Brussels Press, 48 p.

Dunlap, R. E., Gallup, Jr., G. H., and Gallup, A. M. (1992). *Health of the planet: A George H. Gallup Memorial Survey*. Gallup International Institute, Princeton, New Jersey.

Gowda, M. V. R., Fox, J. C., and Magelky, R. D. (1997). Students' understanding of climate change: insights for scientists and educators. *Bulletin of the American Meteorological Society*, 78: 2232-2240.

Groves, F., and Pugh, A. (1999). Elementary pre-service teacher perceptions of the greenhouse effect. *Journal of Science Education and Technology*, 8(1): 75-81.

Hills, G. (1989). Students' "untutored" beliefs about natural phenomena: primitive science or common sense? *Science Education* 73: 155-186.

Jeffries, H and Stanisstreet, M. (2001). Knowledge about the "greenhouse effect" have college students improved, *Research in Science and Technological Education*, 19(2), 205-220.

Jensen, B., and Schnack, K. (1997). The action competence approach in environmental education. *Environmental Education Research*, 3(2): 163-178.

Keller, J. M. (2006). Development of a concept inventory assessing student's beliefs and reasoning difficulties regarding the greenhouse effect (Doctoral dissertation). Available from Dissertations and Theses database. (UMI No. 3237466).

Kempton, W., Boster, J. S., and Hartley, J. A. (1995). *Environmental Values in American Culture*, MIT Press, Cambridge, Massachusetts.

Krause, D. (1993). Environmental consciousness: An empirical study. *Environment and Behavior* 25: 126-142.

Lave, J (1988). *Cognition in Practice*, Cambridge University Press, Cambridge, p. 14-15.

Mohapatra, A. K. (2009). Students' perception of the global warming. *Journal of Indian Education*. Vol.XXXV (1), p. 35-45.

Osterlind, K. (2005). Concept formation in environmental education: 14- years olds' work on the intensified greenhouse effect and the depletion of the ozone layer. *International Journal of Science Education*, 27: 891-908.

Papadimitriou, V. (2004). Prospective primary teachers' understanding of climate change, greenhouse effect and ozone layer depletion. *Journal of Science Education and Technology*, 13, 299-307.

Read, D., Bostrom, A., Morgan, M. G., Fischhoff, B., and Smuts, T. (1994). What do people know about global climate change? 2 Survey studies of educated laypeople. *Risk Analysis*, 14: 971-982.

Rye, J. A., Rubba, P. A., and Wiesenmayer, R. L. (1997). An investigation of middle school students' alternative conceptions of global warming. *International Journal of Science Education*, 19: 527-551.

Sudarmadi, S., Suzuki, S., Kawada, T., Netti, H., Soemantri, S., and Tugaswati, T. A. (2001). A survey of perception, knowledge, awareness and attitude in regard to environmental problems in a sample of two different social groups in Jakarta, Indonesia. *Environment, Development and Sustainability* 3:169-183.

2013, Sustainable Approaches for Environmental Conservation *Pages* **201–211**
Editors: **D.R. Khanna, A.K. Chopra, R. Bhutiani, Gagan Matta & Vikas Singh**
Published by: **BIOTECH BOOKS, NEW DELHI**

Chapter 26

Climate Change: A Hovering Threat on Human Rights

Sakshee Sharma and Saloni Mathur

Rajiv Gandhi National University of Law,
Patiala, Punjab

In the present day world when top countries are aiming to reach new heights of development, many climate change issues have come into the picture, which are directly or indirectly related to the heavy industrialization and other human activities. Although policies are being framed by these countries but somehow it seems that the measures taken by them to stop the resulting climate change is not in consonance with the damage done. This study focuses on the link between the human activities to achieve higher standards of living on one hand and on the other, the resultant infringement of the human rights acquired by the man from nature. The research throws light on how the populations of the underdeveloped countries which have a minimal share in the process of climate change are the most vulnerable lot. The rising sea levels and natural disasters due to climate change have given way to malnutrition, hunger, lack of access to water and adequate housing, exposure to disease, loss of livelihood and permanent displacement. The threat to the pacific island nations and other low lying countries due to the rising sea levels have also jeopardized the civil and political rights of the people living here. Further this research will show how farmers of agrarian countries are victims of the evident climate change.

This paper will bring together the cause of the climate change and its detrimental ramifications on the poor and vulnerable population in the form of human rights violation.

Keywords: *Climate change, Human rights, Violation, Underdeveloped.*

Introduction

Changes in the earth's climate have been naturally occurring since the creation of the planet. Yet a growing body of evidence is attributing some climate changes to human made environmental pollutants (Merrill, 2010). The studies conducted by climatologists concluded that the overall pattern actually demonstrated a steady decrease in temperature over the first 900 years of the last 1000 years, followed by a dramatic increase in the 20th century. The findings suggest that the 1990-2000 was the warmest decade of the whole millennium, indicating that the temperature increase over the last century is not typical of normal climate variability. The researchers also co-related their finding with factors known to affect the climate and determine that solar variability and volcanism were the main influences during the first 900 years, but that human activity contributed too much of the variation in the 20th century (Martinez, 2005). Global warming and ozone depletion are environmental events influenced by human activity (Merrill, 2010). Global warming refers to an average increase in the earth's temperature which in turn causes changes in climate. An increase in the earth's temperature is the result of natural causes and human activities including greenhouse gases, ozone depletion, deforestation, volcanoes, *etc.* (Merrill, 2010). Greenhouse gases (GHGs) are the gases that absorb infrared radiations in the atmosphere (Merrill, 2010). It is important to make the distinction between naturally occurring GHGs and GHGs that are released from industrial plants and through vehicle emissions. Natural GHGs, such as water vapor, carbon dioxide, and other gases, exist in the atmosphere and act to catch some of the energy that is radiated back into space from the heat of the Earth's surface. This natural greenhouse effect is beneficial because it maintains the Earth's average temperature around sixty degrees Fahrenheit, making the planet a hospitable environment. However, excessive amounts of gases emitted from industrial sources and automobiles (CO_2, N_2O, CFCs, PFCs, HFCs, HCFCs) have increased the planet's ability to "trap" this energy, causing an increase in the average global temperature. This rise in global climate, and its impact on the Earth's ecosystem, will likely have far-reaching effects (Martinez, 2005). After the industrial revolution, productivity and efficiency increased dramatically as production of goods shifted from the home to factories. Unfortunately, along with the technological advances came and enormous increase in the amount of atmospheric concentrations of GHGs. Since the Industrial Revolution, concentrations of carbon dioxide have increased by nearly thirty percent, concentrations of methane have more than doubled, and nitrous oxide concentrations have risen by about fifteen percent (Martinez, 2005). Industrialization has given rise to other toxic gases and their adverse consequences are acid rains and ozone hole. Cutting of trees, deforestation, increased building activities are leading to desertification. Global warming is the most crucial and contentious issue of our times. Inter-governmental Panel on Climate Change (IPCC) was created in 1988 by the United Nations Environment Program and the World Meteorological Organization to assess scientific and technical research concerning climate change and to evaluate potential impacts and possible options for adaptation on mitigation (IPCC). In 2001, the IPCC determined that the average global air had temperature had increased one Fahrenheit degree over the last century (Martinez, 2005). The IPCC also predicted that the average global temperature would rise by another 2.5 to 10.4 degrees Fahrenheit by the year 2100 and concluded that the temperature change has been caused primarily by human activities releasing greenhouse gases into the atmosphere (Martinez, 2005).

Clearly, climate change has had and will continue to have adverse impacts on fundamental human rights:

Right to Life

"The anthropogenic climate change violates Right to life in at least two ways. First, climate is projected to result in an increased frequency of severe weather events, such as tornadoes, hurricanes, storm surges and floods and these can lead to a direct loss of life. Storm surges can have a devastating effect. R.F. Mclean and Alla Tsyban write for example, "Skorm surges flooding in Bangladesh has caused very high mortality in the coastal population (*e.g.*, at least 225,000 in November in 1970 and 138000 in April 1991), with the highest mortality among the old and weak. Land that is subject to flooding-at least 15 per cent of the Bangladesh land area-is disproportionately occupied by people living a marginal with few options or resources for adaptation. Climate change will also produce flooding and landslides as these can be devastating. The fourth assessment report of the IPCC reports that in 1999, 30,000 died from storms followed by floods and landslides in Venezuela. In 2000/2001, 1813 died in floods in Mozambique. In addition to severe weather events, climate change will also involve heat waves, and these, too, will lead to loss of life..." (Gardiner *et al.*, 2010).

The right to life is anchored in Article 6 of the International Covenant on Civil and Political Rights (ICCPR). In its general comments nos. 6 and 14, the Human Rights Committee, which monitors the convention, postulated the right to life as the supreme right, which may not be infringed upon even in times of emergency (Rathgeber, 2010).

The quality of the environment affects the ability of people to enjoy the universally held right to life. Direct impacts include the increased incidents of natural disasters, while indirect impacts include poorer standards of health, nutrition, access to clean drinking water and susceptibility to disease. The immediacy of these threats indicates that positive action must be taken to ensure the full enjoyment of the right to life. Given the fact that mitigation remains the only known means to fully prevent the catastrophic impacts of climate change and that the critical decisions do not lie within the control of each state itself, the full protection of the right to life from climate change mostly depends upon the actions of the international community (Blazogiannaki, 2009).

The right to life is protected in both the UDHR (Universal Declaration of Human Rights) and the ICCPR (International Covenant on Civil and Political Rights). Article 3 of the UDHR provides "everyone has the right to life, liberty and security of person". Article 6(1) of the ICCPR provides 'very human being has the inherent right to life. This right shall be protected by law. No one shall be arbitrarily deprived of his life. The right to life of children also receives specific protection in article 6 of the CRC. In its General Comment on the right to life, the UN Human Rights Committee warned against interpreting the right to life in a narrow or restrictive manner. It stated that protection of this right requires the State to take positive measures and that "it would be desirable for state parties to take all possible measures to reduce infant mortality and to increase life expectancy."[1] Another implication of the hovering climate change to the Right to Life is in the form of farmer suicides. Data from the National Crime Records Bureau indicate that More than 17,368 Indian farmers killed themselves in 2009, the worst figure for farm suicides in six years and an increase of 1,172 from the previous year's figure. Loss of government subsidies, international competition, failure of crops due to recent erratic climate patterns are all being blamed for a staggering number of Indian farmers who are resorting to suicide[2]. Global Warming began to affect world agriculture in the late nineteenth century and its increasing intensity could be more devastating to agriculture than any previous events in human

---

1 Available at http://www2.ohchr.org/english/law/ccpr.htm

2 Available at http://www.huffingtonpost.com/2011/01/11/india-farmers-commiting-s_n_807621.html

history. Prolonged droughts and reduced irrigational sources, decreasing water table due to meltdown of glaciers have adversely affected the crop production in most agrarian countries (Tauger, 2011) Successive crop failures have forced many farmers to commit suicide.[3] As articulated by the Deputy High Commissioner for Human Rights, climate change can have both a direct and indirect impact on human life. The effect may be immediate, as in the aftermath of climate-change induced extreme weather, or may appear gradually, as deterioration in health, diminishing access to safe drinking water and susceptibility to disease increases.[4]

Right to Health

The ICESCR recognizes the right to the "highest attainable standard of physical and mental health' and the CESCR considers this right indispensible for the enjoyment of other human rights. The right to health is widely protected in other international and regional instruments and under national constitutions. As interpreted by the CESCR and other authoritative or adjudicatory bodies, the substantive content of this right includes timely and appropriate health care, access to safe and potable water, adequate sanitation, an adequate supply of safe food, nutrition and housing, healthy occupational and environmental conditions and access to health-related education, information. These are considered the basic determinants of health which, in the assessment of World health Organization (WHO), climate change will place at risk. Climate change is expected to have significant health impacts, including by increasing, inter alia, malnutrition; the number of people suffering from death, disease and injury from heat waves, floods, storms, fires and droughts; and cardio-respiratory morbidity associated with ground-level ozone (Mclnerney *et al.*, 2011).

The right to health includes a range of factors necessary to lead a healthy life, such as food, nutrition, housing, access to safe water and a healthy environment. Climate change is likely to increase deaths from malnutrition, heat stress and infectious diseases worldwide. It will also alter traditional sources and access to clean water (Blazogiannaki, 2009).

Climate change is a significant and emerging threat to public health, and changes the way we must look at protecting vulnerable populations.[5] Although global warming may bring some localized benefits, such as fewer winter deaths in temperate climates and increased food production in certain areas, the overall health effects of changing climate are likely to be overwhelmingly negative. Climate change affects the fundamental requirements for health – clean air, safe drinking water, sufficient food and secure shelter.[6] Climate change has serious implications for public health. Catastrophic weather events, variable climates that affect food and water supplies, ecosystem changes are all associated with global warming and pose health risks. Climate and weather already exert strong influences on health: increased deaths in heat waves, and in natural disasters such as floods, as well as changing patterns of life-threatening vector-borne diseases such as malaria and other existing and emerging infectious diseases are observed.[7]

---

3 Available at http://www.greenmuze.com/climate/heat/1036-mass-farmer-suicide-in-india.html

4 Available at http://www2.ohchr.org/english/issues/climatechange/docs/submissions/Australia_HR_Equal_Opportunity_Commission_HR_ClimateChange_4.pdf

5 Available at http://www.who.int/globalchange/en/

6 Available at http://www.who.int/mediacentre/factsheets/fs266/en/index.html

7 Available at http://www.who.int/features/factfiles/climate_change/en/index.html

All populations will be affected by climate change, but some are more vulnerable than others. People living in small island developing states and other coastal regions, megacities, and mountainous and polar regions are particularly vulnerable. Children – in particular living in poor countries – are among the most vulnerable to the resulting health risks and will be exposed longer to the health consequences. The health effects are also expected to be more severe for elderly people and people with infirmities or pre-existing medical conditions. Areas with weak health infrastructure – mostly in developing countries – will be least able to cope without assistance to prepare and respond.[8]

Relying on the IPCC assessment, the OHCHR report on climate change and human rights states, "Climate change is projected to affect the health status of millions of people, through increase in malnutrition, increased diseases and injury due to extreme weather events and increase burden of diarrhoeal, cardio-respiratory and infectious diseases. Global warming may also affect the spread of malaria and other vector borne diseases in some parts of the world. Overall, the negative health effects will disproportionately be felt in Sub-Saharan Africa, South Asia and the Middle East" (Mclnerney *et al.*, 2011). Further, the fourth assessment report of the IPCC notes states that climate change is projected to increase the burden of diarrhoeal diseases in low-income regions by approximately 2-5 per cent in 2020. It adds that dengue, too, will increase dramatically, and it reports research that estimates that in 2080, 5-6 per cent billion people would be at risk of dengue as a result of climate change and population increase, compared with 3.5 billion people if the climate remained unchanged. Human induced climate change thus clearly results in a variety of different threats to the human rights to health (Gardiner *et al.*, 2010).

Further, Global warming can adversely affect by causing disruption in food production and economic performance. It may also increase the occurrence of illness, injury and death because of more extreme temperatures, increased floods and severe storms, droughts, increased frequency and changing patterns in waterborne, food-borne, vector-borne and rodent borne diseases. Ozone depletion leads to higher levels of UV-B radiation reaching the earth's surface, which is associated with skin cancer, skin damage, cataracts, suppression of the body's immune system and interference with the physiological and developmental process of plants. Global warming leads to changes in the environment. Environmental changes associated with global warming include melting of arctic glaciers; increases in floods, hurricanes, heat waves, forest fires, changes in ecosystems such as coral reefs; extinctions of some amphibian species; and increase in the spread of infectious diseases among human populations (Merrill, 2010). Due to increased precipitation in wet months and excessive heat in dry months, many officials fear that the incidence of infectious diseases will increase due to weather conditions optimal for the spread of disease. While casual relationships between global warming and increase in disease transmission are hard to establish, some researchers are beginning to find evidence for this connection among animal populations (Merrill, 2010). In one study, researchers found that 67 per cent of 110 species of harlequin frogs with 78-83 per cent of extinctions occurring during years with unusually warm weather. Specifically, the researchers asserted that temperatures are now optimal for the transmission of *B. dendrobaitidi*, a bacterium known to kill harlequin frogs. Evidence of increased infectious diseases due to climate change is also starting to surface among human populations during 1998, the hottest year of the past century, storms associated with *El Nino* (warming of the ocean surface of the Western coast of South America) and hurricanes were abnormally frequent and intense, causing

---

8 Available at http://www.who.int/features/factfiles/climate_change/en/index.html

massive deforestation, forest fires and flooding. As a result, Indonesia and Brazil experienced an epidemic of respiratory illness and Central America experienced increases in water and insect borne diseases such as Cholera and Malaria. Impacts of climate change on global health also includes malnutrition from changes in crops, heat related illnesses from excessive heat exposure, injury and death from natural disasters, vector borne illnesses from changes in migrating animals and allergies from air pollution and many more. Climate change has the potential to alter distribution of vectors by altering the weather conditions that influenced their breeding places. . The world's leading authority on global warming, the IPCC, has concluded that unchecked global warming will cause a significant increase in human mortality due to extreme weather and infectious diseases. No country, even industrialized nations like the United States, will escape from these impacts. Malaria could become even more common in the U.S. as global warming worsens. IPCC scientists project that as warmer temperatures spread north and south from the tropics, and to higher elevation, malaria-carrying mosquitoes will spread with them. They conclude that global warming will put 65 per cent of the world's population at risk of infection.

Right to Food

Climate change is a defining challenge of our times. Its impact and implications will be global, far-reaching and largely irreversible. Climate change is already increasing the risk of exposure to hunger, malnutrition and food insecurity among the poorest and most vulnerable people. Natural disasters are becoming more frequent and intense, land and water are becoming more scarce and difficult to access, and increase in agricultural productivity is becoming more difficult to achieve.[9] The right to adequate food is a human right, inherent in all people, to have regular, permanent and unrestricted access, either directly or by means of financial purchases, to quantitatively and qualitatively adequate and sufficient food corresponding to the cultural traditions of people to which the consumer belongs and which ensures a physical and mental, individual and collective fulfilling and dignified life free of fear (Ziegler, 2001). The right to food has been recognized since the adoption of the UN Declaration on Human Rights in 1948. The UN Food and Agriculture Organization has defined food security as a situation 'when all people, at all times, have physical, social and economic access to sufficient, safe and nutritious food that meets their dietary needs and food preferences for an active and healthy life'.[10] Thus Right to food is an internationally established human right of the world population. According to the UN ambassador to Maldives, "Until now, the global discourse on climate change has tended to focus on the physical or natural impacts of climate change. The immediate and far-reaching impact of the phenomenon on human beings around the world has been largely neglected." There is no hesitation in denying the fact that climate change will have a wide range of effects on the environment, which could have knock-on consequences for food production. Climate change threatens the ability of entire regions to feed themselves.[11]

At last year's COP in Copenhagen, Olivier De Schutter, the Special Rapporteur on the right to food, said "climate change is a ticking time bomb for global food security". De Schutter went on to say that global warming disproportionately affects some of the poorest countries – especially the most vulnerable, including small-scale farmers and indigenous peoples who depend on the land for their

---

9 Available at http://www.wfp.org/content/climate-change-and-hunger-responding-challenge

10 Available at http://www.grida.no/polar/news/4442.aspx

11 Available at www.srfood.org

livelihoods.[12] The Regional food production is likely to decline because of increased temperatures accelerating grain sterility; shift in rainfall patterns rendering previously productive land infertile, accelerating erosion, desertification and reducing crop and livestock yields; rising sea levels making coastal land unusable and causing fish species to migrate; and an increase in the frequency of extreme weather events disrupting agriculture (Dupont and Pearman, 2006). For example, in Australia, up to 20 per cent more droughts are expected by 2030 and up to 80 per cent more droughts by 2070 in south-Western Australia.[13] It is to be noted that climate change will not only disrupt the set patterns of agriculture and reduce the food production globally but is also a threat to the food security by reducing the availability and supply of food. Oxfam, quoting the International Panel on Climate Change noted that "climate change could halve yields from rain-fed crops in parts of Africa as early as 2020, and could put 50 million more people worldwide at risk of hunger"[14]

Another effect of climate change on the Right to food is in the form of migration. Migration of population, vulnerable to the detrimental effects of climate change poses a great threat to the food security of the local residents and more severely the migration population. What is worst is, many countries have resorted to the use of bio fuels which can reduce greenhouse gas emission by 90 per cent. But this measure to mitigate climate change is in turn violating the Right to food as it is diverting agricultural lands and food crops toward the production of bio fuels. Thus indirectly it result in rising food prices, food shortage, affecting especially the poor and underdeveloped population.[15]

The impacts of climate change on the world's populations are going to worsen. We know that they will be felt disproportionately by some of the poorest countries and the most vulnerable within those countries. And we know that small-scale farmers and indigenous peoples – as well as those, more generally, who depend on land and water resources for their livelihoods and constitute half of the world's hungry – will suffer most.[16] It is the urgent need of the hour to build a food system which is resilient to climate extremes.

Right to property and Right to Self-Determination

"The greatest single impact of climate change could be on human migration-with millions of people displaced by shoreline erosion, coastal flooding and agricultural disruption."- IPCC, 1990 (Corlet,2008). While climate change has been the subject of intense debate and speculation within the scientific community, but insufficient attention has been given to the humanitarian consequences it will generate. Just as the causes of climate change are being analyzed and its consequences projected, it is equally vital to anticipate foreseeable movement scenarios and strengthen the responses to the humanitarian consequences. Climate change is already undermining the livelihoods and security of many people, exacerbating income differentials and deepening inequalities. Over the last two decades the number of recorded natural disasters has doubled from some 200 to over 400 per year. Nine out of every ten natural disasters today are climate-related (UNHCR, 3). On current predictions, as many as 200 million people may be displaced as a consequence of climate change by the year 2050, that is to say, approximately one per cent in every fourteen worldwide. Entire nations, such as Maldives and

12 Available at www.grida.no/polar/news/4442.aspx

13 Available at: http://www.ipcc.ch/pdf/assessment-report/ar4/wg2/ar4-wg2-chapter11.pdf

14 Available at www.srfood.org

15 Available at www.hreoc.gov.au/about/media/papers/hrandclimate_change.html

16 Available at www.momagri.org/uk/focus-on-issues/-climate-change-and-the-right-to-food-_639.html

Tuvalu, are likely, if current GHG emission trends continue, to be lost because of sea level rise, rendering their inhabitants stateless. Tuvalu is negotiating agreements with Australia and New Zealand to move its 1200 strong population and the Maldives has started saving to buy land to resettle its four lakhs strong population to India or Sri Lanka. Australia's previous government (reportedly) refused Tuvalu's request twice while New Zealand agreed to accept 75 immigrants every year (McInerney *et al.*, 2011). Environmental change including climate change presents a new threat to human security. Faced with an unconceivable scale of environmental change, migration may be an adjustment mechanism of first resort, or a survival mechanism of last resort. Migration may be an adaptation mechanism for those with the resources to move early and far enough away from danger. Or, in extreme cases and for those with fewer means to move, migration may be an expression of failed adaptation. Extreme weather as a manifestation of climate change is increasingly problematic for the people of Mozambique. In 2001, 2007 and 2008 heavy rains caused flooding along the Zambezi River in central Mozambique. Flooding in 2007 was then exacerbated by the impact of Cyclone Favio. Many people were made homeless. Droughts, coastal soil erosion and rising sea levels – which may be connected to climate change – also affect a large number of people in Mozambique. The river delta regions and the 2,700 km- long coastline are at particularly high risk of inundation and erosion. In Mozambique, environmental stressors (particularly flooding) contribute to migration and displacement. People are displaced during the flood emergency period; following recurring flooding events, people are relocated on a permanent or semi-permanent basis. Along the Zambezi River valley, temporary mass displacement is taking on permanent characteristics. The field research did not detect large-scale international migration resulting from the Zambezi River flooding or significant rural-urban migration patterns for flood-affected groups. Instead, the research revealed that government-organized resettlement programs dominate the environmentally induced movement pattern for flood-affected areas. Resettlement removes people from the physical danger of extreme floods but can lead to other environmental, social and economic difficulties. Subsistence farmers and fishermen are moved away from fertile lands on riverbanks and to higher, drought-prone areas. Some resettled people attempt to return periodically to work in their fields in low-lying river areas in order to maintain land ownership and their livelihoods as farmers. Resettlement often causes these people to lose their livelihoods, forcing relocated households to depend almost entirely on governmental and international aid. As extreme weather events continue to hit Mozambique, the Government of Mozambique will increasingly face decisions about how to manage people at risk and on the move due to environmental factors (Warner *et al.*, 2008).

Taking a human rights approach to climate change, grounded in the principle of the inherent dignity of the human person, implies that it is not only the total numbers of those displaced that matter. Every single person who is forced to move from their home, against their will, must have a remedy available to them which respects their rights, protects their rights and if necessary, fulfils their rights as recognized under International Human Rights Law.

The Right to Self-Determination is a fundamental principle of international law. Common Article 1, paragraph 1, of the International Covenant on Economic, Social and Cultural Rights and the International Covenant on Civil and Political Rights establishes that "all peoples have the right of self-determination", by virtue of which "they freely determine their political status and freely pursue their economic, social and cultural development". Important aspects of the right to self-determination include the right of a people not to be deprived of its own means of subsistence and the obligation of a State party to promote the realization of the right to self-determination, including for people living outside its territory (Office of the United Nations High Commissioner for Human Rights,14).

If climate change continues unmitigated, sea-level rise is expected to result in the total inundation of small island states like Maldives to Marshall islands *etc.* It may thus destroy one of the hallmarks of statehood: the country's territory. Additionally, the combination of sea- level rise, rising temperatures and extreme weather events threatens to render the islands inhabitable. Therefore, climate change impacts constitute a threat to the enjoyment of the right of certain people to self-determination. The loss of land and State renders all other rights, political and civil as well as economic, cultural and social rights unattainable. Without international cooperation, small island developing states, which as a whole emit however less than 1 per cent of global GHG emissions, have little capacity to reverse the impacts of climate change in extinguishing their territory (Blazogiannaki, 2009).

The rights found within the international human rights legal code which are particularly relevant to the discussion of climate change induced displacement include the right to adequate housing and rights in housing; the right to security of tenure; the right not to be arbitrarily evicted; the right to land and rights in land; the right to property and the peaceful enjoyment of possessions; the right to privacy and respect for the home; the right to security of the person, freedom of movement and choice of residence; and housing, land and property (HLP) restitution and/or compensation following forced displacement. All of these entitlements and obligations are in order that people everywhere are able to live safely and securely on a piece of land, to reside within an adequate and affordable home with access to all basic services and to feel safe in the knowledge that these rights will be fully respected, protected and fulfilled. Indeed, the normative framework enshrining these rights is considerable, constantly evolving and ever expanding.

Conclusion

Earth's increasing temperature and the resultant natural disasters is no longer disputed by any credible scientific agency. The increased greenhouse gas emissions since the industrial revolution have up till now slowly yet evidently altered the climate patterns. The further enhancing of natural greenhouse effect, the damaging effects of human activities could be seen within the next 25 years. Rise in sea levels, increased flooding and droughts, frequent natural disasters *etc.* resulting into loss of life and property, increase in infectious diseases. Loss of livelihood, adverse affects on health and living conditions and forced migration are amongst the major ramifications of the human activities induced climate change.

Climate change and the resultant human rights violations have a created a wave of desertification among many nations of the world as the responsibility for climate change is not equally distributed. Some people and groups have emitted more greenhouse gases than others and continue to do so and hence are more to be blamed. This climate change will affect the poorest and the more vulnerable population of the world. This vulnerability is determined by political-economic processes that benefit some people and disadvantage others. In recent past, scientists have discovered evidences of environmental disruption that coincides with abrupt climate change. May politicians have given more importance to the economic concerns of the country and have dismissed the inevitable dangers.

Nations world wide have taken steps to combat with the increasing greenhouse effects. Through many international conventions and treaties such as, Stockholm Conference, Rio Declaration, Montreal Declaration, Kyoto Protocol, United Nations Framework Convention on Climate Change, many nations have committed themselves for the purpose to reverse the damages already caused by human activities. The Kyoto protocol sets binding targets for 37 industrialized countries and the European community for reducing greenhouse emissions. Thus, every country specially the highly industrialized ones have a fixed amount of carbon credit which they need to comply with. But the major loophole in the protocol lies in the fact that provision of carbon trading has been included. It is argued by many critics that trading of carbon credits does little to solve pollution problems over all, since less industrialized nations that do not pollute sell their conservation to the highest bidder. In this way, the major greenhouse gas emitters continue with their set patterns of emissions or even more. Thus, instead of reduction in greenhouse gas emissions this provision of carbon trading has increased the emissions. Further, though Kyoto Protocol has a detailed compliance regime which includes penalties in case of failure to meet emissions targets, yet a decision has not been taken that whether non-complying countries can be legally prevented from trading in emissions. The above mentioned loopholes must be done away with through an ammendment to the protocol. Many conferences and conventions are viewed as a great disappointment due to the lack of binding nature. The developed and industrialized countries rarely comply with the emission targets whether legally or illegally.

The population throughout the world have adapted to the climate change in diverse ways that are usually unnoticed, uncoordinated and unaided by the national governments. This world is small and round. Climate change is a global issue and is not confined to any one or more nations rather it is an issue which can be solved only when all nations take steps together in harmony to combat this issue. Therefore, there is a need for collection action to be taken by all the members of the international community. It is obligatory for all nations by the principle adopted in Stockholm Conference to assist each other by providing methods and technologies to meet the problem of environment. Therefore, there is a need for international cooperation.

The major human rights violated due to the climate change are interlinked to one another. Thus mitigation from one violation will in turn reduce the risk of violation of another. Hunger, poverty, malnutrition, diseases, starvation deaths *etc.* are all the consequences of violation of right to food. Mitigation and adaptation measures to deal with violation of right to food which will also help us deal to with the violation of right to life. Agro-economics is the safest and the most efficient measure to deal with the issue. Small-scale farmers can double food production within 10 years in critical regions by using ecological methods. Agro ecology applies ecological science to the design of agricultural systems that can help put an end to food crisis and address climate-change and poverty challenges. A fundamental shift towards agro ecology is the best possible way to boost food production and improve the situation of the poorest.

The whole issue of climate change stems out from "human activity induced climate change", it is the humans who are to be blamed for violating the human rights of the others. Though this violation is does not take place conspicuously at the first instance because these human activities have an after effect. Initially, it is the people of underdeveloped nations who are most affected because of their high vulnerability to such changes but in the end, it is the whole human community which is going to and will suffer the detrimental effects of climate change. All conferences, conventions will prove to be a failure until we the humans, do not understand and realize and foresee the damages caused by their activities to the whole environment and human community.

The IPCC report has strengthened the claims of different scientists that climate change has and will bring about many natural disasters thus, violating the human rights bestowed upon to the population by the virtue of being human. The response of many nations to this report is overwhelming. The efforts put in by different countries to reduce their carbon emissions and scientific inventions to replace CFCs are a commendable starting point. The different being held from time to time indicate a general consensus to the fact the matter of climate change and its impact on human rights is of great importance to the human community. Still, much has to done to help the vulnerable population who are the captives of political and social shackles and to elevate them from this vulnerability.

References

Blazogiannaki, Maria. "Human Rights and Climate Change". Convention on the conservation of European wildlife. 2009. https://wcd.coe.int/wcd/viewdoc.jsp

Corlet, David. Stormy Weather: The Challenge of Climate Change and Displacement: Sydney: University of New South Wales Press Ltd. 2008.

Dupont, Alan and Pearman, Graeme. " Heating up the planet: Climate Change and Security". 2006. http://www.lowyinstitute.org/publication.asp

Gardiner, Mark Stephen; Caney, Simon and Jamieson, Dale. Climate Change: Essential Readings, London: Oxford University Press, 2010.

Goyal, Malti. Energy Sources and Global Warming. Bombay: Allied Publishers Ltd, 2005.

Martinez, H. Leah. "Post Industrial Revolution Human Activity and Climate Change: Why the United States must implement mandatory limits on industrial green house gas emissions." Journal of land use 20 (2005): 407-411.

Merrill, M. Ray. Environmental Epidemiology: Principles and Methods, New Delhi: Jones and Bartlett, 2010.

McInerney, Siobhan, Lankford; McOrion and Rajamani, Lavanya (2011). Human Right s and Climate Change: A Review on International Legal Dimensions, Washington, D.C: The World Bank.

Rathgeber, Theodor. "Climate Change Violates Human Rights." 2010

Tauger, B. Mark. Agriculture in World History, Oxon: Routledge, 2011.

Ziegler, Jean. "The Right to Food". Report by Special Rapporteur on the Right to Food Commission on Human Rights. 2001. http://daccessdds.un.org/doc/UNDOC/GEN/GO1/110/PDF/GO111035.pdf

Warner, Koko, Dun Olivia and Stal, Marc (2008). " Climate Change and Displacement". Forced Migration Review 31 (2008): 14, 18

2013, Sustainable Approaches for Environmental Conservation *Pages* ***213–221***
Editors: **D.R. Khanna, A.K. Chopra, R. Bhutiani, Gagan Matta & Vikas Singh**
Published by: **BIOTECH BOOKS, NEW DELHI**

Chapter 27

Khadi: The Environmentally Friendly Fabric

Neeta Tiwade[1] and Pravin Charde[2]

[1]Department of Textile and Clothing, Home Science Faculty, Sevadal Mahila Mahavidyalaya, Nagpur – 4400 09, M.S.
[2]Principal, Sevadal Mahila Mahavidyalaya, Nagpur – 440 009, M.S.

Khadi for one's self is not only self employment but also a way to live with proud. Khadi is a symbol of strength and self-sufficiency and to provide employment for millions, which has continued till this day. Many of men, women and children are compelled to go to bed empty stomach, without investing a penny, they can start weaving and feed their family. Khadi is mainly woven in pure cotton but it can also be woven in a mix of fibers, which is light, soft and comfortable to wear. Researcher studied and observed that Khadi is an eco-fabric and process involves no environmental pollution with a sample of 100 people working in its institution Nag Vidharbha Charkha Sangh, Mul of Chandrapur District, working in Sawali area and nearby places. It has been observed that making of khadi is eco-friendly because it used organic cotton and avoided harmful chemicals from processing and dying of the cloth, which is very harmless to environment. No chemicals or petroleum products are used. Khadi weaving from natural fibers, which is non-synthetic made it fully biodegradable, pure and non-polluting. Mostly spinning and weaving is dominated in rural area hence environment is clean and airy. This khadi institution is providing economical and social platform to rural people.

Keywords: *Eco-friendly, Socio-economical, Environmental, Biodegradable, Wonderful khadi.*

Introduction

Khadi for one's self is not only self employment but also a way to live with proud and has been found to be a unique instrument to promote communal harmony and religious tolerance. Gandhi promoted Khadi for self sustainability but his whole premise was that everyone, rich or poor or in between should have access to food, shelter and clothing in a self-reliant way. This sector generates employment for the poor and it's a material that breathes to make their daily livelihood. Spinning is mostly done by the girls and women in the villages, while weaving is dominated by men. It is primarily a means to provide employment to the unemployed rural people. Through the medium of khadi weaving, the weaver expresses art and designing by the spindle and loom.

Khadi is a symbol of strength and self-sufficiency and to provide employment for the millions, has continued till this day. Many of men, women and children are compelled to go to bed empty stomach without investing, they can start weaving and feed their family. Khadi is very relevant this time because it is not a poor man's fabric although it provides employment to poor man in rural area. It helps in supporting poor, helpless rural women as they can work independently and earn their living.

What is Khadi?

Khadi, a fabric of hand woven and hand-spun fabric, enlarged its attraction for the discerning public due to its unique characteristics-hand crafted, often matchless texture and appearance, eco-friendly, truly heritage product.

Khadi is mainly woven in pure cotton but it can also be woven in mix of fibers, which is light, soft and comfortable to wear. It is the only fabric where the play of texture is so unique that no two fabrics will be absolutely identical, thus lending its uniqueness and matchlessness in terms of feel and texture. In fast moving economy, community suffer from economic instability, this sector is a socio-economically sustainable way.

Aims and Objectives

1. To study the factors which are responsible for economical development of rural people.
2. To study whether Khadi is an eco-fabric and process involves no environmental pollution.

Materials and Methods

The methodology pertaining to this study is presented as follows

Selection of the Area and Sample

A random sample of 100 people working in spinning, weaving, dyeing of its khadi institution in Mul of Chandrapur district, Nag Vidharbha Charkha Sangh, production center in Sawali area and nearby villages of Mul.

Methods of Data Collection

Two types of tools are used *i.e.* interviewed schedule and observation schedule to collect data.

Data Collection

Researcher contacted samples in Sawali and nearby villages and asked various questions about working in different capacities and recorded the answers in questionnaire. It is also observed the process of carding, spinning, dyeing and weaving to have a first hand knowledge of its working.

Analysis of Data

Researcher utilized her ability and skills to obtain required data. After the data collection, such data is transferred to master chart, which helped in preparation of Tables and interpretation of data along with observation of working in spinning, dyeing and weaving are as under.

Results and Discussion

Table 27.1: Place of Working

Sl.No.	Place of Working	No. of Respondents (N=100)	Percentage
1	Own House	45	45 per cent
2	Shed	55	55 per cent
	Total	**100**	**100 per cent**

It indicates that 45 per cent people are working in their own house in rural area that have no agriculture land and no other option of employment and 55 per cent people working in institution shed.

Table 27.2: Employment of Rural Area

Sl.No.	Employment	No. of Respondents (N=100)	Percentage
1	Rural	98	98 per cent
2	Urban	02	02 per cent
	Total	**100**	**100 per cent**

The above table, it indicates that 98 per cent people working in rural areas and 2 per cent people are working in institution.

Table 27.3: Types of Working

Sl.No.	Requirement	No. of Respondents (N=100)	Percentage
1	Spinners	73	73 per cent
2	Weavers	25	25 per cent
3	Dyeing and Printing	02	02 per cent
	Total	**100**	**100 per cent**

From the above table, 73 per cent people are working on six spindle charkha and spinning wheel for spinning, 25 per cent weavers are working on loom and 2 per cent are working in dyeing and printing.

Table 27.4: Spinning of Khadi

Sl.No.	Spinners	No. of Respondents (N=100)	Percentage
1	House	31	42 per cent
2	Shed (institution)	42	58 per cent
	Total	**73**	**100 per cent**

* Out of 100 samples, 73 are spinners.

Figure 27.1 and 27.2: Spinning on Charkha

Figure 27.3: Spinning on Charkha

This table indicates that out of 73 spinners, 42 per cent working in homes and other 58 per cent in shed. It is observed that there is no noise pollution in spinning (Figures 27.1–27.3).

The process of spinning is as follows

1. *Ginning* - process of separating fibers from cotton seeds.
2. *Cardings* - carding machine separates fibers and then passed through moving steel bars that remove short fibers, tangles and then collected as slivers.
3. *Roving* - further slivers are thinned out and twisted to strengthen it, is called roving which is directly spun into produce yarn. Roving is wound on bobbins for spinning.
4. *Spinning* - Mostly spinning is done on new model charkha (NMC). In some villages, traditional and spinning wheel is used. The spinning draw the combed cotton and add twist making it tighter and thinner, until it reaches the yarn thickness needed for weaving.

Dyeing Process in Khadi (Figures 27.4–27.7)

Firstly yarn are dyed to give base for printing, then sent to washing unit, where it goes to washing tanks and finally wash in plain water, which are again dried in sunlight. The dyeing often begins at low temperature and continues heating of dye bath until wanted tone of color, intenseness and evenness is obtained.

Chemicals which meet organic fiber processing standards are used in khadi. Processing of color on yarn and fabric with natural dyes from various fruits, flowers, leaves is completely eco- friendly and avoid use of harmful chemicals at all stages, which are free from chemical fertilizers and pesticides. Finished cloth is completely eco friendly cotton yarn in earthy natural colors, prevents pollution.

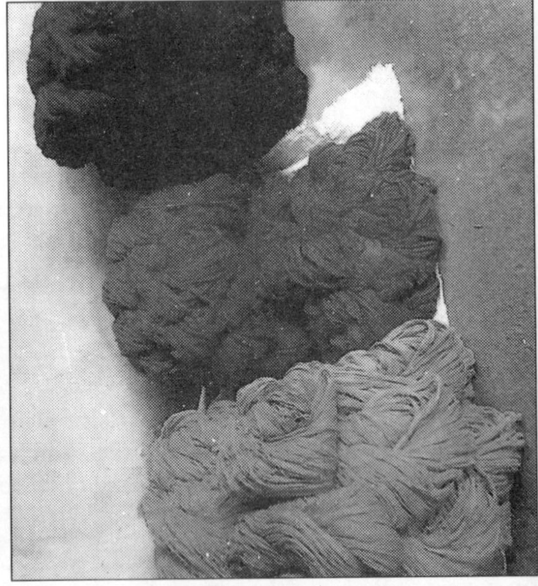

Figure 27.4–27.7: Dyeing Process

Table 27.5: Weaving of Khadi

Sl.No.	Weavers	No. of Respondents (N=100)	Percentage
1	House	25	100
2	Shed	—	—
	Total	**25**	**100**

* Out of 100 samples, 25 are weavers.

Weavers are working on loom in homes in rural area.

Weaving - yarn is supplied by institution to weavers at their doorstep for weaving on loom with a fly shuttle, which involves interlacing a set of longer threads (called the warp) with a set of crossing threads (called the weft).

Making of khadi is eco-friendly since it used organic cotton and avoid harmful chemicals from processing and dying of the cloth, which is very harmless to environment. No chemicals or petroleum products are used in weaving. Khadi weaving from natural fibers is non-synthetic, made it fully biodegradable, pure and non-polluting.

Figure 27.8: Weaving on Handloom

Figure 27.9–27.10: Weaving on Handloom

Conclusion

1. Making of khadi is eco-friendly since it use organic cotton and avoided harmful chemicals from processing and dying of the cloth, which is very harmless to environment.
2. Processing of color on yarn and fabric with natural dyes from various fruits, flowers and leaves is completely eco- friendly and avoided use of harmful chemicals at all stages.
3. Khadi weaving from natural fibers is non-synthetic, made it fully biodegradable, pure and non-polluting.
4. This institution provided employment with facilities to rural people at their doorstep hence they are economically developed.
5. Mostly spinning and weaving is dominated in rural area hence environment is clean and airy.
6. This sector provided economical and social platform, living them proudly, even they are not educated, old aged and no other option of employment and also helping to control the rural migration.

Acknowledgement

Author is thankful to the Hon'ble Shri Badke, President, Nag Vidharbha Charkha Sangh, Mul, District Chandrapur who provided full co-operation in this research work.

References

Agrawal, Meenu, 2008. *Economic Reforms, Unemployment and Poverty: The Indian Experience*. New Century Publications, New Delhi, p. 256.

Bedi, J.S., 2002. *Economic Reforms and Textile Industry: Performance and Prospects*. Commonwealth, New Delhi, p. 510.

Dash, T.R. and Sen, K.K. *Co-operatives and Economic Development*, p. 359.

KVIC, *Annual Report* 2009-10.

KVIC, *Journal 'Jagriti'*, August 2008 to February 2011.

Prasad, Avadh, 1994. *Khadi Technology: A Techno-Social Study*. Rawat Publications, Jaipur, p. 11.

Sahasrabuddhe, P.K., 1989. *Bleaching, Dyeing and Printing of Khadi*. Directorate of Publicity, KVIC, Mumbai, p. 147.

2013, Sustainable Approaches for Environmental Conservation *Pages* ***223–229***
Editors: **D.R. Khanna, A.K. Chopra, R. Bhutiani, Gagan Matta & Vikas Singh**
Published by: **BIOTECH BOOKS, NEW DELHI**

Chapter 28

Nanotechnology and Environment: Boon or Bane

Praveena Chaturvedi and Nipur

Department of Computer Science,
Kanya Gurukul Mahavidyalaya Dehradun, Uttarakhand

Since 1980, the capability of the technology responsible for environmental health and safety has steadily tough. The technologies cannot perform their basic functions now, and they are completely unable to cope with the new challenges they are facing in the 21st century. This paper describes some of these challenges, focusing on next-generation nanotechnology and suggests changes that could revitalize the health based on technologies. When the integration of sciences and the humanities refers to collaboration between sciences/engineering and the humanities/social sciences, it constitutes an essential element of science and technology policy, taking how science and technology are positioned in modern society into consideration. Now that every field of science and technology, and even the entire society, are rapidly being computerized, the need for such science and technology policies is not limited to the software sector. Such policies are needed across many other sectors.

Nanotechnology has potential applications in many sectors of the Indian economy, including consumer products, health care, transportation, energy and agriculture. In addition, nanotechnology presents new opportunities to improve how we measure, monitor, manage and minimize contaminants in the environment. Nano-materials can have different chemical, physical, electrical and biological characteristics. For example, an aluminium can is perfectly safe, but nano-sized aluminium is highly explosive and can be used to make bombs.

In this paper, we will take a short tour around some of the major development areas in nanotechnology and its effect on the environment.

Keywords: *Nanotechnology, Nano-materials, Environment, Healthcare and Energy conversion.*

Introduction

The nanotechnology is not just a new emerging technology it's a way of life in the 21st century. It affects almost every area of human activity in one way or another. To better understand how nanotechnology could revolutionize such diverse areas regarding Environment and Information technology, we need to review a bit of fundamental physics. Two sets of theories relate to this discussion: classical mechanics, which governs the world of our direct perception and quantum mechanics, which governs the world of atoms and molecules. The term "*nanotechnology*" was first defined by Tokyo Science University, NorioTaniguchi in a 1974 paper (Taniguchi, 1974) as follows: 'Nano-technology' mainly consists of the processing of, separation, consolidation, and deformation of materials by one atom or one molecule. Nanotechnology is apprehensive with the world of hidden miniscule particles that are dominated by forces of physics and chemistry that cannot be applied at the macro- or human-scale level. These particles have come to be defined by some as nanomaterials, and these materials possess unusual properties not present in traditional and/or ordinary materials. Technology generally refers to "the system by which a society provides its members with those things needed or desired". In its original sense, "nanotechnology" refers to the projected ability to construct items from the bottom up, using techniques and tools being developed today to make complete, high performance products. The term nanotechnology has come to be defined as those systems or processes that provide goods and/or services that are obtained from matter at the nanometer level, that is, from sizes at or below one billionth of a meter.

Nanotechnology has emerged in the last decade as an exciting new research field. In history, tremendous human progress becomes possible through converging technologies stimulated by advances in four core fields: Nanotechnology, Biotechnology Information technology and new technologies based in Cognitive Science (NBIC). For Information Technology and engineering, nanotechnology is the new micro technology, enabling, for example, faster and better computer chips. For the biosciences, nanotechnology is a window to an entirely new research area. As Cognitive Science, it not only gains more and more opportunities for union with other sciences but also becomes a solid basis for a range of innovative new technologies advancing individual and group creativity (Wolbring *et al.*, 2002) Human intellectual and social performance will be greatly enhanced by nano-enabled, portable information systems and communication devices, by biotechnology treatments for disorders of the mind or memory, and by increased understanding of how the human brain and senses actually function.

Quantum Effects of Nanotechnology

Nanotechnology is the science of the small; the very small. It is the use and manipulation of matter at a tiny scale. At this size, atoms and molecules work differently and provide a variety of surprising and interesting uses. The "nano" in nanotechnology comes from the Greek word *nanos*, which means dwarf. Scientists use this prefix to indicate 10^{-9} or one-billionth. Thus a nanosecond is one-billionth of one second; a nanometer is one-billionth of one meter *etc.* Objects that can be classified as having something to do with nanotechnology are larger than atoms but much smaller than we can perceive directly with our senses. One way to look at this size scale is that one nanometer (nm) is about 100,000 times smaller than the diameter of a single human hair. Hence, Nanotechnology is the engineering of functional systems and entails the measurement, prediction and construction of materials on the scale of atoms and molecules (Danial and Jurgan, 2009).

When particles are created with dimensions of about 1-100 nms, the so called quantum effects rule the behavior and properties of particles. Properties such as melting point, fluorescent, electrical conductivity, magnetic permeability and chemical reactivity change as a function of the size of the particle. For example, nanoscale gold illustrates the unique properties that occur at nanoscale. Nanoscale gold particles are not yellow color with which we are familiar; they appear red or purple. Another fascinating and powerful result of the quantum effects of the nanoscale is the concept of "tunability" of properties. That is, by changing the size of particles, a scientist can literally fine-tune a material's property of interest. Another potent quantum effect of the nanoscale is known as "tunneling". The scanning tunneling microscope and flash memory is based on this technology.

Healthcare and Nanotechnology

"Nanotechnology for Healthcare" offers unique opportunities for radically improved diagnosis, prevention and treatment of diseases. Hospitals will benefit greatly from nanotechnology with faster, cheaper diagnostic equipment. The lab-on-a-chip will be able to analyze the patient's ailments in an instant, providing point of care testing and drug application. Nanotechnology has the potential to modernize healthcare for the next generation. There are three key areas in which it could do this: Diagnosis, Prevention and Treatment.

Healthcare: Diagnosis

The earlier disease is diagnosed, the more likely that treatment will be successful. Advances in bioassays offer the potential to identify the presence disease-causing microbes long before a person shows any symptoms.

Through advances in imaging technologies we can look into a person's body and more accurately determine the presence and location of cancer and damage to the body's tissues and organs. Advanced monitoring ensures that a patient's recovery is complete. Not only do these advances bring about improved quality of life, they also contribute to a more efficient and cost-effective health service.

Healthcare: Treatment

At present, treatment of illness tends to use quite old-fashioned, well-established methods. These methods are obviously not always entirely successful, with general solutions and treatments often being applied to very specific problems.

Furthermore, current healthcare can often cause additional problems such as rejection or a bad reaction to a transplant. Nanotechnology can help to address some of these issues, in a number of ways. It will be possible to provide personalised medicine designed for individual patients and their particular illness.

Nanotechnology will aid in the delivery of just the right amount of medicine to the exact spots of the body that need it most. Nanoshells, approximately 100 nm in diameter, will float through the body, attaching only to the effective cells. When excited by a laser beam, the nanoshells will give off heat in effect, cooking the tumour and destroying it. Nanotechnology will create biocompatible joint replacements and artery stents that will last the life of the patients instead of it having to be replaced every few years.

Healthcare: Prevention

The majority of healthcare is reactive rather than preventative. This is often some time after the initial infection or trauma, which means that tissue damage and suffering has already occurred. In some cases this damage can be irrepairable, in others permanent reminders remain (such as loss of function of that body part or scarring). Therefore, possibly the most important aspect of nanomedicine in the future will be its potential to prevent illness, rather than simply treating it. Nanotechnology will contribute to this through more effective monitoring of individuals health and more sterile hospital environments (limiting the opportunity for bacteria, viruses and other microbes to cause secondary disease).

Energy Conversion and Storage Nanoelectrodes

Nanosized materials are known to take on peculiar properties compared to the bulk material. Their electronic and mechanical properties are known to improve *e.g.* higher electrical conductivity and greater strength. Nanodimensional materials also have an extraordinarily high surface area to volume ratio. All of these properties would bring beneficial effects if they could be retained when the material is assembled into a structure capable of being used as an electrode – nanostructured electrodes.

The production and use of oil and other fossil fuels continues to rise on a yearly basis to record new levels. At the same time, fewer new oil fields are being identified which means that sometime over the next few decades oil production will not be able to satisfy demand. Not only we running out of oil, but our use of fossil fuels is putting an increasing burden on the environment in terms of greenhouse gases and global warming. Alternative energy sources are known, but their efficiency based on cost is poor. Nanotechnology advances can improve this.

Lower Use of Material and Energy

Nanotechnology offers two ways to reduce material and energy use. First, nanomaterials are generally more active than bulk materials, and so either less material is required or the abilities of that material (*e.g.* insulation) are greatly enhanced.

Second, more abundant materials can often be converted into excellent substitutes for rare materials through the application of nanotechnology (*e.g.* replacing platinum in catalytic convertors with nanostructured metal oxides).

These new applications can often be more energy efficient, both in manufacturing and in ultimate use. However, there needs to be a life cycle analysis approach to this to determine whether the new application is sustainable and whether it will actually reduce or increase the burden on the environment. This means determining the total material and energy use from product development through manufacture, use, and final recycling or disposal.

Energy Storage

Not only do we need new ways of generating energy, we need better ways of storing it. Nanotechnology is leading to improved, environmentally-friendly batteries and super capacitors. We also need to reduce damage to the environment. Particularly toxic are those chemicals we use as solvents. Nanotechnology is leading to their eradication through the development of new nano coatings and nano structured surfaces that can effectively repel dirt and other contaminants.

Nanoscale Processes for Environmental Improvement

The subject of "Global Warming" itself is generating a lot of heat now a day. Having struggled with the issue of ozone layer depletion in the last decades, the world is now facing the challenge of global warming and resultant climate change. It is said "environment is not what we inherit from our ancestors; we borrow it from our future generations" We have already started facing the consequences of hasty use of natural resources leading to ecological imbalance. Thus, to hand over to our future generations the same level of environmental quality which we are presently enjoying is becoming difficult day by day. This makes the issue of environmental sustainability more important than it ever was.

Nanotechnology also has the potential to improve the environment, both through direct applications of nanomaterials to detect, prevent and remove pollutants, as well as indirectly by using nanotechnology to design cleaner industrial processes and create environmentally responsible products (Bergeson, 2004).

Environment: Preserving the Planet

The doom of the planet is not a viewer sport "We live in an interconnected world, where all parts of the world are affected by what happens in all other parts" There is no doubt that the pressures we are putting on the planet are leading to potentially catastrophic consequences.

So what can we do to limit the damage and ensure a future for our children? Firstly, the negative aspect is: The fossil fuel that oils our everyday lives is responsible for 44 per cent of the carbon dioxide we emit annually. The good news is that energy from sunlight is sufficient to meet our needs ten thousand times over. Today, more efficient and cheaper solar energy collectors are in the process of being developed using nanotechnology. These collectors could be deployed as small units in our homes. They work particularly well in diffuse light, so would suit even less sunny climates. This would have the benefit of not sterilizing precious land (a diminishing resource for food) and quickly improve the quality of many people's lives, especially in poorer housing or in the less developed world.

Negative Aspects of Nanotechnology

Health

Nanoparticles have been shown to be absorbed in the livers of research animals and even cause brain damage in fish exposed to them after just 48 hours. If they can be taken up by cells, then they can enter our food chain through bacteria and cause a health threat like mercury in fish, pesticides in vegetables or hormones in meat. The increasingly-popular carbon nanotube (20x stronger and lighter than steel) looks very much like an asbestos fiber – what happens if they get released into the air? Being carbon-based, they wouldn't set off the usual alarms in our bodies, making them difficult to detect.

Environmental

There are two thoughts regarding the environmental implications of nanotechnology: One is positive and the other is potentially harmful (Theodore and Kunz, 2005). If nonmaterial really are as strong as diamonds, how decomposable or persistent are they? Will they litter our environment further or present another disposal problem like nuclear waste or space litter? In the distant future, will self-replicating nanobots – necessary to create the trillions of nanoassemblers needed to build any kind of product – run amok, spreading as quickly as a virus, in the infamous "gray goo" scenario?

Privacy

As products shrink in size, eavesdropping devices too can become invisible to the naked eye and more mobile, making it easier to invade our privacy. Small enough to plant into our bodies, mind-controlling nanodevices may be able to affect our thoughts by manipulating brain-processes.

Terrorism

Capabilities of terrorists go hand in hand with military advances, so as weapons become more powerful and portable, these devices can also be turned against us. Nanotech may create new, unimaginable forms of torture – disassembling a person at the molecular level or worse. Radical groups could let loose nanodevices targeting to kill anyone with a certain skin color or even a specific person.

Society

With all the potential abuses of nanotech, many experts advocate a strong system to regulate and monitor nanotech developments. But because nanotech laboratories can be small and mobile, surveillance needs to be practically everywhere – devolving a free society into a Big Brother scenario. Also, what is the impact on the economy? If nations can make anything they want, will they lose all incentive to trade? *What about morality – should we be playing with god-like powers?*

Conclusion

Nanotechnology is already in many of the everyday objects around us, but this is only the start. It will allow limitations in many existing technologies to be overcome and thus has the potential to be part of every industry. Virtually every area of modern life will be touched and improved upon by nanotechnology in the coming decades. Nanotechnology can offer improvements to the durability of construction materials, textiles and the functionality of electronic products, safe and efficient transportation.

Nanotechnology is now beginning to enable scientists to domesticate atoms, but at the same time, this capability is raising health and safety issues. All branches of science and technology may be converging and these are major domains, each with huge power to transform human life also. Nanotechnology and information technology are enablers, as well as creative fields in their own right, giving other branches of science and technology new powers. Biotechnology and Cognitive Science directly concern the human body and mind and have the greatest possible implications for human physical and mental health.

The future of nanotechnology has great potential. However, it also has the potential to change society more than the industrial revolution. It will affect everyone and so should be developed for everyone. All members of society should have a voice in this.

At the same time, there are many other possible impacts that people are worried about. Can we weigh the good versus bad to see whether we should move forward with nanotech or not? The answer might not be that simple.

References

Bergeson, L.L., 2004. "Expect a Busy Year at EPA." *Chemical Processing*, 17 (Feb.).

Hristozov, Danial and Ertel, Jurgan, 2009. Nanotecnology and Sustainability: Benefits and Risk of Nanotecnology and Environmental Sustainability" published in Forum der forshung. No.22. Jahr, pp. 161-168, BTU Cottbus Elgenverlag ISSN No. 0947- 6989.

http://www. aiche.org

http://www. nanoforum.org

http://www.nobel.se/physics/laureates/1986/press.html

http://www.swissre.com

Taniguchi, N., 1974. *Proc. Intl. Conf. Prod. Eng.* Tokyo, Part II, Japan Society of Precision Engineering.

Theodore, L. and Kunz, R.G., 2005. *Nanotechnology: Environmental Implications and Solutions.* John Wiley and Sons, Inc.

Wolbring, G., Roco, M.C. and Bainbridge, W.S., 2002. Science and technology and the triple D (Disease, Disability, Defect). "Converging technologies for improving human performance: *Nanotechnology, Biotechnology, Information Technology and Cognitive Science.* Arlington, VA, National Science Foundation.

2013, Sustainable Approaches for Environmental Conservation *Pages* **231–234**
Editors: **D.R. Khanna, A.K. Chopra, R. Bhutiani, Gagan Matta & Vikas Singh**
Published by: **BIOTECH BOOKS, NEW DELHI**

Chapter 29

Physical, Mental and Spiritual Health: Living a Balance Life

Kalpana Jadhav and Kalpana Kulkarni
Department of Home Science,
Yeshwant Mahavidyalaya, Wardha, M.S.

Health is an important resource for living, but it is very specific to people's lives. To achieve good health it is important to personalize the information presented to make it relevant to your daily life, *e.g.* physical activity, spending time with friends and family, enjoying time to contemplate the world around you and challenging yourself to reach new heights. Mental health is not mere absence of mental illness. A mentally healthy person is one who is free from internal conflicts, well adjusted (*i.e.* able to get along well with others and he accepts criticism and is not easily upset), searches for identity, a firm sense of self esteem, knows himself (his needs, problems and goals), good self control and faces problems and tries solve them intelligently. Spirituality is defined as the direct and personal relationship which we have with a Cosmic or Universal power. This presupposes that the majority of human beings believe in a power greater than themselves. Albert Einstein the genius mathematician and scientist had a profound belief in a Cosmic power which created and constantly energizes this universe.

Keywords: *Physical health, Mental, Spiritual, Life.*

As per World Health Organization (1984)" Health is a state of complete physical and social well being and not merely an absence of disease or infirmity". As said health is not simply the absence of disease; it is something positive, a joyful attitude to life, and a cheerful acceptance of the responsibilities that life puts upon the individual. A healthy individual is a man who is well balanced bodily and mentally and well adjusted to his physical and social environment. There is a close interaction between a healthy mind and a healthy body. In the Hippocratic writings it is said that "a wise man ought to realize that health is his most valuable possession and learn to treat his illnesses by his own judgement".

Health is an important resource for living, but it is very specific to people's lives. To achieve good health it is important to personalize the information presented to make it relevant to your daily life, *e.g.* physical activity, spending time with friends and family, enjoying time to contemplate the world around you and challenging yourself to reach new heights.

Physical Health

The physical dimension involves caring for your physical body - eating the right foods, getting enough rest and relaxation, and exercising on a regular basis. If we don't have a regular exercise program, eventually we will develop health problems. A good program builds your body's endurance, flexibility and strength. The health status is determined primarily by socioeconomic development. Important socio-economic factors are;

1. *Economic status*: The economic progress is the major factor in reducing morbidity, increasing life expectancy and improving quality of life.
2. *Educational status*: Second major factor influencing health is education. Literacy coincides with poverty, malnutrition, ill health, high infant and child mortality rates.
3. *Occupation*: Unemployed people usually show higher incidence of ill health and death. It can cause psychological and social damage.

Apart from above factors, other factors that have strong impact on our outlook on health include family influences, the media, culture or society. It is the perspective on health that will largely guide our lifestyle, habits and behaviors throughout our lives. But the achievement of health imparts strong effect on our outlook and is not a single event; attaining and maintaining health and wellness is an ongoing process, and indeed are source for everyday life.

Mental Health

Mental health is not mere absence of mental illness. A mentally healthy person is one who is free from internal conflicts, well adjusted (*i.e.* able to get along well with others and he accepts criticism and is not easily upset), searches for identity, a firm sense of self esteem, knows himself (his needs, problems and goals), good self control and faces problems and tries solve them intelligently.

The mind needs self-supportive attitudes, positive thoughts and viewpoints and a positive self-image. You also need to give and receive forgiveness, love and compassion; you need to laugh and experience happiness; you need joyful relationships with yourself and others.

Another factor affecting mental health is Stress. It is the body's natural reaction to situations that are not comfortable or pleasing to the person. Stress can be of many things - from the physical reactions that a person's body has, to the mental situations that are caused by having things to do, things to worry about, and things to deal with on a regular basis. The body requires proper stress management. Having a healthy mind is essential to the healing process and emotional healthy living is a core component of achieving a healthy mind. Our brain chemicals run our body. They are the messengers and communicators. If these chemicals are out of balance then our body will be out of balance

It's important to keep your mind sharp by reading, writing, organizing and planning. Read broadly and expose yourself to great minds. Television is the great obstacle to mental renewal.

Spiritual Health

Spirituality is defined as the direct and personal relationship which we have with a Cosmic or Universal power. This presupposes that the majority of human beings believe in a power greater than themselves. Albert Einstein the genius mathematician and scientist had a profound belief in a Cosmic power which created and constantly energizes this universe.

Spiritual dimension refers to that part of individual which reaches out and strives for meaning and purpose in life. The spiritual dimension is your center, your commitment to your value system. It draws upon the sources that inspire and uplift you and tie you to timeless truths of humanity. Spiritual health requires inner calmness, openness to your creativity and trust in your inner knowing. For some it requires having a relationship with a higher power.

The Ways to Improve Physical, Mental and Spiritual Health

Participate in regular physical activity, avoid smoking and follow good nutritional balance diet. It is important to get lots of energy throughout the day without being overly tired. Find time to be vigorously active almost daily. Develop healthy relationships, meet and interact with people and participate in a variety of social activities. Be a good listener, ready to learn from life's challenges. Be passionate about a number of different things in your life and society. Enjoy studying nature and the beauty around you.

Spiritual practices can help us to develop the better parts of ourselves. They can help us to become more creative, patient, persistent, honest, kind, compassionate, wise, calm, hopeful and joyful. These are all part of the best health care.

Meditation can offer inner peace and health. Listen carefully, try reaching back, examine your motives, and write your worries in the sand.

Every day we should commit at least one hour for renewing three dimensions: physical, mental, and spiritual, so that you can work more quickly and effortlessly. In today's busy world, people are rushing around with complex schedules, a variety of activities, leading to long days. In many homes both spouses need to work to support themselves and their family and time is limited for personal relationships and for health. Children have to take additional responsibilities to look after aging parents.

The mind/brain is said to override our emotions, our nervous systems, and most of our physical needs. We can control many functions in our bodies with our minds. Biofeedback is one method used to have the mind control what the body does. But we need to learn to think with our heart as well as our brain. The power of the mind is beyond our imaginations. Since thought always precedes action we come to the conclusion that if we can change our thinking we can change our behavior. The foundations of our thoughts are the beliefs and values we have developed over time. These beliefs and values must be identified, reviewed and changed if we want to change our thought patterns and subsequent actions.

If we want to feel truly alive and open to life's opportunities we need to look carefully at our physical, mental, emotional and spiritual health. We need to identify our shortcomings and make those changes we need so that we are truly vibrating at the highest level and enjoying life to the fullest.

References

http://www.healthy-holistic-living.com/Definition-of-Healthy-Living.html#ixzz1UJRqHdRV

http://www.healthy-holistic-living.com/Emotional-Healthy-Living.html#ixzz1UJSTq0yU

Vipasanna Meditation: Journey to a Healthy Mind

Mental Health: New Understanding, New Hope. The World Health Report. Geneva: World Health Organization; 2001c.

A Report of the World Health Organization, Department of Mental Health and Substance Abuse in Collaboration with the Victorian Health Promotion Foundation and University of Melbourne. Geneva: WHO; 2005.

2013, Sustainable Approaches for Environmental Conservation *Pages* **235–259**
Editors: **D.R. Khanna, A.K. Chopra, R. Bhutiani, Gagan Matta & Vikas Singh**
Published by: **BIOTECH BOOKS, NEW DELHI**

Chapter 30

Ayurvedic Science and Epidemic/ Disaster Due to Climate Change

Sheetal Prakash Vyas, Dinesh Karma, Tushar Bhoyar, Mayank Parsai and Shailendra Mishra

International Institute of Health Management Research, New Delhi

Researchers have found that there is a close link between local climate and the occurrence or severity of some diseases and other threats to human health. It is estimated that climate change contributes to 150,000 deaths and 5 million illnesses each year, and the World Health Organization estimates that a quarter of the world's disease burden is due to the contamination of air, water, soil and food.

' Prevention is better than cure'; this principle has been well recognized and followed by ayurveda since the era B.C. Immunology is included in the swasthavritta, a part of this science and that too with a wider perspective than the present modern medical science. currently available vaccination, which is needless if the rules laid by ayurveda are properly observed, according to our experience in the medical field for last thirty years.

The aim of this Chapter is to relate the causes of climate change and their effects according to ayurveda with the modern aspect.

Keywords: Ayurveda, Disaster, Climate change, Global warming.

Introduction

Climate Change is neither a myth nor a prophecy, but something we need to prepare for. Global climate change has emerged as the greatest threat facing human kind today. It affects society, the economy and most ecosystems. It causes poverty and human insecurity.The effect of Climate Change depends upon a number of local factors, but least developed nations are often the most affected.

Researchers have found that there is a close link between local climate and the occurrence or severity of some diseases and other threats to human health. It is estimated that climate change contributes to 150,000 deaths and 5 million illnesses each year, and the World Health Organization estimates that a quarter of the world's disease burden is due to the contamination of air, water, soil and food.

In the last quarter of the 20th century, the average atmospheric temperature rose by about 1 degree Fahrenheit. By 2000, that increase was responsible for the annual loss of about 160,000 lives and the loss of 5.5 million years of healthy life, according to estimates by the World Health Organization. The toll is expected to double to about 300,000 lives and 11 million years of healthy life by 2020.

The biggest tolls were in Africa, on the Indian subcontinent, and in Southeast Asia. Most of the increased burden of death and disease were from malnutrition, diarrhoea, malaria, heat waves, and floods.

Climate changes also impacts on health include: Increased frequencies of heat waves; more variable precipitation patterns compromising the supply of freshwater, higher risks of water-borne diseases; and a rise in coastal flooding due to rising sea levels *etc.*

The Ayurveda has also described the causes of climate change and their impact in the Great granthas like Charak samhita, Astang Sangraha *etc.*

Ayurveda, as the name implies is truly a science of 'life'. It is not only a healing science, but it's a guide to live an ideal living style for every human on this globe.

Basically, ayurveda is mainly concerned in maintaining health of every individual rather than treating the diseases. According to it, one can enjoy healthy life by observing certain rules laid by the science; categorized under the chapter ' swasthavritta'. Those are divided into dinacharya (daily regimen), ritucharya (rules for change in diet and daily regimen according to the changes in climate) and sadvritta (rules for mental health and social behavior).

The word 'ecology' which has been coined recently by the modern science (at the end of twentieth century); but one may be astonished to know that the age old indian medical science has thought thoroughly about it about 5000 years ago and has laid down the rules and regulations according to seasonal changes.

' Prevention is better than cure'; this principle has been well recognized and followed by ayurveda since the era B.C. Immunology is included in the swasthavritta, a part of this science and that too with a wider perspective than the present modern medical science. currently available vaccination, which is needless if the rules laid by ayurveda are properly observed, according to our experience in the medical field for last thirty years.

Another specialty of this science is that it treats every patient individually, not in parts but as a whole. because the philosophy on which the indian sciences have been based, thinks that every living creature and plants upon this earth is made up of a peculiar combination of fractions of all the basic elements *i.e.* ' panch mahabhutas' present in the universe, so that plant kingdom and animal kingdom both come under the influence of ecological climatic changes and need to change themselves according to the seasonal variation.

According to tridosha theory of ayurveda vata, pitta and kapha regulate the various complex mechanisms going on in every individual. Due to changes in the climate (seasonal changes) the balance of all these three doshas is affected and a particular dosha may get vitiated at a particular period of change in season. To bring the proper balance of these three doshas, ayurveda advises the use of certain plant and animal products for medicinal purpose and purification methods known as 'panchakarmas' *i.e.* vamana, virechana, basti, nasya and rakta – mokshana.

Not a single science available, other than ayurveda, has thought about diet (ahar) part so scrupulously and thoroughly. Indian residents used to taste various delicious recipes when the so-called advanced western world had started change in its diet from raw meat to roasted meat. All the edibles according to ayurveda are divided into six tastes and each taste has its effect over the tridoshas in human individual.

'Prakriti' concept is another unique specialty of ayurveda. According to the predominance of the dosha, three main varieties of prakriti are said to be vata prakriti, pitta prakriti and kapha prakriti. According to this concept two or more off springs of the same parents, even the twins may differ in prakriti. The diet and behavioral part also varies according to prakriti.

Psychiatry is a specialized branch of modern medicinal science, which has been evolved in the 20th century. Ayurveda is the only science that gives an equal importance to the mind and the body and that too from time immemorial. Texts of ayurveda have described various kinds of manas-prakritis as well. In Kottakal in south india, there is a big mental hospital where only ayurvedic treatment is given for various psychological disorders.

Ayurveda emphasizes that one should attend properly to various natural urges (vegas) like sneezing, coughing, urinating, passing motions, laughing, passing gases through anus or mouth, hunger, thirst *etc.* these are never to be stopped or restricted. Not only that but ayurveda has explained that almost all diseases are created due to stopping (not attending) to these urges or by enforcing them.

Aim of the Paper

The aim of this Chapter is to relate the causes of climate change and their effects according to ayurveda with the modern aspect.

Methodology

This was secondary descriptive study, but some issues were collected from Ayurveda grantha regarding causes and effects of climate change.

The Modern Aspect of Climate Change: Causes and their Effects

The climate of a region or city is its typical or average weather. For example, the climate of Hawaii is sunny and warm. But the climate of Antarctica is freezing cold. Earth's climate is the average of all the world's regional-climates.

Climate change, therefore is a change in the typical or average weather of a region or city. This could be a change in a region's average annual rainfall, or it could be a change in a city's average temperature for a given month or season.

Climate change is also a change in Earth's overall climate. This could be a change in Earth's average temperature, for example. Or it could be a change in Earth's typical precipitation patterns.

Earth's Climate is Changing

Earth's climate is always changing. In the past, Earth's climate has gone through warmer and cooler periods, each lasting thousands of years.

Observations show that Earth's climate has been warming. Its average temperature has risen a little more than one degree Fahrenheit during the past 100 years or so. This amount may not seem like much. But small changes in Earth's average temperature can lead to big impacts.

Figure 30.1: Burning Coal, Oil and Gas to Create Energy Releases Gases into the Air
***Source*: NASA**

Climate Change: How do we know?

The Earth's climate has changed throughout history. Just in the last 650,000 years there have been seven cycles of glacial advance and retreat, with the abrupt end of the last ice age about 7,000 years ago marking the beginning of the modern climate era – and of human civilization. Most of these climate changes are attributed to very small variations in Earth's orbit that change the amount of solar energy our planet receives.

The current warming trend is of particular significance because most of it is very likely human-induced and proceeding at a rate that is unprecedented in the past 1,300 years.

Earth-orbiting satellites and other technological advances have enabled scientists to see the big picture, collecting many different types of information about our planet and its climate on a global scale. Studying these climate data collected over many years revealed the signals of a changing climate.

Certain facts about Earth's climate are not in dispute:

- ☆ The heat-trapping nature of carbon dioxide and other gases was demonstrated in the mid-19th century. Their ability to affect the transfer of infrared energy through the atmosphere is the scientific basis of many JPL-designed instruments, such as AIRS. Increased levels of greenhouse gases must cause the Earth to warm in response.
- ☆ Ice cores drawn from Greenland, Antarctica, and tropical mountain glaciers show that the Earth's climate responds to changes in solar output, in the Earth's orbit, and in greenhouse gas levels. They also show that in the past, large changes in climate have happened very quickly, geologically-speaking: in tens of years, not in millions or even thousands.

What is Causing Earth's Climate to Change?

Some causes of climate change are natural. These include changes in Earth's orbit and in the amount of energy coming from the sun. Ocean changes and volcanic eruptions are also natural causes of climate change.

Most scientists think that recent warming can't be explained by nature alone. Most scientists say it's very likely that most of the warming since the mid-1900s is due to the burning of coal, oil and gas. Burning these fuels is how we produce most of the energy that we use every day. This burning adds heat-trapping gases, such as carbon dioxide, into the air. These gases are called greenhouse gases.

Causes of Climate Change

The earth's climate is dynamic and always changing through a natural cycle. What the world is more worried about is that the changes that are occurring today have been speeded up because of man's activities. These changes are being studied by scientists all over the world who are finding evidence from tree rings, pollen samples, ice cores and sea sediments. The causes of climate change can be divided into two categories - those that are due to natural causes and those that are created by man.

Natural Causes

There are number of natural factors responsible for climate change. Some of the more prominent ones are continental drift, volcanoes, ocean currents, the earth's tilt and comets and meteorites. Let's look at them in a little detail.

Continental Drift

You may have noticed something peculiar about South America and Africa on a map of the world - don't they seem to fit into each other like pieces in a jigsaw puzzle?

About 200 million years ago they were joined together! Scientists believe that back then, the earth was not as we see it today, but the continents were all part of one large landmass. Proof of this comes from the similarity between plant and animal fossils and broad belts of rocks found on the eastern coastline of South America and western coastline of Africa, which are now widely separated by the Atlantic Ocean. The discovery of fossils of tropical plants (in the form of coal deposits) in Antarctica has led to the conclusion that this frozen land at some time in the past, must have been situated closer to the equator, where the climate was tropical, with swamps and plenty of lush vegetation.

The continents that we are familiar with today were formed when the landmass began gradually drifting apart, millions of years back. This drift also had an impact on the climate because it changed the physical features of the landmass, their position and the position of water bodies. The separation of the landmasses changed the flow of ocean currents and winds, which affected the climate. This drift of the continents continues even today; the Himalayan range is rising by about 1 mm (millimeter) every year because the Indian land mass is moving towards the Asian land mass, slowly but steadily.

Volcanoes

When a volcano erupts it throws out large volumes of sulphur dioxide (SO_2), water vapour, dust, and ash into the atmosphere. Although the volcanic activity may last only a few days, yet the large volumes of gases and ash can influence climatic patterns for years. Millions of tonnes of sulphur dioxide gas can reach the upper levels of the atmosphere (called the stratosphere) from a major eruption. The gases and dust particles partially block the incoming rays of the sun, leading to cooling. Sulphur dioxide combines with water to form tiny droplets of sulphuric acid. These droplets are so small that many of them can stay aloft for several years. They are efficient reflectors of sunlight, and screen the ground from some of the energy that it would ordinarily receive from the sun. Winds in the upper levels of the atmopshere, called the stratosphere, carry the aerosols rapidly around the globe in either an easterly or westerly direction. Movement of aerosols north and south is always much slower. This should give you some idea of the ways by which cooling can be brought about for a few years after a major volcanic eruption.

Mount Pinatoba, in the Philippine islands erupted in April 1991 emitting thousands of tonnes of gases into the atmosphere. Volcanic eruptions of this magnitude can reduce the amount of solar radiation reaching the Earth's surface, lowering temperatures in the lower levels of the atmosphere (called the troposphere), and changing atmospheric circulation patterns. The extent to which this occurs is an ongoing debate.

Another striking example was in the year 1816, often referred to as "the year without a summer." Significant weather-related disruptions occurred in New England and in Western Europe with killing summer frosts in the United States and Canada. These strange phenomena were attributed to a major eruption of the Tambora volcano in Indonesia, in 1815.

The Earth's Tilt

The earth makes one full orbit around the sun each year. It is tilted at an angle of 23.5° to the perpendicular plane of its orbital path. For one half of the year when it is summer, the northern hemisphere tilts towards the sun. In the other half when it is winter, the earth is tilted away from the sun. If there was no tilt we would not have experienced seasons. Changes in the tilt of the earth can affect the severity of the seasons - more tilt means warmer summers and colder winters; less tilt means cooler summers and milder winters.

The Earth's orbit is somewhat elliptical, which means that the distance between the earth and the Sun varies over the course of a year. We usually think of the earth's axis as being fixed, after all, it always seems to point toward Polaris (also known as the Pole Star and the North Star). Actually, it is not quite constant: the axis does move, at the rate of a little more than a half-degree each century. So Polaris has not always been, and will not always be, the star pointing to the North. When the pyramids were built, around 2500 BC, the pole was near the star Thuban (Alpha Draconis). This gradual change in the direction of the earth's axis, called precession is responsible for changes in the climate.

Ocean Currents

The oceans are a major component of the climate system. They cover about 71 per cent of the Earth and absorb about twice as much of the sun's radiation as the atmosphere or the land surface. Ocean currents move vast amounts of heat across the planet - roughly the same amount as the atmosphere does. But the oceans are surrounded by land masses, so heat transport through the water is through channels.

Winds push horizontally against the sea surface and drive ocean current patterns. Certain parts of the world are influenced by ocean currents more than others. The coast of Peru and other adjoining regions are directly influenced by the Humboldt current that flows along the coastline of Peru. The El Niño event in the Pacific Ocean can affect climatic conditions all over the world.

Another region that is strongly influenced by ocean currents is the North Atlantic. If we compare places at the same latitude in Europe and North America the effect is immediately obvious. Take a closer look at this example - some parts of coastal Norway have an average temperature of -2°C in January and 14°C in July; while places at the same latitude on the Pacific coast of Alaska are far colder: -15°C in January and only 10°C in July. The warm current along the Norewgian coast keeps much of the Greenland-Norwegian Sea free of ice even in winter. The rest of the Arctic Ocean, even though it is much further south, remains frozen.

Ocean currents have been known to change direction or slow down. Much of the heat that escapes from the oceans is in the form of water vapour, the most abundant greenhouse gas on Earth. Yet, water vapor also contributes to the formation of clouds, which shade the surface and have a net cooling effect.

Any or all of these phenomena can have an impact on the climate, as is believed to have happened at the end of the last Ice Age, about 14,000 years ago.

Human Causes

The Industrial Revolution in the 19th century saw the large-scale use of fossil fuels for industrial activities. These industries created jobs and over the years, people moved from rural areas to the cities. This trend is continuing even today. More and more land that was covered with vegetation has been cleared to make way for houses. Natural resources are being used extensively for construction, industries, transport and consumption. Consumerism (our increasing want for material things) has increased by leaps and bounds, creating mountains of waste. Also, our population has increased to an incredible extent.

All this has contributed to a rise in greenhouse gases in the atmosphere. Fossil fuels such as oil, coal and natural gas supply most of the energy needed to run vehicles, generate electricity for industries, households *etc.* The energy sector is responsible for about ¾ of the carbon dioxide emissions, 1/5 of the methane emissions and a large quantity of nitrous oxide. It also produces nitrogen oxides (NOx) and carbon monoxide (CO) which are not greenhouse gases but do have an influence on the chemical cycles in the atmosphere that produce or destroy greenhouse gases.

Greenhouse Gases and their Sources

Carbon dioxide is undoubtedly, the most important greenhouse gas in the atmosphere. Changes in land use pattern, deforestation, land clearing, agriculture and other activities have all led to a rise in the emission of carbon-dioxide.

Methane is another important greenhouse gas in the atmosphere. About ¼ of all methane emissions are said to come from domesticated animals such as dairy cows, goats, pigs, buffaloes, camels, horses, and sheep. These animals produce methane during the cud-chewing process. Methane is also released from rice or paddy fields that are flooded during the sowing and maturing periods. When soil is covered with water it becomes anaerobic or lacking in oxygen. Under such conditions, methane-producing bacteria and other organisms decompose organic matter in the soil to form methane. Nearly 90 per cent of the paddy-growing area in the world is found in Asia, as rice is the staple food there. China and India, between them, have 80-90 per cent of the world's rice-growing areas.

Methane is also emitted from landfills and other waste dumps. If the waste is put into an incinerator or burnt in the open, carbon dioxide is emitted. Methane is also emitted during the process of oil drilling, coal mining and also from leaking gas pipelines (due to accidents and poor maintenance of sites).

A large amount of nitrous oxide emission has been attributed to fertilizer application. This in turn depends on the type of fertilizer that is used, how and when it is used and the methods of tilling that are followed. Contributions are also made by leguminous plants, such as beans and pulses that add nitrogen to the soil.

How We All Contribute Every Day

All of us in our daily lives contribute our bit to this change in the climate. Give these points a good, serious thought:

- ☆ Electricity is the main source of power in urban areas. All our gadgets run on electricity generated mainly from thermal power plants. These thermal power plants run on fossil fuels (mostly coal) and are responsible for the emission of huge amounts of greenhouse gases and other pollutants.
- ☆ Cars, buses, and trucks are the principal ways by which goods and people are transported in most of our cities. These are run mainly on petrol or diesel, both fossil fuels.
- ☆ We generate large quantities of waste in the form of plastics that remain in the environment for many years and cause damage.
- ☆ We use huge quantity of paper in our work at schools and in offices. Have we ever thought about the number of trees that we use in a day?
- ☆ Timber is used in large quantities for construction of houses, which means that large areas of forest have to be cut down.
- ☆ A growing population has meant more and more mouths to feed. Because the land area available for agriculture is limited (and infact, is actually shrinking as a result of ecological degradation!), high-yielding varieties of crop are being grown to increase the agricultural output from a given area of land. However, such high-yielding varieties of crops require large quantities of fertilizers; and more fertilizer means more emissions of nitrous oxide, both from the field into which it is put and the fertilizer industry that makes it. Pollution also results from the run-off of fertilizer into water bodies.

Factors in the Occurences of Natural Disasters

Here are the causes of "Climate Change" and it includes: agriculture and cattle-raising, dams and megaprojects, forest fires, illegal and legal logging, mangroves and shrimp farming, mining, oil and gas production and usage, and plantations. Global warming, and slash-and-burn farming the other aggravating factors in the presence of powerful natural disasters that inflicts great damage to properties, source of livelihood and infrastructures all over the planet especially the South Asia and South East Asia.

Pollution

The pollution emanating from different industrial zones all over the world, vehicles, CFCs from our air-conditioned units and refrigerators, and burnt plastic damages Ozone layer that protects our planet from the heat of the sun and us humans, also adds significantly to the Global Warming we are experiencing now.

Deforestation

An island nation in the Caribbean, Haiti is reported to have a history of deforestation. Plagued with social problems and population, the forests are indeed threatened. Mudslides smothered thousands of lives a testament of the effects brought by massive deforestation in the country. Deforestation due to "kaingin" or slash-and-burn farming, growing populace and illegal logging is stripping our forests of the trees that protect us from floods and serves as a habitat to different animals, causing mudslide, soil erosion and landslide.

The Evidence for Rapid Climate Change is Compelling

Republic of Maldives: Vulnerable to Sea Level Rise

Sea Level Rise

Global sea level rose about 17 centimeters (6.7 inches) in the last century. The rate in the last decade, however, is nearly double that of the last century.[4]

Global Temperature Rise

All three major global surface temperature reconstructions show that Earth has warmed since 1880. Most of this warming has occurred since the 1970s, with the 20 warmest years having occurred since 1981 and with all 10 of the warmest years occurring in the past 12 years. Even though the 2000s witnessed a solar output decline resulting in an unusually deep solar minimum in 2007-2009, surface temperatures continue to increase.

Warming Oceans

The oceans have absorbed much of this increased heat, with the top 700 meters (about 2,300 feet) of ocean showing warming of 0.302 degrees Fahrenheit since 1969.

Flowing Meltwater from the Greenland Ice Sheet

Shrinking Ice Sheets

The Greenland and Antarctic ice sheets have decreased in mass. Data from NASA's Gravity Recovery and Climate Experiment show Greenland lost 150 to 250 cubic kilometers (36 to 60 cubic miles) of ice per year between 2002 and 2006, while Antarctica lost about 152 cubic kilometers (36 cubic miles) of ice between 2002 and 2005.

Visualization of the 2007 Arctic Sea Ice Minimum

Declining Arctic Sea Ice

Both the extent and thickness of Arctic sea ice has declined rapidly over the last several decades.

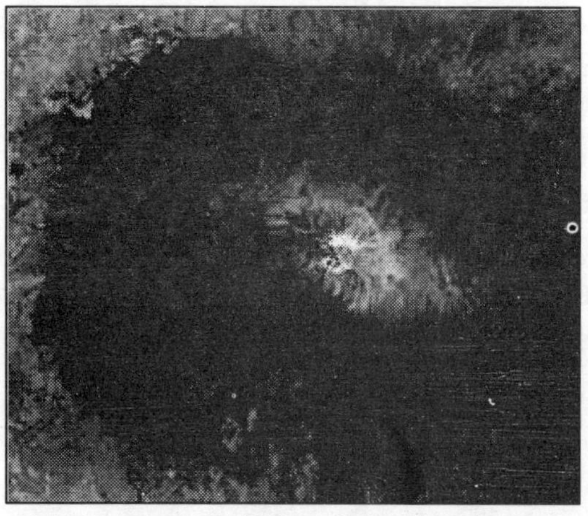

The Disappearing Snowcap of Mount Kilimanjaro, from Space

Glacial Retreat

Glaciers are retreating almost everywhere around the world – including in the Alps, Himalayas, Andes, Rockies, Alaska and Africa.

Extreme Events

The number of record high temperature events in the United States has been increasing, while the number of record low temperature events has been decreasing, since 1950. The U.S. has also witnessed increasing numbers of intense rainfall events.[11]

Ocean Acidification

Since the beginning of the Industrial Revolution, the acidity of surface ocean waters has increased by about 30 percent.[12,13] This increase is the result of humans emitting more carbon dioxide into the atmosphere and hence more being absorbed into the oceans. The amount of carbon dioxide absorbed by the upper layer of the oceans is increasing by about 2 billion tons per year.

The Current and Future Consequences of Global Climate Change

The potential future effects of global climate change include more frequent wildfires, longer periods of drought in some regions and an increase in the number, duration and intensity of tropical storms.

Global climate change has already had observable effects on the environment. Glaciers have shrunk, ice on rivers and lakes is breaking up earlier, plant and animal ranges have shifted and trees are flowering sooner.

Effects that scientists had predicted in the past would result from global climate change are now occuring: loss of sea ice, accelerated sea level rise and longer, more intense heat waves.

Scientists have high confidence that global temperatures will continue to rise for decades to come, largely due to greenhouse gasses produced by human activities. The Intergovernmental Panel on Climate Change (IPCC), which includes more than 1,300 scientists from the United States and other countries, forecasts a temperature rise of 2.5 to 10 degrees Fahrenheit over the next century.

According to the IPCC, the extent of climate change effects on individual regions will vary over time and with the ability of different societal and environmental systems to mitigate or adapt to change.

The effects, or impacts, of climate change may be physical, ecological, social or economic. Evidence of observed climate change includes the instrumental temperature record, rising sea levels, and decreased snow cover in theNorthern Hemisphere. According to the Intergovernmental Panel on Climate Change, "[most] of the observed increase in global average temperatures since the mid-20th century is very likely due to the observed increase in [human greenhouse gas] concentrations". It is predicted that future climate changes will include further global warming (*i.e.*, an upward trend in global mean temperature), sea level rise and a probable increase in the frequency of some extreme weather events. Signatories of the United Nations Framework Convention on Climate Change have agreed to implement policies designed to reduce their emissions of greenhouse gases.

Temperature Changes

Global mean surface temperature difference from the average for 1880–2009 the IPCC's Assessment Reports on climate change showed global warming of around 0.6 °C over the entire 20th century. The future level of global warming is uncertain, but a wide range of estimates (projections) have been made. The IPCC's "SRES" scenarios have been frequently used to make projections of future climate change. Climate models using the six SRES "marker" scenarios suggest future warming of 1.1 to 6.4 °C by the end of the 21st century (above average global temperatures over the 1980 to 1999 time period). The range in temperature projections partly reflects different projections of future social and economic development (*e.g.*,economic growth, population level, energy policies), which in turn affects projections of greenhouse gas (GHG) emissions. The range also reflects uncertainty in the response of the climate system to past and future GHG emissions (measured by the climate sensitivity).

Physical Impacts of Climate Change

Key climate indicators that show global warming.

Working Group I's contribution to the IPCC Fourth Assessment Report, published in 2007, concluded that warming of the climate system was "unequivocal." This was based on the consistency of evidence across a range of observed changes, including increases in global average air and ocean temperatures, widespread melting of snow and ice and rising global average sea level.

Human activities have contributed to number of the observed changes in climate. This contribution has principally been through the burning of fossil fuels, which has led to an increase in the concentration of GHGs in the atmosphere. This increase in GHG concentrations has caused aradiative forcing of the climate in the direction of warming. Human-induced forcing of the climate has likely to contribute to a number of observed changes, including sea level rise, changes in climate extremes (such as warm and cold days), declines in Arctic sea ice extent and to glacier retreat.

Human-induced warming could potentially lead to some impacts that are abrupt or irreversible (see the section on Abrupt or irreversible changes). The probability of warming having unforeseen consequences increases with the rate, magnitude, and duration of climate change.

Effects on Weather

Observations show that there have been changes in weather. As climate changes, the probabilities of certain types of weather events are affected.

Changes have been observed in the amount, intensity, frequency and type of precipitation. Widespread increase in heavy precipitation have occurred, even in places where total rain amounts have decreased. IPCC (2007d) concluded that human influences had, more likely than not (greater than 50 per cent probability, based on expert judgement), contributed to an increase in the frequency of heavy precipitation events. Projections of future changes in precipitation show overall increase in the global average, but with substantial shifts in where and how precipitation falls. Climate models tend to project increasing precipitation at high latitudes and in the tropics (*e.g.*, the southeast monsoon region and over the tropical Pacific) and decreasing precipitation in the sub-tropics (*e.g.*, over much of North Africa and the Northern Sahara).

Evidence suggests that, since the 1970s, there have been substantial increases in the intensity and duration of tropical storms and hurricanes. Models project a general tendency for more intense but fewer storms outside the tropics.

Accumulated cyclone energy in the Atlantic Ocean and the sea surface temperature difference which influences such, measured by the U.S. NOAA.

Extreme Weather

Since the late 20th century, changes have been observed in the trends of some extreme weather and climate events, *e.g.*, heat waves. Human activities have, with varying degrees of confidence, contributed to some of these observed trends. Projections for the 21st century suggest continuing changes in trends for some extreme events. Solomon *et al.* (2007), for example, projected the following likely (greater than 66 per cent probability, based on expert judgement) changes:

- ✰ An increase in the areas affected by drought;
- ✰ Increased tropical cyclone activity;
- ✰ And increased incidence of extreme high sea level (excluding tsunamis).

Projected changes in extreme events will have predominantly adverse impacts on ecosystems and human society.

Glacier Retreat and Disappearance

A map of the change in thickness of mountain glaciers since 1970. Thinning in orange and red, thickening in blue.

IPCC (2007a:5) found that, on average, mountain glaciers and snow cover had decreased in both the northern and southern hemispheres. This widespread decrease in glaciers and ice caps had contributed to observed sea level rise. With very high or high confidence, IPCC (2007b and 2007d:11) made number of projections relating to future changes in glaciers:

- ✰ Mountainous areas in Europe will face glacier retreat
- ✰ In Polar regions, there will be reductions in glacier extent and the thickness of glaciers.
- ✰ More than one-sixth of the world's population are supplied by meltwater from major mountain ranges. Changes in glaciers and snow cover are expected to reduce water availability for these populations.
- ✰ In Latin America, changes in precipitation patterns and the disappearance of glaciers will significantly affect water availability for human consumption, agriculture and energy production.

Oceans

The role of the oceans in global warming is a complex one. The oceans serve as a sink for carbon dioxide, taking up much that would otherwise remain in the atmosphere, but increased levels of CO_2 have led to ocean acidification. Furthermore, as the temperature of the oceans increases, they become less able to absorb excess CO_2. The ocean have also acted as a sink in absorbing extra heat from the atmosphere. This extra heat has been added to the climate system due to the build-up of GHGs.

More than 90 percent of warming that occurred over 1960–2009 is estimated to have gone into the oceans.

Global warming is projected to have a number of effects on the oceans. Ongoing effects include rising sea levels due to thermal expansion and melting of glaciers and ice sheets, and warming of the ocean surface, leading to increased temperature stratification. Other possible effects include large-scale changes in ocean circulation.

Acidification

About one-third of the carbon dioxide emitted by human activity has already been taken up by the oceans. As carbon dioxide dissolves in sea water, carbonic acid is formed, which has the effect of acidifying the ocean, measured as a change in pH. The uptake of human carbon emissions since the year 1750 has led to an average decrease in pH of 0.1 units. Projections using the SRES emissions scenarios suggest a further reduction in average global surface ocean pH of between 0.14 and 0.35 units over the 21st century.

The effects of ocean acidification on the marine biosphere have yet to be documented. Laboratory experiments suggest beneficial effects for few species, with potentially highly detrimental effects for a substantial number of species. With medium confidence, Fischlin *et al.* (2007) projected that future ocean acidification and climate change would impair a wide range of planktonic and shallow benthic marine organisms that use aragonite to make their shells or skeletons, such as corals and marine snails (pteropods), with significant impacts particularly in the Southern Ocean.

Oxygen Depletion

The amount of oxygen dissolved in the oceans may decline, with adverse consequences for ocean life.

Sea Level Rise

Sea level rise during the Holocene.

Sea level has been rising 0.2/cm/yr, based on measurements of sea level rise from 23 long tide gauge records in geologically stable environments.

There is strong evidence that global sea level rose gradually over the 20th century. With high confidence, Bindoff *et al.* (2007) concluded that between the mid-19th and mid-20th centuries, the rate of sea level rise increased. The IPCC (2007d, p. 5) reported that between 1961 and 2003, global average sea level rose at an average rate of 1.8 mm per year (mm/yr), with an uncertainty range of 1.3–2.3/ mm/yr. Between 1993 and 2003, the rate increased above the previous period to 3.1 mm/yr (uncertainty range of 2.4–3.8 mm/yr). The IPCC (2007d) was uncertain whether the increase in rate from 1993 to 2003 was due to natural variations in sea level over the time period, or whether it reflected an increase in the underlying long-term trend.

Ocean Temperature Rise

From 1961 to 2003, the global ocean temperature has risen by 0.10 °C from the surface to a depth of 700 m. There is variability both year-to-year and over longer time scales, with global ocean heat content observations showing high rates of warming for 1991–2003, but some cooling from 2003 to 2007. The temperature of the Antarctic Southern Ocean rose by 0.17 °C (0.31 °F) between the 1950s and the 1980s, nearly twice the rate for the world's oceans as a whole. As well as having effects on ecosystems (*e.g.* by melting sea ice, affecting algae that grow on its underside), warming reduces the ocean's ability to absorb CO_2.

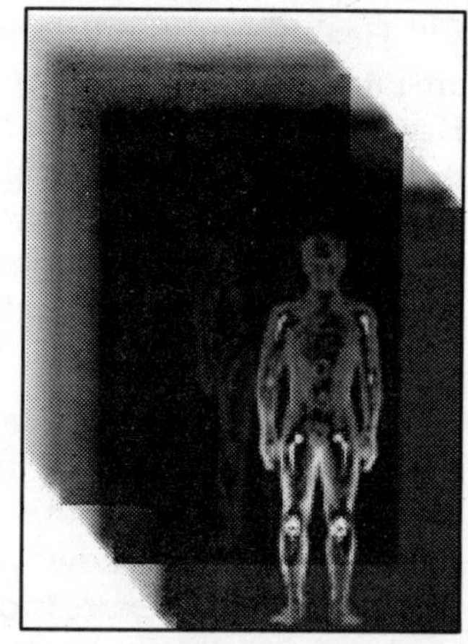

Impact on Human Health

Throughout the world, the prevalence of some diseases and other threats to human health depend largely on local climate. Extreme temperatures can lead directly to loss of life, while climate-related disturbances in ecological systems, such as changes in the range of infective parasites, can indirectly impact the incidence of serious infectious diseases. In addition, warm temperatures can increase air and water pollution, which in turn harm human health.

Human health is strongly affected by social, political, economic, environmental and technological factors, including urbanization, affluence, scientific developments, individual behavior and individual vulnerability (*e.g.*, genetic makeup, nutritional status, emotional well-being, age, gender and economic status). The extent and nature of climate change impacts on human health vary by region, by relative vulnerability of population groups, by the extent and duration of exposure to climate change itself and by society's ability to adapt to or cope with the change.

Climate change poses a wide range of risks to population health - risks that will increase in future decades, often to critical levels, if global climate change continues on its current trajectory.The three main categories of health risks include: (i) direct-acting effects (*e.g.* due to heat waves, amplified air pollution and physical weather disasters), (ii) impacts mediated *via* climate-related changes in ecological systems and relationships (*e.g.* crop yields, mosquito ecology, marine productivity), and (iii) the more diffuse (indirect) consequences relating to impoverishment, displacement, resource conflicts (*e.g.* water) and post-disaster mental health problems.

Climate change thus threatens to slow, halt or reverse international progress towards reducing child under-nutrition, deaths from diarrhoeal diseases and the spread of other infectious diseases. Climate change acts predominantly by exacerbating the existing, often enormous, health problems, especially in the poorer parts of the world. Current variations in weather conditions already have many adverse impacts on the health of poor people in developing nations and these too are likely to be 'multiplied' by the added stresses of climate change.

A changing climate thus affects the prerequisites of population health: clean air and water, sufficient food, natural constraints on infectious disease agents, and the adequacy and security of shelter. A warmer and more variable climate leads to higher levels of some air pollutants and more frequent extreme weather events. It increases the rates and ranges of transmission of infectious diseases through unclean water and contaminated food, and by affecting vector organisms (such as mosquitoes) and intermediate or reservoir host species that harbour the infectious agent (such as cattle, bats and rodents). Changes in temperature, rainfall and seasonality compromise agricultural production in many regions, including some of the least developed countries, thus jeopardising child health and growth and the overall health and functional capacity of adults. As warming proceeds, the severity (and perhaps frequency) of weather-related disasters will increase - and appears to have done so in a number of regions of the world over the past several decades. Therefore, in summary, global warming, together with resultant changes in food and water supplies, can indirectly cause increases in a range of adverse health outcomes, including malnutrition, diarrhoea, injuries, cardiovascular and respiratory diseases and water-borne and insect-transmitted diseases.

Health equity and climate change have a major impact on human health and quality of life, and are interlinked in a number of ways. The report of the WHO Commission on Social Determinants of Health points out that disadvantaged communities are likely to shoulder a disproportionate share of the burden of climate change because of their increased exposure and vulnerability to health threats. Over 90 percent of malaria and diarrhoea deaths are borne by children aged 5 years or younger, mostly in developing countries. Other severely affected population groups include women, the elderly and people living in small island developing states and other coastal regions, mega-cities or mountainous areas.

Climate change can lead to dramatic increases in prevalence of a variety of infectious diseases. Beginning in the mid-70s, there has been an "emergence, resurgence and redistribution of infectious diseases". Reasons for this are likely multicausal, dependent on a variety of social, environmental and climatic factors, however, many argue that the "volatility of infectious disease may be one of the earliest biological expressions of climate instability". Though many infectious diseases are affected by changes in climate. Vector-borne diseases, such as malaria, dengue fever and leishmaniasis, present the strongest causal relationship. Malaria in particular, which kills approximately 300,000 children annually, poses the most imminent threat.

Concept of Ayurveda Science Regarding Climate Change and Recommendations/Solutions to Prevent the Epidemic/Disaster

(This description is based on Ayurveda Grantha- Janpadodhavansha chapter in Vimana Sthana of Charaka Samhita)

Historical/Mythical Cause of Diseases

In early times during destruction of Daksa's sacrifice when the human beings fled in various directions, fast running, swimming, running, jumping and leaping *etc.* which agitated the body gave rise to gulma. Prameha and kustha arose due to intake of fatty material. Insanity arose due to fear, torture and grief. Epilepsy arose due to impure contact of various creatures. Fever arose from the forehead of the great Lord (Rudra). Internal hemorrhage arose from the excessive heat (of fever). Phthisis arose due to excessive sexual intercourse by the king of stars (moon).

Initial Origin of Disorders

During initial age (krtayuga):

In early times, no undesirable consequence arose except from unrighteousness.

During the initial age (krtayaga), people were: having prowess like the sons of gods, exceedingly pure and with vast influence, having perceived the gods, godly sages, virtue, religious sacrifices and method of their performance; with the body compact and stable like the essence of mountains, and complexion and sense organs clear, having strength, speed and valor like the wind, with well-formed buttocks, endowed with appropriate measure (size), physiognomy, cheerfulness and corpulence, were devoted to truthfulness, straightforwardness, un-cruelty, charity, control of the senses, observance of rules, penance, fasting, celibacy and vows; devoid of fear, attachment, aversion, confusion, greed, anger, grief, conceit, illness, sleep, drowsiness, fatigue, exhaustion, lassitude and "holding" and were having immeasurable life-span.

For those having exalted mind, qualities and actions the crops grew endowed with inconceivable rasa, virya, vipaka, prabhava, and other properties due to presence of all qualities in earth *etc.* (the five elements?) In the beginning of Krtayuga.

At the declining of krtayuga: due to over-receiving (see calendar for definition of this term) there arose heaviness in bodies of these wealthy persons; heaviness of the body led to fatigue, lassitude, hoarding, holding and greed in successive order.

In Tretayuga: greed gave rise to malice, speaking lie, passion, anger, conceit, dislike, roughness, violence, fear, infliction, grief, anxiety, excitement *etc.* successively.

In treta, a quarter of righteousness disappeared due to which there was reduction of a quarter in the yearly duration of the yugas (ages) and consequent degradation of quarter in unctuousness, purity, rasa, vipaka, prabhava and other properties of the crops.

Because of this the bodies of the people due to intake of food degraded by a quarter in properties and other behaviors were not resistant as earlier and as such were pervaded by pitta and vata and were attacked first by diseases like fever *etc.* Thus the living beings were gradually affected by decrease in their life-span.

In yuga after yuga a quarter of righteousness is reduced in this order along with similar reduction in the qualities of living beings leading finally to dissolution of the universe. After completion of 100 years, there is a loss of 1 year in the lifespan of the living beings in respective ages. Thus is said the initial origin of disorders.

Causative Factors along with Preparation for Treatment of Epidemics

It has been observed abnormal conditions of stars, planets, moon, sun, air and fire and also of the environment which derange the seasons too. Shortly hereafter the earth too will not provide properly rasa, virya, vipaka an prabhava to herbs, consequently, due to absence of these requisite properties spread of diseases is certain.

Hence before destruction and loss of nutrients in the earth, it was recommended that the Extract the herbs lest they should lose their rasa, virya, vipaka and prabhava.

We shall make use of these properties-rasa, virya, vipaka and prabhava for those who are devoted to us and whom we like, because there will not be any difficulty in counteracting the epidemic disorders if the drugs are well-collected, well-processed and well-administered.

Why Does an Epidemic Effect Everyone in a Community?

Even though people differ in constitution, strength, age *etc.*, they may all succumb to the epidemic, due to 4 factors which become deranged. They are:

1. Air
2. Water
3. Place
4. Time

"They are important in progressive order because of the degree of their indispensability"

Again: Deranged air, water, place and time are the cause of epidemics.

Air

Causing illness; not in accordance with the season, excessively moist, speedy, harsh, cold, hot, rough, blocking, terribly sounding; excessively clashing with each other, whistling, and affected with unsuitable smell, vapor, gravels, dust and smoke.

Water

Water is devoid of merits when it is excessively deranged in respect of smell, color, taste and touch, is too slimy, deserted by aquatic birds, aquatic animals are reduced and is unpleasing.

Place (Land)

It's normal color, smell, taste and touch is too much affected, it contains excessive moisture, is troubled by reptiles, violent animals, mosquitoes, locusts, flied, rats, owls, vultures, jackals *etc.*, has groves of grasses an creepers and abundance of diffusing plants (*i.e.* kudzu? Or *i.e.* ragweed?); has a new look, has fallen, dried and damaged crops, smoky winds; crying out of birds and dogs; bewilderment and painful condition of various animals and birds; the community with abandoned and destroyed virtue, truthfulness, modesty, conduct, behavior and other merits; the rivers constantly agitated and over-flooded, frequent occurrence of meteorites, thunderbolts, and earthquakes, fierce and crying appearance; the sun, the moon and the stars with rough, coppery, reddish, white and cloudy appearance frequently; as if filled with confusion and excitement, torture, crying and darkness with frequent crying sound as if seized by guhyaka.

Time

Time should be known as contrary if it is having signs contrary, excessive or deficient to those of the season (*i.e.* a drought during the rainy season, or and excessively cold winter.)

Even when these 4 epidemic-producing factors are present, a person who has been properly treated with preventative care will be immune against the diseases.

Treatment and Prevention in Epidemics

The people do not suffer, if managed with preventive therapy. For those who have no similarity in either death or previous deeds, pancakarma is the best treatment. Thereafter proper use of rasayana measures and management with the drug collected previously is recommended.

Truthfulness, benevolence, charity, offerings, worship of gods, observance of noble's conduct, calmness, self-protection, residence in healthy places, observance of celibacy and company of those who are observing celibacy, discourse of religious scriptures, narratives of self-controlled great sages, constant company with religious, pure and those regarded by the elders-this is the management for the protection of life

For those whose death is not certain during that difficult period.

The Root Cause of Aggravation of V, P, K, which give Rise to Epidemics and Destroy the Community

The root cause of all V, P, K is unrighteousness. That also arises from the misdeeds of previous life but the source of both is intellectual error. Such as -when the heads of the country, city, guild, and community having transgressed the virtuous path deal unrighteously with the people, their officers and subordinates, people of the city and community, and traders carry this unrighteousness further. Thus this unrighteousness by force makes the righteousness disappear. Then the people with righteousness having disappeared, unrighteousness has the upper hand and the gods have deserted the place, the seasons get affected and because of this it does not rain in time, or at all or there is abnormal rainfall, winds do not blow properly, the land is affected, water reservoirs get fried up and the herbs giving up their natural propertied acquire morbidity. Then epidemics break out due to polluted contacts and edibles.

Unrighteousness is the Cause of the Destruction of the Community by Weapons

Those who have excessively increased greed, anger, attachment and conceit, disregarding the weak attack each other, or their enemies or are attacked by their enemies resulting in loss of themselves, their kinsmen and enemies.

They are also attacked by the raksasas *etc.* or other organisms due to that unrighteousness or other unwholesome act.

Unrighteousness is the Cause of the Disease Arisen Due to Cursing

Those with righteousness disappeared or moved away from righteousness, behave in unwholesome manner disregarding the good advice from preceptor, elders, accomplished ones, sages and other respectable persons. Consequently those people having been cursed by the preceptors *etc.* are reduced to ashes immediately with many families along with the other individuals who are cursed so.

3 Types of Diseases

Innate

Caused by bodily doshas

Exogenous

Caused by Bhuta (spirits and organisms), poisoned air, fire, trauma *etc.*

Psychic

Caused by non-fulfillment of desires and facing of the undesired.

Types of Physiological ('Somatic') Diseases of the Body

V, P and K

Types of (Psychological) Diseases of the Mind

Rajasic, tamasic

([The source of all illness:])

Prajnaparadha: intellectual error (in use of speech, mind, and body)

Excessive use of: too much application.

Negative use of: non-use of.

Perverted use of: holding up or forcing of urges, sleeping, falling and posturing on uneven places, abnormal posturing, heating, pressing, obstructing breath and torturing

Perverted use of speech: words of betrayal, lying, untimely speech, quarrel, unliking irrelevance, indiscipline, harshness *etc.*

Perverted use of mind: fear, grief, anger, greed, confusion, conceit, envy, wrong knowledge

There are 3 Causes of Disorders

Excessive, negative and perverted uses of sense objects, actions and time. Such as- excessive gazing at the over-brilliant objects is excessive use, avoiding looking altogether is negative use and seeing too near, too distant, fierce, frightful, wonderful, disliked, disgusting, deformed and terrifying objects is perverted use of visual objects. Likewise, to hear too much the loud sound of clouds, drums, cries *etc.* is excessive use; not at all hearing is negative use and hearing of harsh and frightful words and those which indicate death of dear ones, loss, humiliation *etc.* is perverted use of auditory objects. Too much smelling of too sharp, intense and congestant odors is excessive use, not at all smelling is negative sue and smelling of fetid, disliked, impure, decomposed, poisoned air, cadaverous odor *etc.* is perverted use of olfactory objects. Likewise, too much intake of rasas (tastes) is excessive use, not at all taking is negative use of gustatory objects. Perverted use of those will be described in the chapter dealing with the methods of eating except the quantity.

Too much indulgence in very hot and very cool objects and also in bath, massage, anointing *etc.* is excessive use of tactile objects; total abstinence from them is negative use and application of tactile objects such as hot and cold bath *etc.* without the usual order and also the touch of uneven surface, injury, dirty objects, organisms *etc.* is perverted use of tactile object.

So the condition of all the sense organs produced by the overall tactile sensation, when they are harmful, is known as unwholesome conjunction of sense organs and it's objects which is of 5 types each having 3 sub-divisions. The objects which are accepted properly are known as wholesome ones.

Action is application of speech, mind and body. Too much application of these is excessive use and their total non-application is negative use. Holding up or forcing of urges, sleeping, falling and posturing on uneven places, abnormal posturing, heating, pressing, obstructing breath and torturing is the perverted use of bodily actions. Words indicating betrayal, lying, untimely speech, quarrel, unliking, irrelevance, indiscipline and harshness *etc.* come under the perverted use of speech. Fear, grief, anger, greed, confusion, conceit, envy, wrong knowledge is the perverted use of mind.

In short, any other harmful action of speech, mind and body which has not been mentioned above, should be taken as their perverted use (Of course if they are excessive or negative uses they should be taken in those respective categories.)

The 3 Aggravating Factors for the Physical and Psychic Doshas

1. Unsuitable contact of objects with sense organs
2. Intellectual error
3. Consequence [of the above two]

Table 30.1: Samprapti Chart = 3 Passages for Disease

Passage	*Components of the Passage*	*Diseases Related to this Passage*
External passage	Dhatus and twak ("skin including rasa dhatu located in that")	Glands, boils, diabetic boils, scrofula, wart, granuloma, moles, leprosy and other skin diseases, freckles, *etc.* and erysipelas, oedema, gaseous tumor, piles and abscess *etc.*
Middle passage	Vital parts (urinary bladder, heart, heat, etc.), bone-joints and bound ligaments and tendons dhatus)	Hemiplegia, stiffness of sides, convulsion, facial paralysis, wasting, tuberculosis, pain inbone-joints, prolapse of rectum *etc.* and also the diseases of head, heart and urinary bladder.
Internal passage	Belly (Great Channel; G.I.tract); stomach, intestines, colon	Diarrhoea, vomiting, alasaka, fever, cholera, cough, dyspnea, hiccup, hardness of bowels, abdominal enlargement, spleen enlargement *etc.* along with erysipelas, oedema, gaseous tumor, piles abscess etc.

National Research Council (NRC), 2001. Climate Change Science: An Analysis of Some Key Questions. National Academy Press, Washington, DC

Schwartz and Randall, 2003. An Abrupt Climate Change Scenario and Its Implications for United States National Security (PDF) (22 pp, 915K, About PDF) October 2003.

World Health Organization (WHO), 2003. Climate change and human health - risks and responses. McMichael, A.J., Campbell-Lendrum, D.H., Corvalán, C.F., Ebi, K.L., Githeko, A., Scheraga, J.D. and Woodward, A. 322pp.

www.UNCR

2013, Sustainable Approaches for Environmental Conservation *Pages* ***261–265***
Editors: **D.R. Khanna, A.K. Chopra, R. Bhutiani, Gagan Matta & Vikas Singh**
Published by: **BIOTECH BOOKS, NEW DELHI**

Chapter 31

High Prevalence of Gallbladder Cancer Cases in the Gangetic Basin of Bihar: A Hospital Based Study

Arun Kumar, Md. Ali, A. Nath, J.K. Singh and Ranjit Kumar
Mahavir Cancer Institute and Research Centre,
Phulwarisharif, Patna, Bihar

Carcinoma of the gallbladder is the third most common malignancy of the gastrointestinal tract in the Eastern Uttar Pradesh and entire Bihar regions of India. The main source of drinking water in this region is the river Ganges, which is heavily polluted with agricultural pesticides. The incidence of gall bladder cancer is 7.5 times more in north India as compared to the south. It is a lethal disease with a wide geographical, ethnic, and cultural variation suggesting major environmental influences such as diet and life style factors in the development of disease. Not only the Gallbladder cancer but also Breast cancer, Ovarian cancer, Liver cancer, Leukemic cancer in this region is very high.

In the present study, Gall bladder cancer cases were taken up for the demographic study. The cases were distributed district wise and especially the cities or towns located near the river basins of Bihar. The age, Sex, Gallstones in cases, Stage of Cancer *etc.* were also accounted for intensive study.

The study shows that the incidence of Gall bladder Cancer cases were very high in river basin areas of Bihar in comparison to Non- river basin zones. Furthermore, in the age and sex study it revealed that people of river basin areas showed higher prevalence of this case at early age and women showed the higher number of cases in comparison to men. The cases of Gallstones and advance stage was also very high among the people of river basin areas in comparison to non- river basin zones. Among all the river basins of Bihar the Gangetic basin showed the maximum number cases.

When the cancer tissues were assayed for the pesticide assessment, there was huge percentage of DDT content in the Cancer tissue as well as blood samples of the cancer patients. The ratio was not only high for Gall bladder cancer cases but also for Breast Cancer, Ovarian Cancer, Liver Cancer *etc.*

From, the entire study it reveals that rivers of Bihar might contain various chemicals (pesticides) which causes this disease. Although many study support that river Ganga and rivers of North Bihar contains high level of mercury, cadmium, Arsenic, DDT and other PoP's. Thus, the study suggests that these pollutants are directly associated with gallbladder carcinogenesis and other cancer cases in the prevalent areas.

Keywords: *Gall bladder cancer, Gangetic zone.*

Introduction

India sees about 1 million new cancer cases every year. Of these, one third die every year. It is projected that by the year 2025, there is going to be a five fold increase in cancer cases in India – 2.8 fold increase due to tobacco and 2.2 fold increase due to ageing. The later is partly contributed by the doubling of life expectancy since independence. Such a dramatic rise in cancer cases is more than double that predicted for developing countries like USA (ICMR, 2001).

Carcinoma of the gallbladder has a very unusual geographical distribution with pockets of high incidence seen in Chile, Poland, India, Japan and Israel; it occurs rarely in the rest of the world. It is a common malignancy in the Western Bihar and Eastern Uttar Pradesh regions of India. Patients present with extremes of clinical symptoms, indicating benign biliary diseases on the one hand and incurable malignant disease on the other (Diehl, 1980 and Pandey, 2003).

Gallbladder cancer is a relatively rare neoplasm that shows, however, high incidence rates in certain world populations. The interplay of genetic susceptibility, lifestyle factors and infections in gallbladder carcinogenesis is still poorly understood. Population-based data reveal that the incidence of gall bladder cancer is very high in northern Indian cities (5-7 per 100,000 women) and low (0-0.7 per 100,000 women) in southern India. The distribution suggests a high-incidence region comprising Uttar Pradesh, Bihar, Orissa, West Bengal and Assam. The cancer is twice more common in women and is the leading cancer among digestive cancers in women in the northern Indian cities of Delhi and Bhopal (Dhir and Mohandas, 1999).

In the present study epidemiological study of Gall bladder cancer cases has been evaluated to know the causative factors.

Materials and Method

A prospective case – control study was carried out at Mahavir Cancer Institute and Research Centre, Patna, India of year 2008 only. 1038 gall bladder cancer cases were taken up for the demographic study. The cases were distributed district wise and especially the cities or towns located near the river basins of Bihar. The age, sex, gallstones in cases, stage of cancer *etc.* were also accounted for intensive study. The study was approved by the institute's ethics committee. Informed consent was obtained from the patients before inclusion in the study. Cases and controls were taken consecutively and worked up as per a predefined proforma designed to include study with patients from the Gangetic basin region or from non – gangetic basin region. The tissue sample was collected from the 50 Gall Bladder patients who were operated at Mahavir Cancer Sansthan, Patna from the Gangetic basin Zone and non- Gangetic basin Zone and were assayed through HPLC.

Results

Table 31.1 shows intensive study of Gall bladder Cancer cases. The study was carried out on 1538 gall bladder cancer cases. The female showed the highest degree of prevalence (73 per cent of cases) in compare to male (27 per cent cases). The most important factor of this study was with the age, as maximum number was accounted with patients above the age of 50. That is higher the age higher the risk of the cancer type. The stage of cancer also showed highest numbers in Stage- III and IV, that means patients are in advance stage coming for the diagnosis and treatment and also have very less awareness about their disease. The Gall stones in the study also showed 82 per cent of the cancer cases. While the Gangetic vs non-gangetic cases 63 per cent of gall bladder cases were accounted from Gangetic basin regions, that means patients coming from the cities located near the river Ganga are more prone to have this cancer.

Table 31.1: Showing the Details of the Epidemiological Study

	Total Patients	*Male Patients*	*Female Patients*
Number of Patients	1538	416	1122
Age (below 30)	63	20	38
Age (31–50)	628	116	510
Age (above 50)	846	280	574
Stage – I	66	18	48
Stage – II	205	56	149
Stage – III	482	130	352
Stage – IV	785	212	573
Gallstones	1261	340	921
Gangetic Zone	968	261	707
Non Gangetic Zone	570	154	416

Table 31.2: The Level of Pesticide (DDT) in the Gall Bladder Tissue (WHO DDT permissible limit – 2 ppb)

Type of Cancer	*DDT (ppb) in Tissue*
Gall Bladder Cancer (Gangetic Zone)	183
Gall Bladder Cancer (Non-Gangetic Zone)	57

Table 31.2, denotes the level pesticides in the tissue of the Gall Bladder cancer patients. The level much higher than the permissible limit.

Discussion

Gall bladder cancer is a lethal disease with a wide geographical, ethnic, and cultural variation. It is a common malignancy among women in India. The aetiology of gallbladder carcinoma is still obscure and although numerous factors have been implicated, none has stood the test of time (Kumar *et al.*, 2006). In the hospital based study of 1538 patients it was found that most of the cancer cases were in the female patients with stage wise cases showing high prevalence of cancer cases from Stage III and Stage – IV. The higher the age, the higher the cancer risk as above the age of 50 maximum cases were reported. Gall stones were also found in 3/4th of the total cases. Maximum cases were reported from the Gangetic basin Zones *i.e.* from the cities located near the River Ganga. To correlate if there are some chemicals which are causing this type of specific cancer in the people, it was found that DDT was responsible for this particular type of cancer as elevated levels of DDT was found in the Gall bladder tissue. The study directly throws light that cause of this type of cancer is due to the pesticides which are reaching the human body through various food chains leading to mutations in the DNA and finally leading to cancer due to disturbed metabolic activity.

Several studies have been carried out by various scientists of the world correlating this type of cancer (Khan *et al.*, 1999, Shukla *et al.*, 1998, 2001 and 2008, Arundhati *et al.*, 2004, Dutta *et al.*, 2005 and Baez *et al.*, 2010)

Thus, from the entire study it can be concluded that there is high prevalence of gall bladder cancer cases in the Gangetic Zone of Bihar due to the pesticide contamination in either drinking water or food which is leading to such type of cancer cases.

Acknowledgement

The authors are thankful to Mahavir Cancer Sansthan for providing all the necessary infrastructure required for this particular study.

References

Arundhati, R., Mohapatra SC, Shukla HS: A review of association of dietary factors in gallbladder cancer. Indian Journal of Cancer, 2004 (4): 41.

Báez S, Tsuchiya Y, Calvo A, Pruyas M, Nakamura K, Kiyohara C, Oyama M, Yamamoto M. Genetic variants involved in gallstone formation and capsaicin metabolism, and the risk of gallbladder cancer in Chilean women. World J Gastroenterol 2010; 16(3): 372-378.

Dhir V and Mohandas KM. 1999. Epidemiology of Digestive Tract Cancers in India IV. Gall bladder and pancreas. Indian J Gastroenterol; 18(1):24-28.

Diehl A. Epidemiology of gallbladder cancer: A synthesis of recent data. J Natl Cancer Int 1980; 65:1209-1214.

Dutta, U; Nagi, B; Garg, P K; Sinha, S K; Singh, K; Tandon, R K. Patients with gallstones develop gallbladder cancer at an earlier age. European Journal of Cancer Prevention: 2005 – (14) 4:381-385.

Khan ZR, Neugut AI, Ahsan H,: Risk factors for biliary tract cancers. Am J Gastroenterol 1999; 94:149-52.

Kumar JR, Tewari M, Rai A, Sinha R, Mohapatra SC, Shukla HS. An objective assessment of demography of gallbladder cancer. J Surg Oncol. 2006 Jun 15; 93(8):610-4.

National Cancer Registry Programme: Consolidated report of the population based cancer registries 1990-1996. Incidence and distribution of cancer. New Delhi: Indian Council of Medical Research, ICMR; 2001. p. 52-3.

Pandey M. Risk factors for gallbladder cancer: a reappraisal. Eur J Cancer Prev. 2003; 12(1):15-24.

Shukla V K, Chauhan V S, Mishra R N, Basu S: Lifestyle, reproductive factors and risk of gallbladder cancer Singapore Med J 2008; 49 (11): 912.

Shukla, V K, A Prakash, B D Tripathi, D C S Reddy, and S Singh. Biliary heavy metal concentrations in carcinoma of the gall bladder: case-control study. BMJ 1998 (317): 1288-1289.

Shukla, V K; Rastogi, A N; Adukia, T K; Raizada, R B; Reddy, D C S; Singh, S, Organochlorine pesticides in carcinoma of the gallbladder: a case-control study. European Journal of Cancer Prevention: 2001 (10) 2:153-156.

2013, Sustainable Approaches for Environmental Conservation *Pages* ***267–272***
Editors: **D.R. Khanna, A.K. Chopra, R. Bhutiani, Gagan Matta & Vikas Singh**
Published by: **BIOTECH BOOKS, NEW DELHI**

Chapter 32

Evaluation of Energy Consumption Approach among the Countries

G.C. Mishra

*Department of Civil Engineering,
Lingaya's University, Faridabad, Haryana*

Correlation of development and energy consumption has existed with the advent of civilization. Many attempts have been made in that direction and the most obvious of them being the use of Gross Domestic Product (GDP) per capita as an indicator. The shortcomings of such approach are well known and for this reason the HDI (Human Development Index) has been conceived as a composite of longevity, knowledge and standard of living. The existing energy consumption patterns are both physically and socially unsustainable. While the industrialized world faces sweeping energy transitions imposed by an impending decline of petroleum production, much of the non-industrialized world already face signicant energy shortages. It seems unlikely that any single source will succeed in claiming a market share comparable to that currently owned by petroleum, let alone the collective fossil fuels. The transitions can be expected to be no more uniform than the current energy consumption patterns and probably less so.

A correlation is established between the UN Human Development Index and per capita energy consumption for world nations highlighting the case of relationship of these parameters among various countries with special emphasis on Asian countries. A strong relationship between index values and energy consumption is observed for the majority of the world. Additionally, a distinct secondary trend emerges from the dataset, representing heavy energy exporters such as Organization of the Petroleum Exporting Countries and some Former Soviet Union nations, among others. The preliminary observation is made that these two trends closely resemble saturation curves, exhibited by a variety of natural phenomena. For the primary trend, three regions are isolated: a steep rise in human development relative to energy consumption for energy-poor nations; a moderate rise for transitioning nations; and essentially no rise in human development for energy-advantaged nations, consuming large amounts of modern energy. Therefore, it is the need of the hour to re-think of current pattern of development which widens the gap between rich and poor by exploiting natural resources in un-sustainable manner. The best approach for achieving sustainable development among countries of the world will be to have rational approach of energy distribution.

Keywords: Human development index, United Nations, Poverty, Energy, Economy

Introduction

Human Development Index

The UN Human Development Index (HDI) is a comparative measure of poverty, literacy, education, life expectancy, childbirth and other factors for countries worldwide. It is a standard means of measuring well-being. The index was developed in 1990 by the Pakistani economist Mahbub ul Haq and Amartya Sen which has been used since 1993 by the United Nations Development Programme in its annual report. The HDI measures the average achievements in a country in three basic dimensions of human development:

1. A long and healthy life, as measured by life expectancy at birth.
2. Knowledge, as measured by the adult literacy rate (with two-thirds weight) and the combined primary, secondary and tertiary gross enrolment ratio (with one-third weight).
3. A decent standard of living, as measured by gross domestic product (GDP) per capita at purchasing power parity (PPP) in USD.

Energy an Imperative for Economic Development

Access to modern energy services is fundamental fulfilling basic social needs, driving economic growth and fueling human development. This is because energy services have an effect on productivity, health, education, safe water and communication services. Modern services such as electricity, natural gas, modern cooking fuel and mechanical power are necessary for improved health and education, better access to information and agricultural productivity. There are wide variations between energy consumption of developed and developing countries and between the rich and poor within countries, with attendant variations in human development.

Energy is a prime mover of economic growth and development. This is critically important for the developing countries like India where economic development is on the rise. Simultaneously providing adequate and equitable access to basic amenities and services is the immediate priority of the policymakers of the all developing countries. Energy will also be required to meet the targets set up by these countries under the MDGs (Millennium Development Goals) adopted at the UN Millennium Summit held in Johannesburg in September, 2000 for improving the condition of the world's poorest by 2015. Therefore, only economic development can provide a lasting solution to address the problems of the country (ADB, 2007).

Energy Consumption in 2010: Over 5 per cent Growths

Energy markets have combined crisis recovery and strong industry dynamism. Energy consumption in the G20 soared by more than 5 per cent in 2010, after the slight decrease of 2009. This strong increase is the result of two converging trends. On the one-hand, industrialized countries, which experienced sharp decrease in energy demand in 2009, recovered firmly in 2010, almost coming back to historical trends. Oil, gas, coal and electricity markets followed the same trend. On the other hand, China and India, which showed no signs of slowing down in 2009, continued their intense demand for all forms of energy.

Objective of this paper is to attempt and examine the linkages between energy services and human development in developing and developed countries. It does so by comparing modern energy use in developed and developing countries and argues that a threshold of modern energy is required to achieve growth and improvement in human development.

Country with an electric consumption below 4,000 kWh has an HDI above 0.9 and, barring four cases (South Africa, Saudi Arabia, Russia and South Korea), all countries consuming more than this value per capita have an HDI greater than 0.9. This observation is then used to assess the consequences of growth scenarios assuming that, over the 2000 – 2020 time frame, an increasing number of countries reach or exceed the 4,000 kWh thresholds. This assumption is combined with the universally accepted assumption that world population will keep growing over the same period.

Results and Discussion

Energy and Development

The Figure 32.1 displays the relationship between 'Human Development Index' and 'Per Capita Energy Consumption'. "No country in modern times has substantially reduced poverty without a massive increase in its use of energy and/or a shift to efficient energy sources". Conversely, this diagram indicates that it is possible for developed countries to reduce energy consumption and use energy more efficiently without impacting on quality of life.

It is apparent from this figure that, for an energy consumption above 1 ton of oil equivalent (toe)/ capita per year, the value of HDI is higher than 0.8 and essentially constant for all countries. One toe/ capita/year" seems, therefore, the minimum energy needed to guarantee an acceptable level of living as measured by the HDI, despite many variations of consumption patterns and lifestyles across countries. Figure 32.1 HDI versus annual primary energy consumption (commercial + non-commercial)

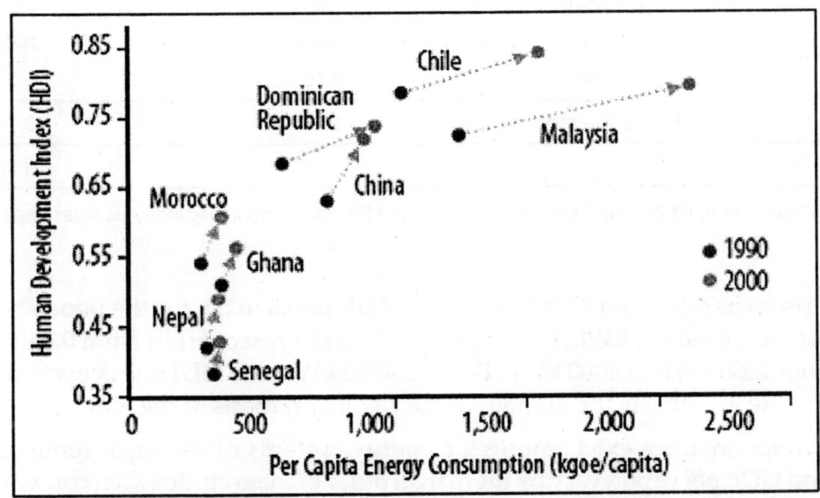

Figure 32.1: Relationship Between HDI and Energy Consumption
***Source*: International Energy Agency UNDP Analysis**

per capita. It also shows that the influence of per capita energy consumption on the HDI begins to decline somewhere between 1 and 3 toe per inhabitant. Thereafter, even with a tripling in energy consumption, HDI does not increase. For the poorest nations, the crisis is one of survival; of finding an effective method out of poverty and for this, energy is essential. For the majority of the world's nations, in the knee between the abyss of desperate poverty and the Olympus of multi-thousand kWh affluence, the energy crisis is a challenge to the stability of their societies and their efforts to arrive at a broadly equitable level of human development, HDI = 0.9, maintained efficiently at about 4000 kWh per capita.

Electricity and the HDI

The United Nations indicated that the electricity consumption per capita needed in order to experience a society with a medium level of human development was just over 1000 kilowatt-hours.

Table 32.1: Selected National HDI and kWh/c

HDI Rank	*Country*	*kWh/capita*	*HDI Value*
1	Norway	26,640	0.963
8	Ireland	6560	0.946
10	U.S.A.	13,456	0.944
11	Japan	8612	0.943
16	France	8123	0.938
20	Germany	6989	0.930
30	Barbados	3193	0.878
40	Qatar	17,489	0.849
52	Cuba	1395	0.817
62	Russian Federation	6062	0.795
85	China	1484	0.755
99	Iran	2075	0.736
108	Viet Nam	392	0.704
127	India	569	0.602
177	Niger	40	0.281
—	World Average	2465	0.741
—	—	(19)	(17)

For a selection of 60 countries, 84 per cent correlation between HDI and annual electricity use per capita in kilowatt-hours (kWh) was found.

This curve rises sharply from 0 kWh, through HDIs of 0.3 to 0.6 for the poorest countries of the world with tens to hundreds of kWh; then it goes through a knee of HDIs from 0.6 to 0.9 with energy consumption from 1000 kWh to 4000 kWh. Beyond 4000 kWh, the HDI rises very gradually from 0.9 toward its asymptote at 1; this is the domain of the wealthy nations of the world.

Important commonalities exist among the energy systems of rather different countries, since energy use (E) and GDP per capita vary by more than order of magnitude when comparing developing to industrialized countries, while energy intensity does not change by more than a factor of 2. In addition, as far developing countries are concerned, this probably reflects the fact that "modern

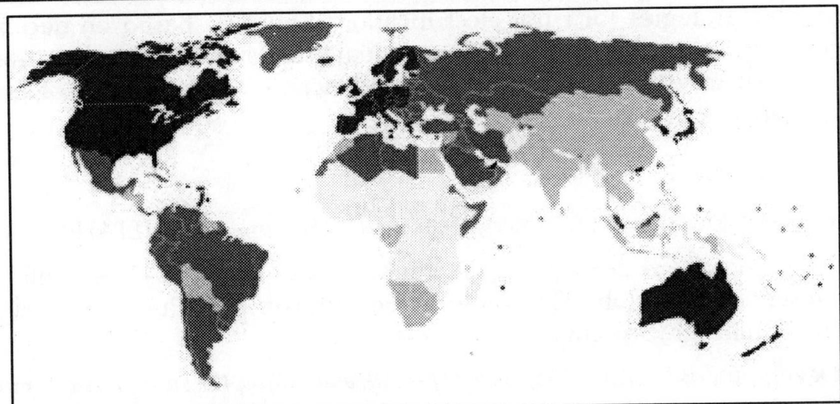

Very High (Developed Country)
High (Developing country)
Medium (Developing country)
Low (Developing country)
Data unavailable

Figure 32.2: World Map Indicating the Human Development Index Category by Country (2010)

sector" of the economy dominates both E and GDP, while the "traditional sector" contributes little to both.

Countries fall into four broad human development categories, each of which comprises 42 countries (except for the second category, comprising 43 countries). The divisions are:

Division	*Comprising*
Very high	42 countries
High	43 countries
Medium	42 countries
Low	42 countries

Conclusion

The advanced nations of the world, have the major responsibility for solving the world's energy crisis because they use the lion's share of the world's energy and because they have the infrastructure, resources and sophistication to both dramatically improve their efficiencies – the largest untapped source of energy today – and to invent new energy cycles that overcome the deficiencies of existing technology.

It is evident from the analysis that addressing the energy needs of developing countries requires a combination of strategies and actions, including the following:

- ☆ Firstly, developing country governments should commit to expanding access to modern energy services by making it a national development priority.

- Secondly, strategies for rural electrification should be based on decentralized power generation. Decentralization has the potential to assist with technology transfer, increased equity in distribution and consumption and increased participation of local people in the supply of energy services.

References

"International Energy Outlook 2005," *Energy Information Administration* (*EIA*) *USDOE*, (5 April 2006)

"The Outlook For Energy - A 2030 View," Exxon Mobil Corporation, 15 September 2004, (5 April 2006), "Presentation by John Constable, senior advisor, UK Public Affairs, at the Scottish Parliament, Edinburgh, Scotland."

ADB (Asian Development Bank). 2007, *Energy for All: addressing the energy, environment, and poverty nexus in Asia*, Manila: ADB.

Census of India. 2001, *Final population: 2001 Census*, New Delhi: Office of the Registrar General.

Electricity Consumption Per Capita (UNDR 2005), *United Nations Development Programme*, (22 March 2006)

Global Energy Futures and Human Development: A Framework for Analysis, October, 2000. World Energy Outlook 2005 [executive summary], *International Energy Agency*, (5 April 2006).

MoF (Ministry of Finance). 2007, *Economic Survey 2006/07*, New Delhi: MoF, government of India.

Pasternak, Alan D., 2006. "Global Energy Futures and Human Development: A Framework for Analysis," UCRL-ID-140773, *Lawrence Livermore National Laboratory* (USDOE), October 2000.

Planning Commission. 2006a, *Integrated Energy Policy – report of the expert committee*, New Delhi: planning Commission, Government of India.

Price, David, 2006. "Energy and Human Evolution," from *Population and Environment: A Journal of Interdisciplinary Studies*, Volume 16, Number 4, March 1995, pp. 301-19.

Tainter, Joseph A., 2006. "Complexity, Problem Solving and Sustainable Societies," from *Getting Down To Earth: Practical Applications of Ecological Economics*, 1996, Island Press, ISBN 1-55963-503-7; [excerpt].

2013, Sustainable Approaches for Environmental Conservation *Pages* **273–280**
Editors: **D.R. Khanna, A.K. Chopra, R. Bhutiani, Gagan Matta & Vikas Singh**
Published by: **BIOTECH BOOKS, NEW DELHI**

Chapter 33

Child Led Disaster Risk Reduction and Climate Change Intervention in Coastal Andhra Pradesh

Meda Gurudutt Prasad

CADME (Coastal Area Disaster Mitigation Efforts)

1002 students died in the 2001 Bhuj Earthquake. 1884 school buildings- collapsed loss of 5950 classrooms, 11761 school buildings suffered major to minor damages. Intervention on child led disaster risk reduction has been initiated in 2008 and it is still going on in all the schools and communities. In 1996 cyclone, as many as 217 children, 62 aged people, 8 pregnant women, 28 physically challenged persons were died at Magasanitippa Village which is located nearby to sea, because of lack of preparedness measures especially lack of pre planned evacuation plan. CADME/ ACTION has initiated Community Based Disaster Preparedness Program at vulnerable villages and child led disaster risk reduction at vulnerable schools of east Godavari district in the year 2008 and as a result of these interventions as many as 30 taskforce groups at community level and 25 child led taskforce groups at vulnerable schools level have been formed.

Keywords: *Disaster, Climate change, Cyclones, Safety.*

Background

In 1977 at Divi Seema cyclone innumerable children died and it was a tragic event in the history of disasters. The family members who were able to swim kissed their family children saying goodbye and left them in the water abruptly as the children are obstacles to swim to survive their lives. Mr.Veerayya who is the resident of Ramakhandam village revealed his situation in that disaster. He also left his 5 years male child named Sambhasivarao in the water and started swimming to save his life at least!! After moving two yards in the water he turned his head to have last look on his son. Sambhasivarao was drowning in the water moving his hands and the father also saw a thatched roof floating on the water at the same time. He went back to his son and lifted his son's hair and put him on the thatched roof along with him.

They floated in the water for 6 hours and find a tree in the water. They (both) climbed the tree and stayed at a tree branch for 15 hours. After receding the water flow, they got down and walked for 25 kilometers to reach the cyclone shelter. This live story is still available in Ramakhandam village near machilipatnem Krushna district. The son name Mr. Sambhasivarao has been renamed it as 'Mruthyunjayarao" (Conqueror of death in English) after this incident.

1002 students died in the 2001 Bhuj Earthquake. 1884 school buildings- collapsed loss of 5950 classrooms, 11761 school buildings suffered major to minor damages.

Kumbakonam fire tragedy .... A deadly fire raged through Lord Krishna School killed 93 children, all below the age of 11 years. Let us learn lessons from earlier tragedies and make our schools a safer place for children.

Intervention on child led disaster risk reduction has been initiated in 2008 and it is still going on in all the schools and communities.

The intervention has been initiated in 25 vulnerable schools and 25 communities in East Godavari district of Andhra Pradesh India. East Godavari District, one of the nine coastal districts of Andhra Pradesh, is a regular victim of multiple disasters and this district was badly affected by all major natural disasters time to time since independent era. Fishing community is the main victim of all disasters and all the families dwell nearby to "Bay Of Bengal" at one or two kilometers distance from the sea. Their main occupation is fishing and it is a routine responsibility of a fisherman for livelihood that goes to fishing at early hours and comes back to home afternoon at 2 pm/3 pm. They do not have RCC buildings which can resist the cyclone wind velocity and live in thatched houses. These conditions make them more vulnerable to natural disasters.

Preparedness measures at community level are nil. And vulnerable communities *i.e.* Children, Aged, Women, Physically Challenged, Mentally Retorted, Pregnant Women and Infants mostly the worst victims of any disaster.

In 1996 cyclone, as many as 217 children, 62 aged people, 8 pregnant women, 28 physically challenged persons were died at Magasanitippa Village which is located nearby to sea, because of lack of preparedness measures especially lack of pre planned evacuation plan.

Who all the actors involved (accurate rendering of people, institutions, *etc.*)?

- ☆ 28,412 children and 36,116 adults from 25 disaster prone villages and 25 vulnerable schools.
- ☆ District Education Department
- ☆ District Fire department
- ☆ Mandal Revenue Office
- ☆ Mandal Development Office
- ☆ District Fisheries Office
- ☆ 150 Teachers from 25 vulnerable Schools
- ☆ State disaster management authority
- ☆ District women and child welfare department

Figure 33.1: Children Practicing to Climb the Rope Bridge to Rescue

Figure 33.2: School Going Child Explaining the Disaster Cycle

Why is this Good Practice?

- ☆ The project is a model to all the vulnerable communities in the country and children studying in vulnerable schools.
- ☆ Capacities of vulnerable communities are increased and confident of combating the disaster situations.
- ☆ Dependency on outside people, like Police, Medical teams, Military for rescue and Medical aid has been reduced drastically.
- ☆ Relief measures to be taken up in the event of disaster are taken care of by Taskforce groups, not depending on outside help.
- ☆ Able Bodied persons in vulnerable villages have specific responsibilities and actions in pre-during-post disaster situations.
- ☆ Women are very active and 50 per cent of women are involved in each taskforce team.

What have been the Key Success/Failure Factors of this Initiative?

CADME/ACTION has initiated Community Based Disaster Preparedness Program at vulnerable villages and child led disaster risk reduction at vulnerable schools of east Godavari district in the year 2008 and as a result of these interventions as many as 30 taskforce groups at community level and 25 child led taskforce groups at vulnerable schools level have been formed.

- ☆ All the 30 taskforce groups at village level are capacitated to combat the disaster situations (pre, during and post) with specific roles and responsibilities in each situation.
- ☆ Perfect planning is done before any disaster, predicted and warning given by metrological department.
- ☆ Disaster drills are organized at regular intervals to familiarize them with their taken tasks and responsibilities.
- ☆ Contingency plans are updated every two months.
- ☆ Local bodies of each village have taken into consideration of the work done by task force group at their respective villages.
- ☆ Contingency plans developed by taskforce groups have been approved at local, mandal and district levels for spontaneous actions in collaboration with government officials for quick response.
- ☆ A joint taskforce committee consists of local task force and government officials at panchayat level, mandal level and district level have been formed to address the problems and to make spontaneous actions in the event of disaster. Selective government officials based on their interest and commitment have also undergone for Disaster Preparedness trainings conducted by CADME/ACTION.
- ☆ Five important teams are formed:
  1. Emergency medical care team
  2. Emergency rescue team
  3. Relief camp management team
  4. Warning and evacuation team
  5. Water and sanitation team

What are the Innovative Elements and Results?

Innovative Aspects

- ✩ Horizontal trainings by trained taskforce members in other vulnerable villages are conducted on their own initiation.
- ✩ School safety maps are developed by trained children in other vulnerable schools in consultation with the respective children and children in other schools are trained by trained children.

Cost Effectiveness

- ✩ There is no need to purchase anything form outside to implement this initiative at vulnerable villages.
- ✩ Task force groups are well versed through the training and how to prepare the floating raft, using the local material like, gunny bags, damaged wooden flanks, waste glass bottles, and torned saris and so on.
- ✩ They are also trained to use waste cloth to be used as bandages for chin, knee head injuries and chest arm and leg fractures.
- ✩ They also construct a floating aid which can rescue two drowning persons with two plastic pots and a rope.
- ✩ The entire training can be organized with very less expenditure and amount that we are investing on this, will give a great result in their community.

What have been the City/Local Government Contribution to Reducing Disaster Risks and Vulnerabilities?

Local Panchayaths of 30 villages have been extending hundred percent cooperation in all aspects. *i.e.* negotiating with all government line departments, financial assistance to provide the food to resource persons, resolutions on involvement of local people in disaster times and so on.

School based DRR

Disaster Preparedness Club in Schools

Before we started our work we got the formal permission from the Education Officer in the District.

We started disaster preparedness clubs in 15 schools in the project villages. Each Club comprises of 100 children from 7-10 standard. Regular classes on DRR are conducted by the Cluster Coordinators outside the school hours (4.30 – 5.30 pm twice a week). The selection of children and leaders among them was done in participatory method involving children.

We have been using a training manual for the training purposes. The trainings in each school are tailor made to address vulnerability of that particular school. The training focuses on peace time issues or emergencies like harms of pesticides and how to protect from it, types of fire and their response, responding to any accidents or children falling *etc.*

To help in better information transaction we have placed a drop in box in each school so that students or club members are using it to ask doubts or clarification and to give their suggestions. These children are an asset to the village community too as they have basic knowledge on disaster preparedness. The emphasis is on clarity on concepts, role of each member in the group and on horizontal training or information dissemination to children in school and in their villages. The clubs are now involved in risk mapping keeping in mind, the both natural and man made hazards

Importance is given on the need for horizontal transfer of knowledge so as to cover many children. To reach out to other children in the school, the trained children perform disaster drills in their own schools every quarter and also during the school assembly or any special school programs, on disaster preparedness. The trained children with the support of the teachers organise training to other students to organize disaster preparedness drills and also help them in increasing their knowledge on disasters by organizing quiz and essay writing competitions.

Child protection in the schools to build awareness on child protection issues is done on regular intervals among the club members.

Formation of Cultural Troupes

Cultural troupes are formed in schools with 3 children from each school trained intensively. The cultural troupe consisting of club members use songs, skits and stories to effectively disseminate information in schools and nearby villages.

Trainings for Teachers from 100 Schools

One of the keys to success of school level program is trained and motivated set of teachers in each school. 200 teachers 100 schools have undergone for CLDRR and child rights and child participation.

Communmty DRR

Training of Trainers

75 people from 25 villages have undergone for training on CLDRR and have been supporting the staff at the village level for training the task force groups.

Module for Trainings

- ☆ Development and Disaster preparedness
- ☆ Disaster history of each vulnerable village
- ☆ Emergency medical care
- ☆ Emergency rescue
- ☆ Relief camp management
- ☆ Contingency plan development
- ☆ Training skills
- ☆ Children's participation and Child protection

Task Force Groups

The task force groups are formed in 25 villages. Each group consists of 50 people *i.e.* 20 children and 30 adults. The groups are now started activities in their respective villages like vulnerability mapping of the village, regular skill trainings in groups.

Grain Banks

As part of CLDRR program we have introduced Grain bank program in all the project villages. During cyclones, floods and other disasters people lose assets and livelihood options and need to stay without food for several days. This was seen widely during the recent floods in the project villages. The village community during the last year has been discussing a plan like this which will help them to take care of themselves during the initial days of disasters. Keeping this point in view grain banks have been established in all the 25 villages.

Advocacy

1. *Advocacy and media training for children*: 30 children underwent for media training and these children who are part of the training program are playing an active role in advocacy efforts.
2. *Consultation meeting with I/NGOs at the state level*: In order to influence policies and practices of the Government and to build constituency to have a shared understanding on the inclusion of children's needs and concerns in disaster management policies, trained children have organised a consultation meeting with I/NGOs at the state level. The participants were primarily be agencies involved in emergency response and Government officials. The consultation the participants to have a shared understanding on the inclusion of children needs and concerns into the state disaster management plans.

Workshop with Education Department Officials

Workshops with the department and regular one to one meetings have been organised at regular intervals in getting the officials to understand and support the need for inclusion of disaster preparedness in school curriculum and for disaster preparedness clubs in schools.

Lessons Learned

- ✰ It is very important to be flexible and pay patience in under to survive and grow the structures (Task groups) developed at community and school levels.
- ✰ Admit to mistakes and correct them.
- ✰ Let the Taskforce take their own decisions in the event disaster whether it is right or wrong.
- ✰ Be flexible with Taskforce members and learn what they say.

Challenges

- ✰ Initially there was no response from the government as well as from the local communities.
- ✰ Having approved the contingency plans and school safety maps has been a challenge.
- ✰ Motivating the community and children towards the intervention is another challenge.
- ✰ Owning of the intervention by the community and children is another challenge.

How to Improve Similar Initiatives in the Future?

The similar intervention can be implemented in other villages improving the above mentioned challenges involving the trained children and Task force groups of this project. District officials and other stakeholders of disaster management are quite impressed of this intervention and started replicating it in other vulnerable villages and schools. Hence the above mentioned challenges are now opportunities to future initiatives.

How Easy Would it be to Replicate this Practice Elsewhere?

As mentioned above this intervention can be replicated any part of the world at vulnerable village or school based on the type of disaster but same methodology can be adopted for any type of disaster.

How Could the Practice be Replicated in a Different Context?

This can be replicated to any context either for Tsunami prone or cyclone prone or flood prone or flash flood prone or fire accident zone or in a peace time.

What Would be the Economic and Political Constraints to Scaling-Up?

There is no political constraint to implement the project but financial constraints are there. As the governments do not have funds to initiate this kind of interventions, it inevitable to search for funding elsewhere to scale up the project.

2013, Sustainable Approaches for Environmental Conservation *Pages* **281–289**
Editors: **D.R. Khanna, A.K. Chopra, R. Bhutiani, Gagan Matta & Vikas Singh**
Published by: **BIOTECH BOOKS, NEW DELHI**

Chapter 34

Planning on Health Management Strategies for Handling Natural Disasters on Mobile

Arpit Bhatnagar and Neha Dang
Tapovan Enclave, Dehradun, Uttarakhand

Natural disasters are often frightening and difficult for us to predict, because we have no control over when and where they happen. What we can control is how prepared we are as communities and governments to deal with the health related issues that natural disasters bring Places that are more likely to have natural disasters, such as the earthquake-prone Pacific Ring of Fire, or coastal areas vulnerable to hurricanes, require accurate methods of predicting disasters and warning the public quickly. Once the people have been informed, evacuation routes must be provided so that they can all leave quickly and safely, even if they travel by foot. Emergency warnings and evacuation plans are not enough, though. Since the use of smart phones is gaining interest in people, the disaster management system was implemented as a smart phone application using Google's Android operating system. The disaster management system Android application known as MyDisasterDroid determines the optimum route along different geographical locations that the volunteers and rescuers need to take in order to serve the most number of people and provide maximum coverage of the area in the shortest possible time. Along with all this it also includes rescue operations,first aids,managing certain conditions in such emergency situations, mapping to nearest healthcare facilities and disposal of bodies in case of death.

Keywords: *Evacuation plan, Emergency warnings, Android operating system, MyDisasterDroid.*

Introduction

Natural disasters have become so commonplace that they hardly receive passing notice on the news unless there have been a large number of casualties. Volcanoes, mudslides, tsunamis and floods are just a few of the ways nature strikes on a daily basis, leaving behind destruction and heartache. Humans have learned to prepare for the possibility of tornadoes, earthquakes, hurricanes and wildfires, but no amount of preparation can lessen the impact that natural disasters have on every aspect of society. The ill effects of disasters are well known and they lead to:

Physical Destruction

The biggest visible effect of natural disasters is the physical ruin they leave behind. Homes, vehicles and personal possessions are often destroyed within a short period of time, leaving families homeless and shutting some businesses down permanently. Tornadoes destroy structures at whim, earthquakes can cause structural damage that might not be apparent at first glance and tsunamis and floods sweep homes off their foundations.

Emotional Toll

Possessions are not hard to replace, as many people keep insurance on their property and tangible goods. The emotional toll of natural disasters is much more devastating. The death of a loved one may be the worst-case scenario but it's not the only lasting emotional effect victims experience. Whole communities may be displaced, separating friends and neighbors; victims face anxiety and depression as they wonder if it could happen again. In extreme cases, they may experience post traumatic stress disorder (PTSD).

Economic Concerns

According to the National Hurricane Center, Hurricane Katrina cost the U.S. $75 billion when it slammed into New Orleans in August 2005. That doesn't include the damage caused in the Florida counties of Miami-Dade and Broward by the same storm. While destruction of this magnitude is not commonplace, even a minor storm can cause considerable damage. At the very least, the local economy must be able to absorb the cost of cleanup and repairs.

Indirect Effects

While the visible effects of natural disasters are immediate and strongly felt, communities that surround ground zero can be indirectly affected by them as well. Natural disasters almost always lead to a disruption in utility services around the area impacted. This can mean life or death for those who rely on dialysis or oxygen to live. Medical assistance is also often slowed, as emergency crews must focus on the victims of the disaster. Banks and other businesses might be closed, affecting a family's ability to withdraw money to pay bills and buy groceries.

The most efficient and effective disaster preparedness systems and capabilities for post-disaster response are usually provided through volunteer contributions and local authority actions at the neighborhood level. These can operate independently, irrespective of reduced, damaged or destroyed infrastructure or capacity elsewhere. Specific actions are also required at the appropriate levels of government, including local authorities, in partnership with the private sector and in close coordination with all community groups, to put into place disaster preparedness and response capacities that are coordinated in their planning but flexible in their implementation. The reduction of vulnerability, as well as the capacity to respond, to disasters is directly related to the degree of decentralized access to information, communication and decision-making and the control of resources. National and international cooperation networks can facilitate rapid access to specialist expertise, which can help to build capacities for disaster reduction, to provide early warning of impending disasters and to mitigate their effects. Women and children are most affected in situations of disaster and their needs should be considered at all stages of disaster management.

Disaster Management

The process of disaster management involves four phases: mitigation, preparedness, response, and recovery. The mitigation phase is the attempt to reduce disaster risks by focusing on long-term measures of eliminating disasters. The preparedness phase is the development of an action plan for an upcoming disaster. The response phase is the mobilization of services and relief when disaster strikes and the recovery phase is the restoration of the affected area to its previous state, the nature of communications need differs at different phases of disasters: early warning, disaster impact, immediate aftermath and recovery. Each of these stages is best served by a different set of methods of communication, and in each of them mobile is of course joining an existing menu of information channels. Mobiles are therefore serving different needs in each case, which is one of the important lessons to be learnt from the examples given here. The distinctive contribution of mobile lies in its unique capacity to disseminate information quickly and informally from individual to individual, which has greater value during some phases of a disaster rather than others.

Role of Technology in Disaster Management

In many countries, as the annual World Disasters Report from the International Federation of Red Cross and Red Crescent Societies makes clear, there are effective warning systems in place which do not depend on high technology at all: for example, hurricane warnings in the Caribbean disseminate meteorological warnings via radio broadcasts, but more importantly by drills taught each year in schools, so that people understand the warnings they are given, and by community-based teams who watch out for local flooding and travel from street to street shouting warnings by megaphone. Improving the effectiveness of early warnings is unlikely to ride on one single technology or communications channel, but rather will involve a mixture of prior surveillance, public education, disaster-preparedness, and a range of communications technologies including the broadcast media and the internet as well as mobiles. On the impact of a disaster, in the chaotic hours that follow, emergency service communications will always be the top priority. In addition, the human value of mobile phones lies in people's desire to contact their family and friends as soon as possible.

Using Wireless Mobile Technologies

More than two billion in the world are mobile users, because of the wide coverage of the mobile phone network, it may be used in the implementation of a disaster management system making it available in everyone's pocket. A study in Bangladesh established that wireless mobile technologies can be used in disaster information management. Results showed that mobile technology may be used to disseminate pre-disaster warnings and post-disaster announcements, to receive information about relief needs, and to exchange information about health hazard. Also, in disaster information management, geographic locations of those in need are important. Using their mobile phones, their locations can either be determined using the mobile network system or through the use of an integrated Global Positioning System (GPS) included in their phone.

Public information campaigns should emphasize the role mobile phones can play in helping recover from a disaster. In the United States, for example, both the Federal Communications Commission and the local operators have posted consumer advisories, telling customers to ensure their handset batteries are charged ahead of an emergency, to have a back-up battery, to keep their phones dry, and to expect the network to be busy in the aftermath of an event such as a hurricane.

Mobile phones tend to play a supplementary role in early warning systems, where prior surveillance, public education and a range of news media, such as broadcast television, radio and the Internet are generally the best way to prepare people for an impending disaster. Mobile phones are not currently an efficient way to broadcast information to large numbers of people, but they can be a useful mechanism for individuals to relay that information on to friends and family who may have missed the initial broadcast. In many countries, there are low-tech, but effective warning systems in place. For example, in the Caribbean, people are warned of approaching hurricanes via radio broadcasts, backed up by annual drills taught each year in schools so that people understand the warnings they are given.

The immediate aftermath of a disaster the contribution of mobile is substantial thanks to the speed with which cellular networks can generally recover from damage, usually within hours or at most a few days. It is typically much easier to repair a wireless base station than hundreds of fixed-line connections. This was strikingly demonstrated in the speed of restoration of mobile services to customers in the Gulf region of the United States after Hurricane Katrina. New mobile networks can be set up relatively quickly in places where there was either no network to begin with or the original network was damaged.

During these chaotic situations, mobile phones can help in the process of recovery. They uniquely give affected people and aid agencies the means to find and receive information specific to their needs. After the Tsunami, for example, a Sri Lankan television employee sent out text messages containing 'on-the-ground' assessments of what was needed and where. The experience of many people in the hours and days after the Tsunami demonstrates that decentralized communication is vital for the efficient matching of resources with needs, and reconnecting dispersed people.

Aid agencies are finding text messages and mobile calls are an increasingly important means of fund-raising, giving them rapid access to funds donated by members of the public. However, some countries impose taxes on these calls and text messages. Governments should review this policy.

Mobile use has spread rapidly, especially in some developing countries. Some poor countries are very vulnerable to natural disasters, and many are badly affected when disaster does strike because their infrastructure and emergency response capabilities are often weak. This report looks at the contribution of mobiles using network data and other evidence. The case studies included here show that the timely spread of reliable information is a vital part of the response to and recovery from a disaster. The nature of communications needs differs at different stages of a disaster, however. The contribution of mobile (and other forms of communication) varies correspondingly. Policy debates should recognize these distinctions.

Current Use of Technology in Disaster Management

UNICEF–Utilizing the Power of SMS

Children's agency UNICEF (United Nations Children's Fund) uses SMS technology to raise funds for disaster response and relief and is a strong advocate of the role of mobile services in emergency situations. "SMS is an important tool to help emergency response, and it's going to become increasingly important, "says Per Stenbeck, International Fundraising Director at UNICEF."It's a definite complement to traditional fundraising techniques and provides a potentially immediate method of donation." The success of SMS technology in fundraising is proven. 27 million EURO was donated in just a couple of weeks in Italy following the December 2004 tsunami emergency.SMS fundraising initiatives have now been launched in most Western European countries. Such services have raised millions of Euros for emergency relief, aiding victims of the Asian tsunami and the Iraq conflict, amongst many others. UNICEF believes that the use of SMS technology has given it the opportunity to increase levels of fundraising from certain segments of the market, for example, the youth sector. In future, the agency is also hopeful that donators will be able to pledge varying amounts, rather than the current fixed donation of 1 EURO per SMS. Despite the advantages, that SMS donations are treated as standard text messages by tax authorities, meaning a percentage of the donation can be lost as tax."Some countries have waived VAT on SMS donations but not all. The amount of VAT charged also varies from country to country."

Ericsson–Responding to Disaster

Established in 1999, Ericsson Response is a global initiative aimed at developing a better and faster response to human suffering caused by disaster. "Mobile technology can definitely help save lives," says Dag Nielsen, Director, Ericsson Response. "We are giving tools so that the humanitarian relief effort can be improved. If you have better tools, the whole relief effort can be enhanced." In partnership with the United Nations World Food Programme, the Office for the Coordination of Humanitarian Affairs and the International Federation of Red Cross and Red Crescent Societies, Ericsson Response provides mobile equipment, as well as people to implement and operate it, to areas affected by both natural disasters and complex disasters. Ericsson Response' services were first called into action following earthquakes in El Salvador and India in January 2001. Since then the initiative has been involved in all major natural and man-made disasters. "We work with the humanitarian organizations in each affected country to help them with their relief efforts, "comments Nielsen."We're helping to bring the power of mobile technology and resources directly to the service of humanity, and, in particular, humanitarian relief operations. The future development of new mobile services will also help speed up the process of recovery."

The Red Cross–Mobilizing the Power of Humanity

As the world's largest humanitarian organization, the International Federation of Red Cross and Red Cross Movement is well placed to highlight the beneficial effects mobile technology can have on areas affected by natural disasters. The Federation has partnered with Ericsson for the provision of mobile phone technology to assist in disaster response and disaster preparation."For major disasters we mobilize staff and volunteers to fly into disaster zones and set up IT and telecommunications services, "says Hugh Peterken, head of IT at the International Federation of Red Cross."We are closely involved with Ericsson Response. Ericsson provides a number of services to us, ranging from donation of mobile phones to providing people to help with training." Over the past few years, the partnership has provided communications equipment and services during operations in Afghanistan, Algeria, Belize, the Dominican Republic, El Salvador, India, Peru, Southern Africa, Tajikistan, Tanzania and Vietnam.

"Mobile technology can certainly help in the response phase," explains Peterken."We have used mobile phones to allow affected families to re-establish contact or confirm to relatives that they are safe. That is a very useful part of the response. The next part is recovery. In that phase, mobile technology is very much used as a business tool. Its value there is well known. There is a general efficiency gain from having people connected through mobile phones." As well as aiding in the response to natural disasters, Peterken adds that the Federation is well placed to help susceptible regions prepare for any likely future events."We are about to pre-position IT and telecoms emergency response kits around the world. We have identified the most at risk areas of the world in terms of disasters, and will target our communications stock for these regions. We know that if we have kit available we can get communications up and running."This programme is being funded by ECHO (European Commission Humanitarian Aid department).

To sum up, the importance of mobiles in both immediate aftermath and recovery phases lies in:

- ☆ The speed of network recovery;
- ☆ The inter-operability of mobile communications;
- ☆ The decentralized, person-to-person flow of information; and
- ☆ The accessibility of mobiles (especially when people are displaced from home).

Challenges in Using Wireless Technology

In the immediate aftermath of a disaster, it is vital for operators and governments to work together to restore mobile networks as quickly and effectively as possible. Mobile networks, like other important parts of the essential infrastructure, are vulnerable to being damaged, and will inescapably face very heavy demands in the immediate aftermath of a disaster, as documented here. But they are capable of much faster recovery than other means of communication.

Emergency response teams from the mobile industry play an important role in either installing a temporary network for as long as needed, or restoring the mobile network extremely quickly. Mobile operators all have emergency plans in place, including measures such as reducing call quality to accommodate higher call volumes as well as planning for repairs and emergency generators. It may also be possible for the public authorities to enhance the capacity of the network to accommodate increased demand after an emergency by providing additional spectrum.

In the recovery phase mobiles also make a completely distinctive contribution. In the case of developing countries in particular, mobiles are likely to be the dominant means of communication for affected members of the public. This flow of information – not mediated by broadcasting agencies or public authorities – ensures that people elsewhere quickly come to know what is happening and what help is needed. In the chaotic aftermath of a disaster, when people are displaced, buildings and infrastructure are destroyed, no central authority can possibly hold all the necessary information and allocate resources to the place of greatest need. Indeed, as the 2005 World Disasters Report firmly pointed out, there can be real drawbacks in outside agencies trying to set themselves up as repositories of information.

Smart Phones and Disaster Response

Based on research released by Smartphone Summit, smart phones account for 10 per cent of all cell phone sales and it is still growing and driving more interest among people. One of the reasons for its continued growth is that it provides information valuable to the users. Moreover, in times of disaster, the more people that have information with them all the time, the more they will be self-reliant allowing rescuers or responders to concentrate on those in the greatest need of help. There are also a number of mobile applications available in smart phones that are beneficial in disaster response. Among these are GPS technology, which can be used in the tracking of rescuers and resources, the translator, which can be used for communication, and the field examiner, which can be used to send information to headquarters for assessment of damages. Indeed, the use of a smart phone in a disaster management system is advantageous.

Android Mobile Development Environment

There are a number of mobile development environments in the market. One of which is Android created by the Open Handset Alliance. Android is an open and comprehensive platform for mobile devices. It is designed to be more open than other mobile operating systems so that developers, wireless operators, and handset manufacturers will be able to make new products faster and at a much lower cost. The end result will be a more personal and more flexible experience to the user]. For this reason, the mobile development environment was used in the implementation of the disaster management system. Android-based disaster management system named MyDisasterDroid (MDD) was implemented.

Geolocation

First, geographic locations of people in need are set. There are two ways that geographic locations can be set: using the application installed in MyDisasterDroid or sending location via text or short message service (SMS) to MyDisasterDroid. Geographic locations or geolocations are described in latitude and longitude. These calculated distances were then used in the genetic algorithm implementation in MyDisasterDroid.

There have been numerous reports and anecdotes about the uses of mobiles during and after each of the disasters in the news headlines since last year's tsunami. In this report we aim to make a more systematic assessment of the role of mobiles than was possible in the immediate aftermath of those events. The report uses network data and other evidence to try to understand how people used mobiles and what their communications needs were in such extreme circumstances. Our research on the impact of mobiles makes it clear that the nature of communications needs differs at different stages of events, which can be characterized as follows:

- Early warnings
- Disaster impact
- Immediate aftermath
- Recovery and rebuilding

SAHANA FOSS

SAHANA is a free and open source disaster management system. It is a web-based collaboration tool that addresses common coordination problems during a disaster. It is a set of pluggable, web-based disaster management solution that provides solutions to problems caused by the disaster and it is designed to help during the relief phase of a disaster. It has already been deployed in different disaster areas including the Philippines during the Southern Leyte Mudslide Disaster in 2006. At present, more tools are being developed however; the current disaster management system does not include a tool that will aid in the response phase during the disaster. The response phase includes search and rescue operations as well as the provision of emergency relief. In this phase, efficiency is important because during this kind of situation, time is of the essence. A second delay may cause someone's life. Thus, a system that determines the most optimum route for the volunteers and rescuers to take in order to serve the most number of people and provide maximum coverage of the affected area in the shortest possible time is beneficial.

Challenges in Using IT Solutions Especially with GPS

Travelling Salesman Problem (TSP)

Determining the most optimum route along different geographical locations is similar to the travelling salesman problem wherein geographic locations represent city coordinates and the rescuers or volunteers represent the travelling salesman. The travelling salesman problem is stated as follows: "Given a finite number of cities and the distance (or cost) of travel between each pair of them, find the shortest (or cheapest) way of visiting all the cities and returning to the starting point." The travelling salesman problem is formally described as a permutation problem with the objective of finding the path of the shortest length (or the minimum cost) on an undirected graph that represents cities or nodes to be visited.

Solution: Genetic Algorithms

A number of algorithms have been developed to solve the TSP. Some of which are linear, dynamic, Monte Carlo, and heuristic search methods. However, results showed that genetic algorithms produced the lowest distance solution among the mentioned optimization methods. Genetic algorithms are computational models inspired by evolution that provides a potential solution to a specific problem. It has a wide range of applications from optimization, test pattern generation, voice recognition, and image processing. It solves problems by mimicking the same processes Mother Nature uses. Usually, when we want to solve a particular problem, we are looking for some solution, which will be the best among others. The space of all possible solutions is called a search space (state space). Each point in the search space represents one feasible solution. Each feasible solution can be "marked" by its value or fitness for he problem. Genetic algorithm starts with a set of solutions (represented by chromosomes) called population. Solutions from one population are taken and used to form a new population will be better than the old population. Solutions which are selected to form new solutions (offspring) are selected according to their fitness – the more suitable solution has a higher chance to reproduce. However, this depends on the selector operator used. The next phase is the crossover phase wherein the selected individuals are mated pair by pair to form a new offspring. Finally, some of the offspring are mutated. This is repeated until some condition (*e.g.* no. of populations or improvement of the best solution) is satisfied.

Conclusion

The disaster management system that facilitates the logistics for the rescue and relief operations during a disaster known as MyDisasterDroid was implemented in an Android based mobile phone. Geographic locations of the people in need were sent via SMS or inputted directly to MyDisasterDroid. Determining the optimum route along the different geographic locations is similar to solving the travelling salesman problem wherein the geographic locations correspond to the cities and the rescuers or volunteers correspond to the travelling salesman. Using genetic algorithm, an optimum route along the given geographic locations was determined. Different genetic algorithm parameters were varied and based on the results, the best combination of genetic algorithm operators to use are the tournament selector operator with a tournament size of 20 and a swapping mutation operator with a mutation rate of 20 for 40 evolutions. Moreover, this system is also flexible because it allows the prioritization to be changed or extended with minimal effort. Indeed, MyDisasterDroid is an application that can be used during the response phase in a disaster especially when time is crucial. The examples described in this paper show the importance and value of mobile communications in the aftermath of disasters.

References

Asia Pacific Telecom Research Ltd., Telecommunications in the Philippines, 1 June 2009.

Coppola, D.P., 2007. *Introduction to International Disaster Management*. Elsevier Inc., USA.

Han, Dongso. In: *Information and Communications University, Daejon, Korea "An Evolving Mobile E-Health Service Platform"*, Youngko and Sungjoonpark.

Mitchell, M., 1996. *An Introduction to Genetic Algorithms*. MIT Press, USA.

United Nations International Strategy for Disaster Reduction Secratariat (UNISDR), Mortality Risk Index, 15 June 2009.

www.pubmed.com

2013, Sustainable Approaches for Environmental Conservation *Pages* **291–294**
Editors: **D.R. Khanna, A.K. Chopra, R. Bhutiani, Gagan Matta & Vikas Singh**
Published by: **BIOTECH BOOKS, NEW DELHI**

Chapter 35
Environmental Education: Have a Concern to It

Anamika

N.A.G. (P.G.) College, Amroha, Uttar Pradesh

Environmental education as an independent field of study arrived on the world scenario in the early seventies and has always been a subject of global concern since the deliberation at Fournex in 1971 and later at Stockholm in 1972 followed by the "Workshop on Environmental Education" at Belgrade in 1975 and then at an Inter-Governmental Conference on the subject at Tbilisi in 1977 organized by the UNEP and UNESCO. It is both theoretical and practical as it is learning how to manage and improve the relationship between human society and the environment in an integrated and sustainable way. Since EVS have been introduced as compulsory subject at each level of formal education system, it implicit greater importance but the reality is different. These attempts have been created Environmental Awareness but Environmental Education is still not been achieved. Why? The answer lies in the complex multidisciplinary nature of subject and the interpretation of its various interweaved aspects. To bridge the gap between environmental education and awareness, in-depth understanding, holistic and naturalistic approach, comprehensive curriculum with effective delivery methods and essentially role models are needed from the beginning of formal education to the higher level of that. Everyone knows that it's a matter of concern but how much we concern? It's a big-big question to answer. Journey for sustainable development is continued from seventies and will be continue, if we would be able to achieve the ultimate aim of environmental education.

Keywords: *Environmental awareness and concern, Environmental education, Collaborative action research, Sustainable development.*

Introduction

Environmental education as an independent field of study arrived on the world scenario in the early seventies and has always been a subject of global concern since the deliberation at Fournex in

1971 and later at Stockholm in 1972 followed by the "Workshop on Environmental Education" at Belgrade in 1975 and then at an Inter-Governmental Conference on the subject at Tbilisi in 1977 organized by the UNEP and UNESCO. It is both theoretical and practical as it is learning how to manage and improve the relationship between human society and the environment in an integrated and sustainable way. Among students, EE create awareness, promote concern, provide competence with opportunity to action and hence prepare better citizens who act in eco-friendly manner.

Environmental Education (EE) is a crucial attempt to cause learners to appreciate, understand and have a basic knowledge of the environment. It entails the development of values, attitudes and skills regarding responsible behaviour on the part of the present generation, bearing future generations in mind. It also aims at the judicious utilization of existing natural resources so that coming generations can enjoy a comfortable way of life as is currently experienced in most parts of the world. In past, it was equated with the environmental sciences, a field which is dominated by the principles and customs of the scientific method but now it's a holistic field of study including social and natural sciences both. Currently it is favored to speak of EE as education in about and for the environment.

Environmental Education is not only a subject like any other and also not merely a collection of data concerning the environment. It is about the way we all should live and also a shared one responsibility. Endow with the educational objectives of cognitive, affective and psychomotor domain about the environment it leads us toward awareness, concern, competence and positive attitude that we can do something for the protection of our environment we all live in. This is the major objective of this course and the syllabus is a framework on which we all must realign our lives.

Although above mentioned lines clearly indicate the need and significance of environmental education in one's life and also justify its compulsion from class 1 to graduation level, yet there are some questions which often knock more or less at the door of everyone's mind as given below-

- Why Environmental Education of everyone's concern?
- Why Environmental Education is still not being achieved?
- What are the hindrances in its way?
- Who are and who will be accountable?
- When it will be achieved?

These or many more concerning issues have been argued in this paper along with suggested guidelines for betterment. It's a beginning effort not the final end because environmental education has no dying end. It was, it is and it will with the existence of human life on earth. If we wish to save human being, we have to create a concern about environmental education because it is the only way which goes towards sustainable development of our environment.

Suggested Guidelines

Let's begin with the prescriptions of the Agenda 21, as finalized by UNCED: "A major priority is to reorient education towards sustainable development by improving each country's capacity to address environment and development in its environmental programmes, particularly in basic learning. This is indispensable for enabling people to adapt to a swiftly changing world and to develop an ethical awareness consistent with the sustainable use of natural resources. Education should, in all disciplines, address the dynamics of the physical/biological and socio-economic environment and human development, including spiritual development." Undeniably education is the key tool to promote such values and improve people's capacity to tackle environmental issues at local, national and international level. The environmental awareness is already there. What required is a comprehensive educational strategy weaved with new set of ethical values and approaches for sustainable development of environment.

Collaborative Action Research Based Environmental Education (CARBEE)

The process by which any educational practitioners (teacher/student/head/principal/administrator) may solve their small scale day-to-day problems scientifically so that they could be able to guide, correct and evaluate their decisions and actions is known as Action Research. When these small action researches are undertaken jointly in group by educational practitioners, they termed as collaborative action researches. It is one of the best approaches to reorient education and to save teachers from the critic they usually face that they are responsible for not achieving the objectives of Environmental Education till yet because they are the centre axis of education.

Yes teachers are the axis but only they could not be blamed for this failure because the curriculum is also a key factor which can't be ignored. In this condition, CARBEE bridge the gap between theory and practice as it provides solution to daily problems associated with teachers or curriculum. Although it consumes more time than routine teaching practices yet it's a small step on the way of sustainability and it should be raise by all teachers as well as teacher educators. No doubt, one day this small step will add big success. The main steps which may be undertaken in CARBEE are as given below-

Identifying/sensing a local environmental problem

↓

Preparing two or more blue prints for solution
(Mention reasons, evidences, resources &time required to act)

↓

Implementing blue prints one by one
(Stop, if problem get solution at first one)

↓

Analysis of the solution

↓

Follow up

Within the current structure of environmental education curriculum and the burden of routine labor on teachers, CARBEE may be a tricky task for many teachers. In fact, their efforts may be small wins not big one. Finally, a shared effort is required. It is not only teachers who need to be educated about environmental education instead of that everyone needs to be educated and to understand the complications which teachers face. Consequently it is also the right time to expand theoretical framework of EE in harmony with huge challenges of sustainability to create new horizon for action and transformation.

Conclusion

Since environmental studies have been introduced as compulsory subject at each level of formal education system, it implicit greater importance but the reality is different. The efforts have been created environmental awareness but Environmental Education is still not been achieved. Why? Everyone knows that it's a matter of concern but how much we concern? These are the some critical issues still waiting for answer. The answer perhaps lies in the complex multidisciplinary nature of subject and the interpretation of its various interweaved aspects. To bridge the gap between environmental awareness and education or between theory and practice; in-depth understanding, holistic and naturalistic approach, comprehensive curriculum with effective delivery methods and essentially role models are needed from the beginning of formal education to the higher level and also at teacher training programmes.

Journey for sustainable development is continue from seventies and will be continue, if we would be able to achieve the ultimate aim of environmental education otherwise one day our environment will destroyed to a extent that we or our forthcoming generation will deadly suffer due to the disaster of nature. At last, neither I submitting any final conclusion here nor I am finishing as Kappeler comments:

I do not wish to conclude or sum up, rounding off the argument so as to dump it in a nut shell for the reader. A lot more can be said about any of the topics I have touched upon..the point is not a set of answers but making possible a different practice. (Cited in Lather, 1991, p. 159)

References

Davis, Julie Margaret, 2003. *Innovation through Action Research in Environmental Education: From Project to Praxis*, Ph.D. Thesis, School of Australian Environmental Studies, Faculty of Environmental Sciences, Griffith University.

Guha, Mohua and Chattopdhyay, Aparajita, 2004. *Environmental Education: A Pathway for Sustainable Development*, Research Paper, IIPS, Mumbai, India.

Thomas, Glyan, 2004. *Skills and Thrills in Outdoor Environmental Education: A contradiction or Beautiful Tension*? Research Paper, La Trobe University in Bendigo, Australia.

Vishvasrao, Arundhati, 2004. *Environmental Education: A Paradigm Shift*. Research Paper, Department of Environmental Sciences, University of Pune, India.

2013, Sustainable Approaches for Environmental Conservation *Pages* ***295–300***
Editors: **D.R. Khanna, A.K. Chopra, R. Bhutiani, Gagan Matta & Vikas Singh**
Published by: **BIOTECH BOOKS, NEW DELHI**

Chapter 36
Eco Safe Facets of Home Science

Amita Bhargawa[1] *and Garima Tyagi*[2]
[1]*V.M.L.G. (P.G.) College, Ghaziabad, Uttar Pradesh*
[2]*N.A.G. (P.G.) College, Amroha, Uttar Pradesh*

Each and every change can be progressive. Progress can also be called modernization. It is not a biological or cultural matter; first of all, it is a social matter. Modernization should be associated with the comfort of life. But in comfortable life, human being does not remember the side effects of modern lifestyle, material goods, their production, utilization and disposal in account of the degradation of environment. The relationship between modernization and eco safe environment is a major and constantly growing social problem. The survival of human beings depends on the intelligence and moral actions in respect to these problems. In this manner Home Science is a subject for good optimism. Home Science may be defined as the study required in developing personal and professional life within a changing society by taking the help of scientific knowledge and methods. It acts as a wonderful learning tool to attain objectives of better-living and success in one's life. Home science is an interdisciplinary field of knowledge with focus on five branches: Apparel and textile, Human development, Food science, Extension education, Home management.

All these branches help us to obtain recent scientific information to cope with the day to day problems like environmental problems. Now a day, researchers are taking interest in eco-safe aspects of Home Science such as organic farming, eco-safe kitchens, eco friendly textile, environmental programmes in extension education and so many more.

Keywords: *Human development, Environment education, Eco friendly, Organic farming.*

Introduction

Each and every change can be progressive. Progress can also be called modernization. It is not a biological or cultural matter; first of all, it is a social matter. Modernization should be associated with the comfort of life. But in comfortable life, human being does not remember the side effects of modern lifestyle, material goods, their production, utilization and disposal in account of the degradation of environment. The relationship between modernization and eco safe environment is a major and constantly growing social problem. The survival of human beings depends on the intelligence and moral actions in respect to these problems. In this manner Home Science is a subject for good optimism.

Home Science may be defined as the study required in developing personal and professional life within a changing society by taking the help of scientific knowledge and methods. It acts as a wonderful learning tool to attain objectives of better-living and success in one's life. Home Science Education provides the opportunities to understand the problems and identify solutions in favour of eco friendly environment in our family, community and nation through identifying their responsibilities, roles and resources. Home science is an interdisciplinary field of knowledge with focus on five branches: Apparel and textile, Human development, Food science, Extension education, Home management.

Extension of all these branches refers to a series of activities, which are applicable and help us to obtain recent scientific information to cope with the environmental problems. The eco safe facets of home Science education are as followed under these sub categories-

Deeds Useful for Eco Safe Environment

Under Food and Nutrition

Organic Farming

Organic agriculture is one among the production methods that are supportive of the environment. The demand for organic food is steadily increasing both in the developed and developing countries. In context of dynamic changes, organic farming can currently be seen as a role model for creating an environmental friendly farming system. Organic agriculture is a production system that sustains the health of soils, ecosystems and people. It relies on ecological processes, biodiversity and cycles adapted to local conditions, rather than the use of inputs with adverse effects. Organic agriculture combines tradition, innovation and science to benefit the shared environment and promote fair relationships and a good quality of life for all involved.

Create an Eco Safe Kitchen

Kitchen can be eco friendly with the material such as counters made up of 'Ice Stone', 100 percent recycled glass and concrete that comes in many colors. While selecting kitchen flooring, pick products like bamboo, cork, recycled glass tiles, recycled metal tile, recycled clay tile, stone tiles and mosaics are available as eco friendly. Seek out a sink made with recycled content, either in stainless steel. For cabinets, formaldehyde-free units made of solid wood responsibly harvested from managed forests or made of an alternative material like wheat board, a composite that incorporates wheat straw can be used.

Use of Environment Friendly Equipments

Energy efficient electronic equipments should be used in the kitchen *i.e.* star rated equipments. When plastic utensils are thrown in the trash, they eventually end up in area landfills. Biodegradable utensils do not need to go the landfill. Manufacturers use petroleum by-products like propane and natural gas to produce plastic equipments. According to the Plastic Planet, manufacturers use 331 million barrels of oil for plastic production each year. This represents 4.6 percent of all United States oil consumption. Instead of using limited oil supplies to makes utensils, biodegradable utensil manufacturers rely on rapidly renewable crops like sugar cane, wheat fiber and corn. These utensils reduce world dependence on oil and help preserve remaining supplies for future generations. Food service items such as trays, disposable cutlery, cups, carryout boxes plates, bowl, platters, compartment trays and even gloves, aprons and hairnets are now being made with biodegradable materials.

Recycle of Waste

Kitchen waste will eventually smell very bad (especially old meat and dairy products) and attract insects and scavengers. So, the quick and simple solution is to bury your kitchen waste in your yard. Since most of us do not have sufficient amount of land to continuously do this, you will want to consider buying or building a composting container. A composting container is basically a wooden or plastic box that speeds up the decomposing of organic matter in the box by increasing the temperature while still allowing plenty of ventilation for bacteria to thrive. This bacteria will then digest organic waste much more quickly than naturally in your yard. These composting containers will also keep out unwanted animals, minimize odors, and even kill some types of pathogens because of the heat. Very quickly nutrient rich and free compost can easily spread around in yard and vegetable garden. This is especially beneficial with keeping the weeds to a minimum just like mulch. Therefore, recycling kitchen waste is a simple way to help save money and give a healthier and more manageable garden that just happens to be good for the environment too.

Under Apparel and Textile

Eco-friendly Fibers

Clothes that are made from the following fibers are eco-friendly-

1. *Hemp*: An amazing natural fiber. Some say hemp could have 25,000 uses. Hemp provides enormous benefit to the natural environment. This is true when used in products and when growing the hemp plant.
2. *Jute*: Similar to hemp, jute is a type of vegetable fiber used for thousands of years, with outstanding potential for the future.
3. *Ingeo*: Trademark for a man-made fiber derived from corn.
4. *Calico*: Fabric made from unbleached cotton. Also referred to as muslin.
5. *Hessian Cloth*: Coarse woven fabric made from jute or hemp.
6. *Organic cotton*: Cotton grown organically (without pesticides etc)
7. *Bamboo Fiber*: Bamboo fabric is very comfortable and 100 per cent biodegradable.
8. *Tencel*: Brand name for a biodegradable fabric made from wood pulp cellulose.
9. *Ramie*: Ramie fibers are one of the strongest natural fibers. Ramie can be up to 8 times stronger than cotton, and is even stronger when wet.
10. *Organic Wool*: Organic wool is wool that has been produced in a way that is less harmful to the environment than non-organic wool.
11. *Organic Linen*: Linen that is made from flax fiber. Could also refer to be linen made from other organically grown plant fibers.
12. *Fortrel ecospun*: Fiber made from plastic containers
13. *Milk Silk*: Silk made from milk
14. *Soy Silk*: Silk made from soybeans
15. *Nettle fiber*: Made from stinging nettle (commonly known as a weed)

Eco Friendly Fabric

Environmental awareness in the textile industry, specializing in providing drapes, curtains and valances (length of fabric fitted over the base of the bed under the mattress) that are well known for high quality craftsmanship, long lasting durability and eco-friendly practices and products. Eco-Friendly Drapery Fabric collection offers natural fabrics like organic 100 per cent pure cotton, hemp and recycled fabrics. When we purchase any eco friendly product from the company, they will voluntarily plant a tree on the person's behalf through the Arbor Day Foundation.

Eco Wears

Beautiful and affordable eco-fashion benefits both us and the planet. Green clothes are always fair trade certified, made from organic cotton, bamboo, hemp and soy and priced to enjoy ultra-soft, durable eco-style for living. Eco Wears men's and women's apparel includes jeans, t-shirts, sweaters, shirts, dresses, skirts, coats, yoga wear and much more are available in market.

Under Home Management

Eco-Friendly Homes

Eco-friendly homes with energy efficient doors and windows, herbal gardens, rainwater harvesting and solar heating are now as fashionable as they are affordable High-energy efficiency and low maintenance will be ideal for such homes. Natural resources, such as sunlight, should be used to cut down on the use of electricity. Windows should be located at strategic places to enable sunlight in to save energy. Rainwater harvesting, solar water heating systems and sewage treatment plants are being promoted by builders in most of the projects. Eco-friendly homes are not just money-saving, but even have an aesthetic beauty. Architect Laurie Baker started off the green architecture, which keep houses cool in summer, warm in winter, recycle water, and harness sunlight. Green architecture is known to cut overall construction costs by close to 25 per cent. Most eco-friendly constructions also save valuable time.

Tips for Environmental Friendly Homes

1. *Install skylights in your roof.* This will allow natural sunlight to illuminate your house for free, especially over rooms that are the darkest.
2. *Use low wattage light bulbs.* These are a great way to save energy easily. They plug straight into any lighting unit and use less energy. Some are even brighter than standard light bulbs but are consuming a lot less energy.
3. *Consider using motion sensor light bulbs for both indoors and outdoors.* These will keep you from leaving a room without turning the lights off. Outdoors, they can save a lot of money on wasted electricity but will illuminate your pathway to the door when needed.
4. *Choose low flow toilet mechanisms.* These use less water while still doing the same job as regular toilets. Obviously, you will save water and money on water bills.

Under Human Development

Human development is psychology based branch of home science in which physical, social, mental and emotional development is focused for all round development of children. Beside these with the use of morality, creativity, reasoning and intelligence we can encourage children to do activities which are in favour of environment. To do so many activities can be done such as-

1. Incorporating eco-friendly concepts into everyday lessons, activities and field trips.
2. Put on a puppet show or play on the environment at your school, church or camp.

3. Encourage children to continue thinking about issues outside of the classroom.
4. Write articles about environmental issues for a school or community newspaper.
5. Make club posters using environmental facts
6. Create art projects about the environment and host an art fair. Reuse material to show that junk can be beautiful if creativity is used.
7. Take photos, make posters or paint pictures reflecting environmental problems.
8. Design an environmental calendar.
9. Give the funds to raise a good environmental cause like Kids for Saving Earth.
10. Publish an environmental newsletter.
11. Visit a wind farm to see how energy is created without depleting natural resources.
12. Share things like books, magazines, movies, games, and newspapers between friends and neighbors instead of purchasing.
13. create and use note pads from once-used paper and make your own cards/letters from once-used products or handmade paper.

Under Extension Education

Extension Education is 'helping people to help themselves' in changing their behaviour (Knowledge, attitude and skill) in a desirable direction in order to bring overall development in an individual, in family, in community and thereby in nation. Through this education, environmental aware citizens can be prepared, many environment related researches can be executed, in extension education various programmes can be implement for the upliftment of environmental awareness *e.g.*-

- ☆ *Craft article programme*: In which various craft articles should be made through children by use of eco friendly materials or waste materials such as- decorative envelops by old invitation cards, beautiful and colourful vase by bottles and useful pot to plant flowers by plastic milk jugs, attractive magazine holders with old cartoons *etc.*
- ☆ *Awareness programmes*: awareness about "Reduce-Reuse-Recycle mantra (3Rs)" should be given in this type of program. These are very helpful for preservation of environment.
- ☆ Various campaigns, seminars and conferences can be organized on environmental education., by non-formal system environment education should spread among different sections of the society;
- ☆ Different media including films, audio, visual and print, theatre, drama, advertisements, hoarding, posters, seminars, workshops, competitions, meetings *etc.* for spreading messages concerning environment and awareness.

Conclusion

Philosophy of Home Science based on the development of the individual, who is the most important component of the community and nation. Home science helps to achieve these purposes in a unique way. People do not know the importance and impact of environment, which can change life in a second. There are two kinds of people in Modern India - those who know what to do, when and in the right manner; while the others who know how to do things, but do not know the right manner. Home Science education deals with both kinds of human being and has many more facets advantageous to environment. So, we can conclude that Home Science education is an effective tool which can surely helpful in eco friendly journey.

References

Coverage of Home Science Information In Popular Dailies; Priyanka A. Kausadikar

How to Save the Environment at home by Danny *et al.*

Tips for Eco Friendly Homes December 23, 2009, www. BharatEstates.com

What are the Benefits of Organic Cleaning Products?, Jomana Papillo-Krupinski, Apr 26, 2011.

www.ask.com

www.google.com

www.kidsforsavingearth.org

2013, Sustainable Approaches for Environmental Conservation *Pages 301–308*
Editors: **D.R. Khanna, A.K. Chopra, R. Bhutiani, Gagan Matta & Vikas Singh**
Published by: **BIOTECH BOOKS, NEW DELHI**

Chapter 37

Effect of Static Magnetic Field on Blood

P.P. Pathak and Archna Jha

Department of Physics,
Gurukula Kangri University, Haridwar, Uttarakhand

We have investigated the effects of static magnetic field on blood sample when exposed to non uniform static magnetic field of 0.35 T. We have plotted the curves between wavelength and transmittance- T per cent with the help of spectrophotometer. Three set of curves are obtained *viz.* before placing in magnetic field, after placing in magnetic field of 0.35 T for 20 min and then after 20 min of removing field. The change in nature of curves showed that there is an effect of magnetic field on blood. We concluded that the most probable reasons for these changes are magnetization of blood and increase in energy of blood due to magnetic field.

Keywords: *Transmittance, Spectrophotometer, Blood, Magnetic field.*

Introduction

Blood is a biomagnetic fluid that exists in a living creature and its flow is influenced by the presence of a magnetic field. It behaves as a magnetic fluid, due to the complex interaction of the intercellular protein, cell membrane and the hemoglobin, a form of iron oxides which is present at a uniquely high concentration in the mature red blood cells, while its magnetic property is affected by factors such as the state of orientation.

According to Pauling and Coryell (1936) the blood possesses two types of properties diamagnetic and paramagnetic. When blood is oxygenated it is diamagnetic and when deoxygenated it is paramagnetic. Haik *et al.* (1999) and Motta *et al.* (1998) estimated the magnetic susceptibility of blood which was found to be 3.5×10^{-6} and -6.6×10^{-7} for the venous and arterial blood, respectively.

Effects of Static Magnetic Field on Erythrocytes

The paramagnetic properties of erythrocytes have been used by Heidelberger *et al.* (1946) and Melville (1975) in efforts to separate Malaria infected cells from healthy cells using a strong magnetic field gradient. On the same basis, the paramagnetic attraction of the erythrocytes, under the effect of inhomogeneous magnetic fields, has been studied by Shiga *et al.* (1996) and Chen (1985) to better understand the effect of the static magnetic field on the blood circulation. These experiments confirm the inhomogeneous magnetic field – induced attraction of paramagnetic erythrocytes.

The Effect of Strong SMF on the Mechanism of Blood Coagulation

The main blood coagulant is fibrinogen. These molecules are changed to fibrin monomers under the influence of the protease thrombin. They are then polymerized to long fibrin molecules that are diamagnetic and can interact with magnetic field (Ueno *et al.*, 1993). In the absence of magnetic fields, the fibrin fiber appears to form a gummed up web. A magnetic field of 10 to 20T is believed to be able to orient the fibrin fibers parallel to the field. Iwasaka *et al.* (1994) performed fibrin degradation product test which showed that under a magnetic field of 8 T, fibrinolytic processes produce on the average 15 per cent more fibrin degradation products than the unexposed samples. If one considers the 190 per cent higher fibrin degradation products in a 400 T^2/m field gradient at 8 T compared to the control sample. This suggests that the magnetic field gradient is the main cause of this effect.

Fibrin gels have demonstrated a tendency to migrate to a region of lower field and fibrin had higher dissolution in a specific direction. Ueno *et al.* (1993) have shown that Fibrin gels formed at 8 T is more soluble in water containing plasma than those formed in the absence of field. Iwasaka *et al.* (1994) studied the homogenous magnetic field effect on the enzymatic activities of plasmin conducted at a $dB/dy < 4$ T/m. It showed no changes in the maximum velocity and the Michaelis constant of plasmin at 8 T. It may be possible that exposure to high magnetic fields will alter the blood coagulation process.

Effect of SMF on Proteins and Lipids of Blood

Proteins and lipids constitute another group of magnetically sensitive biological structures. Ignoring the small set of proteins that have paramagnetic atoms at their centers, all the other proteins interact with magnetic field through their diamagnetic properties and it is the main cause of the orientation of the retinal rods and other membranous organelles in magnetic field. These are the finds of Hong *et al.* (1971). A strong magnetic field can orient biological polymers such as sickle cell hemoglobin (Costa Ribiro *et al.*, 1981), actin (Torbet and Ronzeiro, 1984). De-oxyhemoglobin S, a natural crystalline fiber, forms spontaneously in erythrocytes of the sickle cell erythrocytes is caused by these packed fibers a strong magnetic fields could orient such cells (Hong *et al.*, 1971). Under these conditions, the diamagnetic anisotropy is mainly caused by the heme group of hemoglobin. Accordingly it is unlikely that exposure to high magnetic fields has real in- vivo implications for proteins and lipid structures

Magnetic Susceptibility and Orientation Interaction in the SMF

The magnetic susceptibility χ of a tissue defines the extents of its induced elementary moment. For a molecule the nature of electronic orbital determines the electronic, optical and magnetic properties. Also the non-symmetric characteristic of the susceptibility tensor for a sample is determined by the anisotropic chemical bond. If the values of the susceptibilities are considerably different along one direction (χ_1) compared to other direction (χ_2) then a torque proportional to the difference in susceptibility ($\Delta \chi = \chi_1 - \chi_2$) could be experienced by the molecule. For diamagnetic material χ is negative; that is induced magnetic dipole μ is opposite to the applied field H. Ionic and covalent molecules are diamagnetic, since they have atoms or ions with complete shells.

A torque will act on the anisotropic molecular aggregates in a uniform magnetic field.Biological molecules that orient themselves in a magnetic field are basically anisotropically shaped super molecules. Their internal structure is the cause of the anisotropic susceptibility. The typical dimensions of these molecules are in micron range.

The behavior of biological macromolecules in strong magnetic fields has been studied extensively. These studies have focused on magnetic orientation as a result of anisotropic diamagnetic susceptibility. This is because such reorientation at high fields creates large effects. For example the study of Beaugnon and Tournier (1991) on water, ethanol and acetone as diamagnetic liquids have exhibited magnetic levitation in a magnetic field of more than 20T. The surface of liquid Helium has also been distorted at a magnetic field of 6.5T. Ueno *et al.* (1996) showed the dynamic properties of water at a magnetic field of 8T and magnetic field gradient of 400 T^2/m. In this field, a bulk splitting of water was observed by the strong magnetic field and magnetic field gradient exerts a force on a diamagnetic medium such as water. As blood contains maximum percentage of water which is diamagnetic in nature, so there is probability that the applied magnetic field has effect on it.

Magnetic Field Energy

When an object such as a human body, a red blood cell is placed in a magnetic field, it experiences forces that cause it to tend to move relative to the field and torques that tend to rotate it with respect to the direction of the field. These forces and torques depend on the strength of the field and can range from absolutely negligible to potentially lethal values. To understand whether these forces and torques will be at a significant level in a given situation it is necessary to have mathematical expressions that can be used to calculate them.

Once a magnetic potential energy function 'U' is available to relate the magnetic energy of the object to its location, orientation and material properties, standard techniques of physics can be used to generate the necessary expressions for the forces and torques.

The dipole moment is the integral of the magnetization $\vec{M}$ over the volume V of the object. If magnetization is uniform over the object, m=MV. If a material with a permanent dipole moment m is brought to a point P within a magnetic field $\vec{B_0}$, it acquires energy

$$U = \vec{m} \cdot \vec{B_0}$$

If an object has a magnetic moment proportional to the applied field is brought to P and thereby acquires an induced moment m, its energy is

$$U = \frac{1}{2} mB_0$$

In both cases B_0 is the field existing at P prior to the introduction of the material and it is assumed that the sources of this field are kept constant when the material is put into the field *i.e.* filed in SMF. The factor ½ accounts for the fact that, in the second case, as the material is brought into the magnetic field, its moment gradually increases from zero to m, rather than being at the value m along the entire path.

Schenck (2000) show that the magnetic energy is determined by the strength of the dipole moment, the strength of the magnetic field and the angle between these two vector quantities. The magnetic field exerts forces and torques on the object that have the effect of increasing the magnetic energy. The effect of the forces is to attract paramagnetic materials toward regions of stronger field strength and to push diamagnetic materials towards regions of weaker field strength. The effects of torques are to turn the object such that m is brought into alignment with B_0.

If the force is $\vec{F}$ and torque is $\vec{T}$, we have

$$\vec{F} = \vec{\nabla} U$$

$$\vec{T} = \frac{\partial U}{\partial \theta}\hat{u} \text{ and } U = \vec{m} \cdot \vec{B}_0$$

Where θ is the angle between $\vec{M}$ and $\vec{B}$ and $\hat{u}$ is the unit vector perpendicular to the plane of $\vec{M}$ and $\vec{B}$.

If an object has volume V and susceptibility χ

$$\vec{m} = \vec{M}V = \chi V(\vec{H}_0) = \frac{\chi}{\mu_c} V\vec{B}_0$$

and

$$U = \frac{1}{2}\vec{M}.\vec{B}_0 = \frac{\chi^V}{2\mu_c}\vec{B_0^2}$$

This formula assumes that the absolute value of the susceptibility is much less than 1 and that the particle is sufficiently small that B_0 does not change significantly over it.

Experiment

The aim of the present study was to find the effects of static magnetic field on blood. We know that blood consist of various types of proteins, RBC (which contains hemoglobin), WBC, platelets, lipids, water *etc.* If we consider blood constituent separately according to their magnetic nature we found that deoxygenated-hemoglobin, erythrocytes and few proteins are paramagnetic in nature while most of proteins, lipids, fibrinogen and water are diamagnetic.

If magnetic field is applied all molecules behaves according to their magnetic nature. Haik *et al.* (1999) and Tzirtzilakis (2005) have theoretically shown that there was a decrease in the velocity of blood when it is considered as bio-fluid, flowing in a tube and static magnetic field is applied perpendicular to the flow. They have considered force due to magnetization and Lorentz force, as the cause for this effect. We have performed an experiment to verify that there is an effect of magnetic field on the blood.

Method

We have taken 2 ml venous (deoxygenated) blood from a healthy human.

For analyzing the magnetic effect we have used Shimadzu (Japan) UV – 3600 Spectrophotometer with wavelength range 280 nm – 2000 nm.

The UV-3600 Spectrophotometer can handle measurement with highly precise transmittance and reflectance and uses three detectors to handle a range going from the ultraviolet region to the near-infrared region. The level of sensitivity for the near infrared region has been increased significantly by using an InGaAs detector and a cooled PbS detector for this region. Spectra can be obtained for the

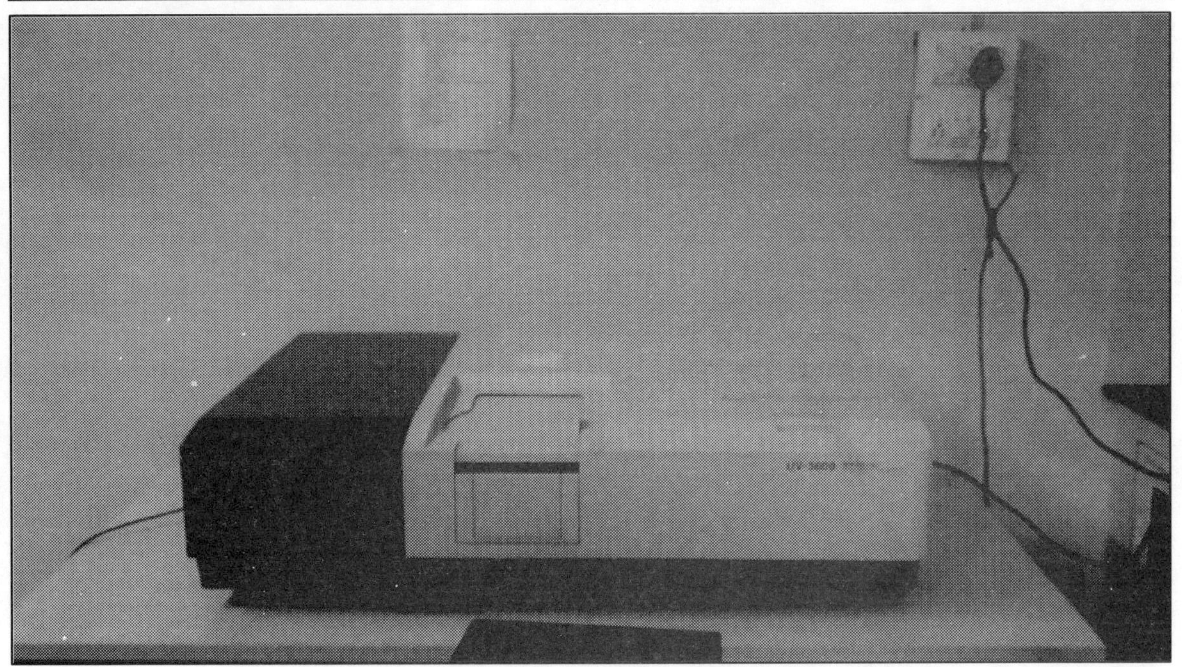

Figure 38.1: Shimadzu *UV* – 3600 Spectrophotometer

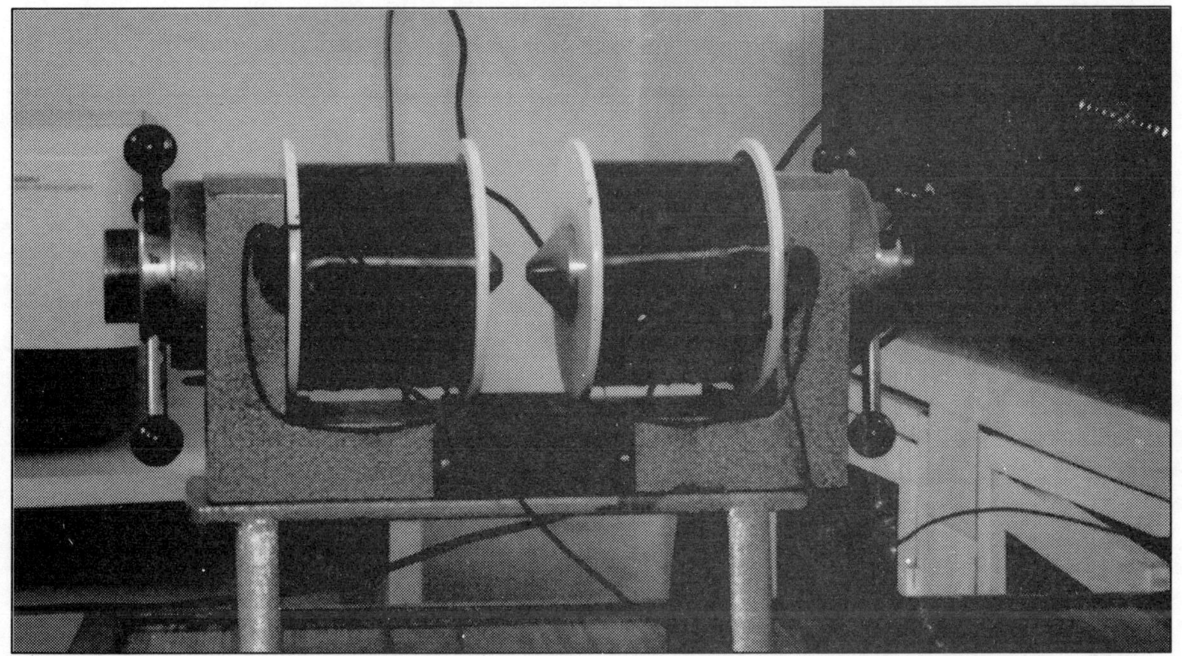

Figure 38.2: Electromagnet Used in Experiment for Producing Non-Uniform Static Magnetic Field

entire range, from the ultraviolet region to the near- infrared region, with a high level of sensitivity and precision.

The 2 ml blood sample was placed in spectrophotometer and transmittance curves are plotted. Then same sample was placed in magnetic field of 3500 Gauss or 0.35 T for 20 minutes, then it was safely placed in spectrophotometer, and again transmittance curves were plotted. The sample was than stirred and left for 20 min. then again placed in spectrophotometer for post exposure curve.

The pre exposure, after exposure and 20 min after removing magnetic field, curves between wavelength and transmittance are shown in Figures 38.3–38.4 respectively.

Result and Discussion

The transmittance curves for blood before and after placing in magnetic field of 0.35 T are obtained in wavelength range 280 nm to 2000 nm.

We compare the trends of the curves and found that the curves of blood before placing in magnetic field are convex, where as the curves obtained after placing in magnetic field are concave in nature.

The study of these curves shows that there is an effect of magnetic field on blood, and the most probable reason for this change is magnetization and change in the energy of blood sample. We know that when a diamagnetic or a paramagnetic molecule placed in a magnetic field it experienced a force that tends to orient them according to field direction and increase the magnetic energy of these molecules. The effect of the force is to push diamagnetic molecules to a region of low magnetic field, so the fibrin fiber orient themselves parallel to the field where as before exposure to field, fibrin fiber form gummed web. The red blood cell or erythrocytes are paramagnetic so orient in the direction of magnetic field as reported by Shiga *et al.* (1996) and Chen *et al.* (1985) in their study. Thus magnetic field shifts these molecules and changes the energy of the blood molecules. Thus blood constituents are same after application of magnetic field but its homogenous property is disturbed. The most important fact

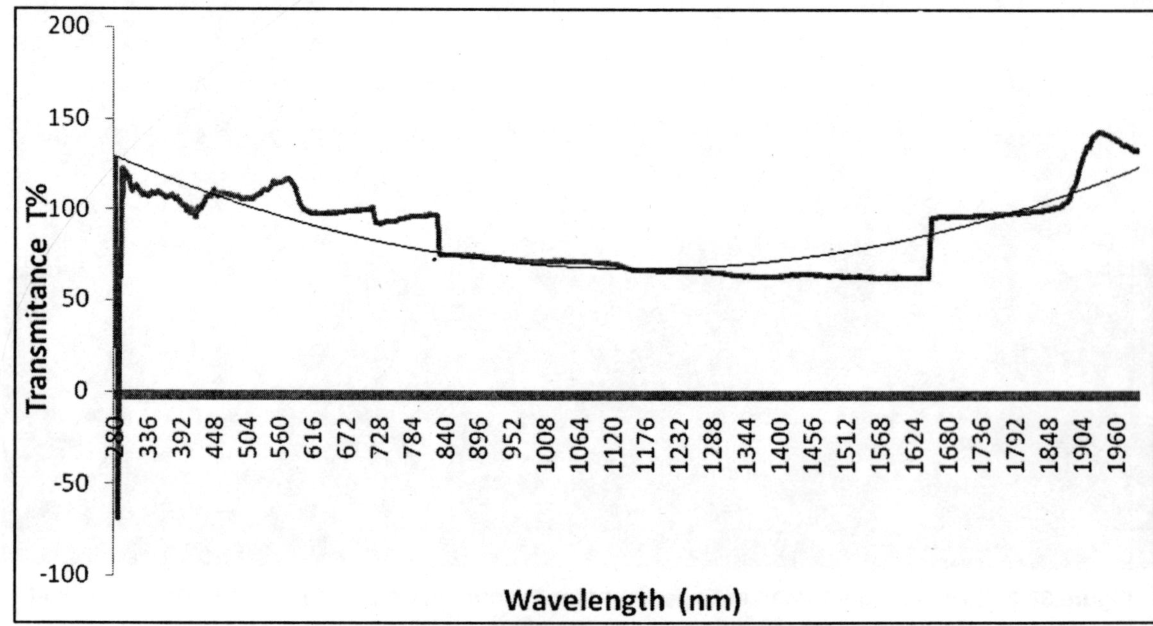

Figure 38.3: Pre Exposure to Static Magnetic Field

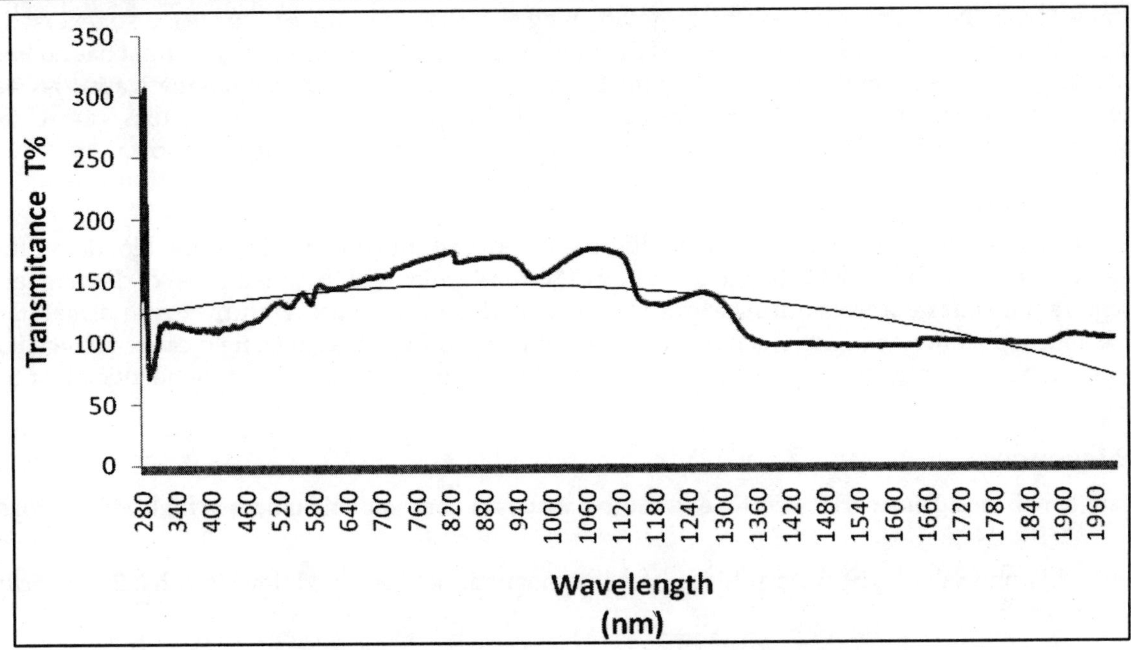

Figure 38.4: Just Afer Exposure to Static Magnetic Field

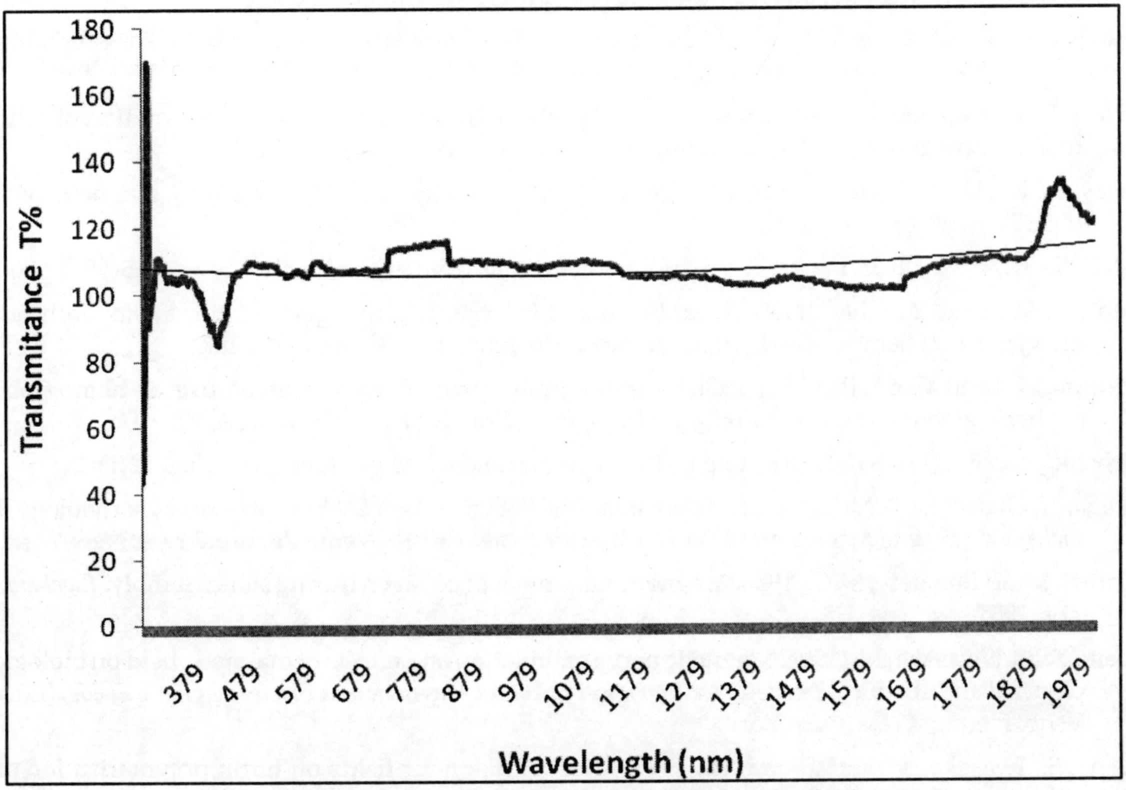

Figure 38.5: 20 Min After Removing Magnetic Field

obtained from these curves is that after removing magnetic field the nature of curve is also reversed it is again convex it conclude that the effect of magnetic field is only during exposure and it has no long last effect. The reason for this is that when field is removed the energy due to magnetization is over come by the thermal energy of these molecules so the random motion of molecules gained the homogenous property of the blood and it again shows the same trend in transmittance curve.

Conclusion

Our experiment clearly indicates that there is an effect of magnetic field on the blood and this effect is most probably due to change in energy of the molecules and the orientation of diamagnetic and paramagnetic molecules during exposure of field and after exposure it returns to its normal state *i.e.* before exposure state. Thus it can be concluded that exposure to magnetic field caused effect. It is however, not possible, for the present to conclude whether this effect is harmful or beneficial for the body.

References

Beaugnon, E. and Tournier, R., 1991. Levitation of water and organic substances in high SMF. *J. Phys.*, 111(10): 1423.

Chen, I.I.H. and Saha, S., 1985. Analysis of intensive magnetic field on blood flow: Part 2. *J. Bioelectricity*, 4: 55.

Costa Ribeiro, P.C., Davidovich, M.A., Waynberg, E., Benski, G. and Kischinevski, M., 1981. Rotation of sickle cells in homogeneous magnetic fields. *Biophysics J.*, 36: 443.

Heidleberger, H., Meyer, M.M. and Damarest, C.R., 1946. Studies in human malaria: The preparation of vaccines and suspensions containing plasmodia. *J. Immunol.*, 52: 325.

Hong, F.H., Mauzerral, D. and Mauso, A., 1971. Magnetic anisotropy and the orientation of retinal rods in homogeneous magnetic field. *Proc. Nat. Acad. Sci., USA*, 68: 1283.

Iwasaka, M., Ueno, S. and Tsuda, H., 1994. Enzymatic activity of plasmin in strong magnetic fields. *IEEE Trans Magnet*, 30(6): 4701.

Melville, D., 1975. Direct magnetic separation of red cells from whole blood. *Nature*, 255(5511): 706.

Motta, M., Haik, Y., Ghandhari, A. and Chen, C.J., 1998. High magnetic field effects on human deoxygenated hemoglobin light absorption. *Bioelectrochem. Bioenerg.*, 47: 297.

Pauling, L. and Coryell, C.D., 1936. The magnetic properties and structure of hemoglobin, oxyhemoglobin and carbonmonoxy hemoglobin. *Proc. Natl. Acad. Sci., USA*, 22: 210.

Schenk, John F., 2000. Safety of strong static magnetic fields. *J. Magn. Reson. Imaging*, 12(1): 2.

Shiga, T., Okazaki, M., Maida, N. and Seiyama, A., 1996. Effects of SMFs on erythrocyte rheology. In: *Biological Effects of Magnetic and Electromagnetic Fields*, (Ed.) S. Veno. Plenum Press, New York.

Torbet, J. and Ronziere, M.C., 1984. Magnetic alignment of collagen during self-assembly. *Biochem. J.*, 219: 1057.

Ueno, S. and Iwasaka, M., 1996. Magnetic nerve stimulation and effects of magnetic field on biological physical and chemical process. In: *Biological Effects of Magnetic and Electromagnetic Fields*, (Ed.) S. Veno. Plenum Press, New York.

Ueno, S., Iwasaka, M. and Tsuda, H., 1993. Effect of magnetic fields on fibrin polymerization and fibinolysis. *IEEE Trans Magn*, 29: 3352.

2013, Sustainable Approaches for Environmental Conservation *Pages* ***309–313***
Editors: **D.R. Khanna, A.K. Chopra, R. Bhutiani, Gagan Matta & Vikas Singh**
Published by: **BIOTECH BOOKS, NEW DELHI**

Chapter 38
Risk on Exposure to Biomedical Waste

Suporna Mukherjee and Sutirtha Mazumdar
International Institute of Health Management Research, New Delhi

Most countries of the world, especially the developing nations, are facing the grim situation arising out of environmental pollution due to pathological waste arising from increasing populations and the consequent rapid growth in the number of health care centers. India is no exception to this and it is estimated that there are more than 15,000 small and private hospitals and nursing homes in the country. This is apart from clinics and pathological labs, which also generate sizeable amounts of medical waste. A few large private hospitals in metros, none of the other smaller hospitals and nursing homes have any effective system to safely dispose of their wastes. With no care or caution, these health establishments have been dumping waste in local municipal bins or even worse, out in the open. In the community the most vulnerable population in terms of exposure to bio medical wastes are health care workers and especially those who dispose bio medical waste and those people who reside near landfill sites. We observed that these health care worker belong to low economic strata and this is a strong confounder to poor health of the health worker and his family.

Keywords: *Biomedical waste, Risk, Exposure, Poor, Hospital.*

Introduction

As per WHO (World Health Organization) norms the healthcare waste includes all the waste generated by healthcare establishments, research facilities and laboratories. In addition, it includes the waste originating from minor or scattered sources such as that produced in the course of healthcare undertaken in the home (dialysis, insulin injections *etc.*)

Most countries of the world, especially the developing nations, are facing the grim situation arising out of environmental pollution due to pathological waste arising from increasing populations and the consequent rapid growth in the number of health care centers. India is no exception to this and it is estimated that there are more than 15,000 small and private hospitals and nursing homes in the country. This is apart from clinics and pathological labs, which also generate sizeable amounts of medical waste. India generates around three million tonnes of medical wastes every year and the amount is expected to grow at eight per cent annually. Creating large dumping grounds and incinerators is the first step and some progressive states such as Maharashtra, Karnataka and Tamil Nadu are making efforts despite opposition.

A few large private hospitals in metros, none of the other smaller hospitals and nursing homes have any effective system to safely dispose of their wastes. With no care or caution, these health establishments have been dumping waste in local municipal bins or even worse, out in the open. Such irresponsible dumping has been promoting unauthorized reuse of medical waste by the rag pickers for some years now.

Surveys carried out by various agencies show that the health care establishments in India are not giving due attention to their waste management. After the notification of the Bio-medical Waste (Handling and Management) Rules, 1998, these establishments are slowly streamlining the process of waste segregation, collection, treatment, and disposal. Many of the larger hospitals have either installed the treatment facilities or are in the process doing so.

The World Health Organisation (WHO) has classified medical wastes according to their weight, density and constituents into different categories. These are:

- ✰ *Infectious*: material-containing pathogens in sufficient concentrations or quantities that, if exposed, can cause diseases. This includes waste from surgery and autopsies on patients with infectious diseases, sharps, disposable needles, syringes, saws, blades, broken glasses, nails or any other item that could cause a cut;
- ✰ *Pathological*: tissues, organs, body parts, human flesh, foetuse, blood and body fluids, drugs and chemicals that are returned from wards, spilled, outdated, contaminated, or are no longer required;
- ✰ *Radioactive*: solids, liquids and gaseous waste contaminated with radioactive substances used in diagnosis and treatment of diseases like toxic goiter; and
- ✰ *Others*: waste from the offices, kitchens, rooms, including bed linen, utensils, paper, *etc.*

Objectives of the Study

General Objective

To identify the adverse health effects of bio medical waste and suggest preventive measures.

Specific Objectives

1. To identify the type of waste generated by the health care organizations.
2. To identify the vulnerable population who are exposed to bio medical waste.

Methodology

This study is a descriptive study based on secondary data, In this study we reviewed various literatures and found out very interesting facts regarding bio-medical waste. In this study we have interacted with some health workers also regarding disposal of bio medical waste, to get an idea what is the scenario in India.

Type of Data

Secondary data we have collected from different literatures including studies from WHO (World Health Organization).

Methods

We have done extensive literature review based on various literatures. After reviewing different literatures we came to know about different diseases causes due to bio medical waste and most vulnerable population who are exposed to bio medical waste.

We had done semi structured interview with the health workers and that helped us to understand the waste disposal scenario in India.

Results and Findings

During our study we observed that a huge amount of bio medical waste is generated worldwide annually and especially in India the number is huge. For example each year, an estimated 16 billion injections, both preventive as well as curative, are administered worldwide. This amounts to almost 44 million injections per day2 out of which 95 per cent are therapeutic in nature. Unsafe injections are reported to have the potency of transmitting infections from patient to patient, patient to health workers and, more rarely, from health workers to patients and to the community at large. Sharps from immunization injections are found to be unsafe with almost 30 per cent of the sharps being either re-used or recycled as documented by WHO (WHO Bulletin, October 1992). In the community the most vulnerable population in terms of exposure to bio medical wastes are health care workers and especially those who dispose bio medical waste and those people who reside near landfill sites. We observed that these health care worker belong to low economic strata and this is a strong confounder to poor health of the health worker and his family. The commonly found diseases are HIV/AIDS, Musculoskeletal diseases, poor nutrition status of the health workers and their family members and various developmental diseases.

Health Hazards

According to the WHO, the global life expectancy is increasing year after year. However, deaths due to infectious disease are also increasing. A study conducted by the WHO reveals that more than 50,000 people die everyday from infectious diseases. One of the causes for the increase in infectious diseases is improper waste management. Blood, body fluids and body secretions which are constituents of bio-medical waste harbour most of the viruses, bacteria and parasites that cause infection.

This passes via a number of human contacts, all of whom are potential 'recipients' of the infection. Human Immunodeficiency Virus (HIV) and hepatitis viruses spearhead an extensive list of infections and diseases documented to have spread through bio-medical waste. Tuberculosis, pneumonia, diarrhoea diseases, tetanus, whooping cough *etc.*, are other common diseases spread due to improper waste management.

Occupational Health Hazards

The health hazards due to improper waste management can affect

- ☆ The occupants in institutions and spread in the vicinity of the institutions
- ☆ People happened to be in contact with the institution like laundry workers, nurses, emergency medical personnel and refuse workers.
- ☆ Risks of infections outside hospital for waste handlers, scavengers and (eventually) the general public
- ☆ Risks associated with hazardous chemicals, drugs, being handled by persons handling wastes at all levels

Discussion

Until recent times, hospital waste in India was not segregated before disposal to the dump or incinerator. Traditionally, recycling in India is conducted from the dumping grounds of immense waste where freelance workers or rag pickers scour the waste manually and sort for recyclable material. These workers then contact relevant industries, which acquire the waste from them. Since most of these rag pickers are women and children from the lowest socio-economic strata, awareness of health risks in general is poor. As a result many of them contact diseases from syringes and needles and other biomedical waste and become carriers of great health risk to the general populace. Since their traditional rejection from mainstream society (as 'untouchables') and their re-emergence as a political front in recent times, it is germane to call for further legislation to ensure education, awareness and health care facilities for their special status, in the context of health hazards in the recycling industry.

Conclusion

We need innovative and radical measures to clean up the distressing picture of lack of civic concern on the part of hospitals and slackness in government implementation of bare minimum of rules, as waste generation particularly biomedical waste imposes increasing direct and indirect costs on society. The challenge before us, therefore, is to scientifically manage growing quantities of biomedical waste that go beyond past practices. If we want to protect our environment and health of community we must sensitize our selves to this important issue not only in the interest of health managers but also in the interest of community

Recommendations

All health care facilities need to follow the government guidelines. Segregation policies must be aimed at reducing the incinerable waste. Only category 1, 2 (human anatomical and animal parts) and 5 (discarded medicines and cytotoxic drugs) are required to go to the incinerator in yellow coloured bags for Mumbai. The rest infectious waste can go for Autoclaving and shredding at the CTF site.

- ☆ All non-infectious wastes like IV containers, glass vials, *etc.*, can be effectively recycled.
- ☆ All non-infectious organic wastes can be recycled through organic composting/vermin composting.
- ☆ We must work towards creating an environmentally responsible health care community and touch upon issues on environment friendly options like mercury and PVC elimination.
- ☆ More effective training tools like producing training videos for project sustainability should also increase the outreach to even remote/rural places.
- ☆ We must take care that polluting technologies like incinerators are not set up within densely populated city limits.
- ☆ Spruce up the biomedical waste transportation system.
- ☆ Guide hospitals and health care institutions in the city to keep back-up plans and also build capacity on environmentally sustainable solutions to send minimum waste for treatment to outside agencies.
- ☆ A grey area that can be closely studied is looking at EPR in medical waste and chalking out an incentive-based programme for the stakeholders for better sustainability.
- ☆ Civil society and citizens in all sections must make a conscious effort to act as watchdogs.

- ☆ The legislative authorities including the pollution control boards and the government must take adequate action for omission/commission of the state law.
- ☆ Other facets of waste management must also include a deeper study relating to elimination of toxics from the health care stream and greening the supply chain.
- ☆ All health care institutions must join in this movement and involvement of NGOs is very much necessary

Thus medical waste with all its complexities can still be effectively handled. Lets work together to create and environment of care and healing.

References

A Hand Book on Health Care Waste Management – a guide for Developing Countries. A WHO Publication.

Bio-Medical Wastes (Management and Handling) Rules, 1997-98 notified on 16th October and Gazetted on 27 October 1997, Ministry of Environment and Forests, New Delhi (20th July 1998).

http://kspcb.kar.nic.in/BMW/

http://parisara.kar.nic.in/pdf/WasteMgmt.pdf

http://www.expresshealthcaremgmt.com

http://www.findarticles.com

http://www.keralapcb.org/

http://www.who.ch/

Implementing Hospital Waste management

Indian Standards 'Solid Wastes-Hospitals-Guidelines for Management: IS 12625:1989. Bureau of Indian Standards, New Delhi.

Report – National Workshop on the Management of Hospital Wastes, 16-18 April, 1998, IIRD, Jaipur (India).

Report of the High Power Committee – Urban Solid Waste Management in India, Planning Commission, Government of India, 1995.

Saurabh Sikka- Mercury Recovery from Biomedical Waste. Congress Proceedings, R'99. Vol. IV, 356-361, Geneva, Switzerland.

2013, Sustainable Approaches for Environmental Conservation *Pages* ***315–322***
Editors: **D.R. Khanna, A.K. Chopra, R. Bhutiani, Gagan Matta & Vikas Singh**
Published by: **BIOTECH BOOKS, NEW DELHI**

Chapter 39

Environmental Science Embeded in Vedic Ruchas

Kalpana Patankar

FRM, Department of Home Science,
Sevadal Mahila Mahavidyalaya, Nagpur, M.S.

Vedas emphasize that the 100 years of human life is guaranteed only when man can respire pure air, drink pure water, eat unadulterated food, can exercise on a land which is in a good condition. But because of pollution of the so called industrial revolution man has become devoid of his basic needs which he requires in an unpolluted condition. Industrial pollution has reached such levels that many places globally are affected by acid rains causing havoc to drinking water resources by contaminating these such as lakes. Vedas pronounce that if man can keep air, water, land and different components of nature (Prakruti) pure and clean then the environment will remain pure and will be beneficial to the mankind.

Keywords: *Veda, Lite, Environment, Ayurveda.*

Vedas deal with every aspect of human life. There is nothing that remains untouched by Vedas. Environment is no exception to this. Air(Vaayu), plants and shrubs, land(Prithvi), Ether(Aakaash) Ravi(Agni/Teja) *etc.*, these are all the components of environment. Vedas identify all these ingredients(Tatwas) by name Deva. Deva means: *DIVYA GUNAM SAMANVITAM*. These and some others donate their heavenly qualities which come together in a symbiotic relationship to maintain the environment in a pure state. Vedas emphasize that the 100 years of human life is guaranteed only when man can respire pure air, drink pure water, eat unadulterated food, can exercise on a land which is in a good condition. But because of pollution of the so called industrial revolution man has become devoid of his basic needs which he requires in an unpolluted condition. Such as water to drink, air to breathe, food to eat, land to cultivate. However, the fact is nothing is easy to get in a pure form or condition in our times.

The smoke, gas and industrial waste coming out of the factories, population explosion, reduction in the forest cover these are all the reasons for pollution. Industrial pollution has reached such levels that many places globally are affected by acid rains causing havoc to drinking water resources by contaminating these such as lakes. Biorhythm of such places is completely disturbed. The pace at which this pollution is growing, a day may not be far away, when this planet will become inhabitable.

Many countries the world over have come together to discuss this serious matter of pollution. Many things are being done at different levels of administration globally as a damage control exercise and to create general awareness amongst people at large. A positive outcome of this has been in creation of a new field, an entirely a new subject called Environmental Engineering. Incidentally the National Environmental Engineering Institute *i.e.* NEERI is situated at Nagpur, a place to which I belong to.

Lets us come to the subject. The message from Vedas is loud and clear. Vedas pronounce that if man can keep air, water, land and different components of nature (Prakruti) pure and clean then the environment will remain pure and will be beneficial to the mankind.

Let us first understand a few important points about vedic times. Vedic seers(Rishis) had an authority to perform Yadnya. This was a method/process which helped them to achieve Riddhi, Siddhi and Sammrudhi for society at large. Which basically meant well being of the people. This they could achieve by having a friendly symbiotic relationship with the forces of nature/Prakruti. This had an all inclusive approach which included living as well as non living. They always tried to maintain and protect the natural habitat of all living organisms and were able to achieve high levels of Biodiversity. All this Rishis were able to achieve by performing Yadnyas. The effective tool to perform such Yadnyas was the effective use of the Vedic Ruchas/Hymns/mantras.

According to Vedas the Prakruti or the nature is composed of Pancha Mahabhootas (Tatwas). These are:

- ✰ Prithvi
- ✰ Aapa
- ✰ Teja
- ✰ Vaayu
- ✰ Aakaash

Vedas show us a way, as to how mankind should use these Tatwas and how to protect these Tatwas. Following pages are devoted to reveal how our Vedic seers(Rishis) had progressed. This fact is stated in the Rucha/Hymn/mantra: Mantras functioned at two levels simultaneously.

- ✰ At one level these mantras are a praise of the different deities.
- ✰ At the same time they function as certain sound vibrations which can also produce energy packets that can be used to achieve a specific objective that was pronounced as a Sankalp at the beginning of these Yadnyas.

Impact of Pure Atmosphere (Vaayumandal)

Vaayu(Air) is a Praana Shakti(Life Force), without which we cannot imagine life. Vedas are vocal, loud and clear that not Vaayu, but pure Vaayu is necessary and beneficial for living organisms. This fact is said in the following Rucha/Hymn/mantra:

1. Mantra

"वात आ वातु भेषजं शुभु मयोधु नो हृदे ।
प्राण आयूंषि तारिषत् ।"

(ऋग्वेद १०.१८६)

"VAAT AA VAATU BHESHJAM SHUBU MAYODHU NO HRUDE I
PRAANA AAYUNSHI TAARISHAT I"

(Rigveda:10.186)

Meaning: " Oh! Vaayu(air), please provide us such a medicine that will keep our heart calm and healthy."

2. Mantra

यददो वात ते गृहे मृतस्य निधिर्हितः ।
ततो नो देहि जीवसे ।"

(ऋग्वेद १. १८६ ३.)

"YADDO VAAT TE GRUHE MRUTASYA NIDHIRHITAAH I
TATO NO DEHI JEEVASE I"

(Rigveda:10.186. 3)

Meaning: Here the Rishi points towards the Amrut (Elixir) Tatwa and says: " Oh Vaayu you are having a reservoir of Amrut, please give us a tiny spec of this, so that we will be long lived.

It is very clear that the Amrut(Elixir) which Vaayu holds within itself is nothing but Oxygen. Oxygen is the Praana, which moves the things within living body and helps removing the excretory waste. This makes man long lived. Reverse of this will show malfunctioning of the body and hence will make him weak and unhealthy, and hence short lived. The following mantra highlights as to how we benefit by breathing in pure and clean air:

3. Mantra

"द्वाविमौ वातौ वात आ सिन्धोरा परावतः ।
दक्षं ते अन्य आ वातु परान्यो वातु यद् रवः ॥"
"आ बात वाहि भेषजं वि वात वाहि यद् रवः ।
त्वं हि विश्वभेषजो देवानां दूत ईयसे ॥"

(ऋ० १० १३७. २-३, अथर्व ४. १३२२-३)

" DWAVIMAU VAATAU VAAT AA SINDHORA PARAAVATAAH I
DAKSHAM TE ANYA AA VAATU PARAANYO VAATU YAD RAVAH I"

(Rigveda:10.137. 2-3)

"AA BAAT VAAHI BHESHAJAM VI VAAT VAAHI YAD RAVAH I
TWAM HI VISHWABHESHAJO DEWANAAM DOOT EEYASE I"

(Atharvaveda:4. 1322-3)

Meaning: In these two mantras two types of Vaayu are indicated. Here Praana Vaayu is requested to bring strength to the body and Apaan Vaayu to throw the excretory waste out of the body.

But if we keep on inhaling polluted air/Vaayu then the above requests to these two Vaayu can never be fulfilled, so the result is our ill health. That again is the highlight of an unpolluted atmosphere which contains Vaayu/air.

Purification of Vaayu (Air) by Plants

If percentage of Carbon-di-oxide (CO_2) increases beyond required specified limits, then atmospheric air becomes polluted. It is the plant kingdom that continuously absorbs the Carbon-di-oxide (CO_2) from the atmosphere. They retain the Carbon and release Oxygen back again in the atmosphere, for us to breathe.

Vedas are saying: if you want to save yourself from air pollution, then carry out plantation of green plants.

4. Mantra

"VANAM AASTHAAPYADHWAM I"

(Rigveda:10. 101. 11)

Meaning: forests should be grown. Increase forest cover.

Man for his benefit if he keeps cutting forest, if he doesn't replant the forests then the percentage of Carbon-di-oxide will increase, this will destroy the plants and crops will get destroyed. Ice on the mountain peaks will start melting and will lead us to deluge. Under such conditions to protect human race Vedas say:

5. Mantra

"अयं हि त्वा स्वधितिस्तेतिजानः प्रणिनाय महते सौभाग्याय ।
अतस्त्वं देव वनस्पते शान्तवल्शो विरोह, सहस्त्रवल्शा वि वयं रूहेम ।
- यजुर्वेद (५ ४३)

"AYAM HI TWAA SWAADHITISTETIJAANAAH PRAANINAAYA MAHATE SAUBHAAGYA AYA I
ATASATWAM DEVA VANASPATE SHAANTAVALSHO VIROH,
SAHASRAVALSHAA VI VAYAM RUHEM I"

(Yjurveda:5. 43)

Meaning: Even if we have to axe a plant, let us do it in such a way that the plant will grow thousand branches more and will germinate thousand times. In the above lines to reduce the pollution plant germination many more times is anticipated. More the merrier, that seems to be the way of thinking. At same time necessity of the replantation of forests is emphasized.

Regarding plants and their utility following mantra says:

6. Mantra

"AOUSHADHI PRATIMODHWAM PUSHPAWATI PRASOOVATIHI I"

(Rigveda:10. 97. 3)

Meaning: Let those medicinal plants continue to enjoy the flowering forever.

At some other place Vedas have to say following:

7. Mantra

"यावतीः कियतीश्चेमाः पृथिव्यामध्योषधीः ।
ता मा सहस्रपर्ण्यो मृत्योर्मुन्चन्त्वहसः ॥
(अथर्व० ६. १० ६)

"YAAVATIHI KIYATISHCHEMAAH PRUTHIVYAAMADHYOSHADHIHIH I
TAA MAA SAHASRAPARANYO MRUTYORMUNCHaTWAHASAH I"

(Atharvaveda: 6. 10. 6)

Meaning: Rishi wishes and prays - Let on this Earth(Prithvi), all those plants who hold 1000 medicinal leaves unto themselves, let them save us from deaths originating from pollution. "Sahasraparni" is a plant which has the power to remove pollution and make the environment pollution free.

Similarly Vedas say:

8. Mantra

"ASHWATTHA WO NISHADANAM PARNE WO WASTISHKRUTA I"

(Rigveda: 10. 97. 5, Yjurveda: 12. 79, 35 4, Taiteerya Samhita: 4. 2 6. 2)

Meaning: One should sit below the Peepal tree and should reside near Palaash trees. Of course for their medicinal value.

Atharvaveda says:

9. Mantra

आयने ते परायणे दूर्वा रोहतु पुष्पिणीः ।
उत्सो वा तत्र जायतां ह्रदो वा पुण्डरीकवान् ॥
(अथर्व० दे०१ ९ ६)

"AAYANE TE PARAAYANE DOORVAA ROHATU PUSHPINIHIHI I
UTSO VAA TATRA JAAYANTAAM HRUDO VAA PUNDAREEKAWAAN I"

(Atharvaveda De.: 196)

Meaning: The road that takes us out of the house should be planted with flowering grass and the grounds should converted into a lake full of Lotuses.

10, 11, 12. Mantras

(10) *"VANAANAAM PATAYE NAMAAH* I"

(11) *"VRUKSHAANAAM PATAYE NAMAAH* I"

(12) *"AAKHYAANAAM PATAYE NAMAAH* I"

(Yjurveda: 16th Adhyaaya.)

Meaning: Forest officials should be recruited to protect forests as a national responsibility in order to save medicinal plants in particular and forest in general.

Now-a-days because of air and land pollution as well as pollution due to synthetic chemicals such as usage of chemical fodders like compost, even food is getting contaminated and polluted.

Vedas have given a call regarding this also, in following way:

13. Mantra

"ANNAPATE NNASYA NO DEHI ANAMIVASYA SHUSHMINAAH I"

(Yjurveda: 11. 83)

Meaning: Vedas wish that our food should be free from any contamination and should not become cause disease. On the contrary it should be giving us health.

AAP/JALA/WATER

Importance of Pure Water

14. Mantra

अथर्ववेद के निम्न मन्त्रों -

शं त आवो हैमवतीः शमु ते सन्तूत्स्याः ।
शं ते सनिष्यदा आपः शमु ते सन्तु वर्ष्याः ॥
शं ते खनित्रमा आपः शं याः कुम्भे भिराभृताः

(१९ २. १-२)

Meaning: Atharvaveda has given these mantras. In this Vedic Rishi explains the water from different sources. Such as: water from Himalaya, eternal water currents, rain water, oasis, thick forest water, water coming out after digging the land, water coming from such other sources. Then he describes water kept in various types of pots such as: earthen pot, pots made out of iron, copper, silver, gold *etc.*

What he wants to say is this: That the water coming from different sources will also have the properties of those soil ores into it. So such water will have high medicinal value and will also add to the taste and quality. Because of these things in it, it brings down the pollution too.

Purification: Vedas give few clues to the purification of contaminated water:

15. Mantra

यासु राजा वरुणो यासु सोमशै विश्वेदेवा यासूर्ज मदन्ति ।
वैश्वानरो यास्वग्निः प्रविष्टस्ता आपो देवीरिह मामवन्तु ॥

(ऋ० ७ ४९)

Meaning: Let those divine waters bring happiness to us. Here Rishis tell us about "Varuna Vishwedewaah" and "Vaishwanar" Agni/fire. Varuna here may be some pure gas(Vaayu) or some water purifying gas. Vaishwanar can be normal fire or electricity. In real terms it means that water contamination can be overcome with these types of air/gases or fire or electricity.

16. Mantra

सवितुर्वः प्रसव उत्पुनाभ्यच्छिद्रेण पवित्रेण सूर्यस्य रश्मिभिः ।

(यजु १. १२. ३१)

Meaning: Yjurveda in particular tells us about a method of using Yantra(Vedic Machines) which suggests using natural sources of heat/energy of Sun to purify water/huge water bodies. Here a specific grass named "Kusha" is also mentioned as water purifier.

If river water becomes impure, then on a very large scale purification of the water by specific methods found references in Rigveda 7. 50. 3.

Purification of Environment with the Help of Agnihotra

Many a times in Vedas Agni or the fire has been glorified with adjectives like Paavaka, Abhivaachaatan, Paavakshochisha, Sapatnadumbhana

(Rigveda: 1. 12. 18-19, Atharvaveda: 19.58.01, Rigveda: 3.9.08, Yjurveda: 3.18)

Agnihotra is performed giving Aahuti(These are all give a ways to Agni/fire) of substances like Ghruta(Ghee), scented substances which are air purifiers and disease-remover.

Charak, bruhannighanturatnakar, yogratnayar, gadnigrha etc Granthas mention many yogas in which by giving certain Aahutis Environment can be brought back to normalcy

According to Atharvaveda the Agnihotra performed in the evening can have its impact lasting till the break of dawn, the next day.

17. Mantra

न तं यक्ष्मा अरुन्धते नैनं शपथो अश्नुते ।

यं भेषजस्य गुल्गुलोः सुरभिर्गन्धो अश्नुते ॥

(१. ३८.)

Meaning: One who is bestowed with the scent of medicine like "Guggle" will always remain disease free. In Agnihotra substances like Ghruta, Guggle, Kesar, Kasturi, Aagaru, Tagar, Shweta-chandan, Fruits, Kanda, Annasomlata, Giloya *etc.* are used as "Havi" in order to purify atmosphere and make the Environment Pollution free.

18, 19 and 20. Mantra

"शग्मा भवन्तु मरुतो नः स्योनाः" (अथर्व ४.२७.३)

"अनवद्यासः शुचयः पावकाः" (ऋ.७५७.५)

"गृभायतः रक्षसः सं पिनष्टन" (ऋ- ७.१०४.१८)

Meaning: Oh Marut/Storm, we feel highly obliged and protected as you have taken away pollution along with your winds. We feel happy as you have completely tied down and grinded the demonical germs of diseases to a halt.

Similar type of following mantras are seen in Atharvaveda:

21. Mantra

आ पर्यन्यस्य वृष्टयोदस्थामामृता व्यम ।
व्यहं सर्वेण पाप्मना वि यक्ष्मेण समायुषा ॥ (३.१ १)

Meaning: When clouds rain they bring a feeling of exaltation all around. The rains clean mountains as well the forest cover. The entire earth is cleaned. All kinds of diseases are washed away. The life expectancy of all living organisms improves.

In the following mantra/Rucha/Hymn the Sun is visualised as the one who makes environment Pollution free.

22. Mantra

"उभाभ्यां देव सवितः पवित्रेण सवेन व ।
मां पुनीहि विश्वतः ॥ (१९.४३)

Meaning: Sun through the influx of Rashmi/rays and rains spreads purity into the atmosphere surrounding the Earth. This is a scientific fact also that Sun rays carry an energy which can destroy the micro organisms resulting in a pure state of environment.

In short one can say that Vedas propogate maintaining environment pollution free by keeping its components(Panchamhaabhootas) Prithvi, Aapa, Teja, Vaayu and Aakaash or Pancha Tatwas clean. These Tatwas express themselves through living as well as non living things. Vedas make us aware of this fact and inspire us to keep the Prakruti/atmosphere neat and clean. Because only then mankind can lead peaceful and meaningful life by enjoying physical, mental and psychological fitness.

This article is only a tiny attempt to exhibit the wisdom lies hidden in the ancient scriptures of Indi.

But even this generates a genuine optimism for many of us like me that if we delve deep then Vedas can surely gives us a deep insight and understanding of Environmental Sciences and provide brilliant solutions to modern maladies.

References

Paryavaran Vidnyaan – Author: Dr. Vidyaadhar Sharma Guleri. Book: Sanskrit Mein Vidnyaan. Publication: Sanskrit Bharti, Delhi.

Rigvedaanchaa Marathi Shaastreeya Anuvaad Author – Dr. V.S.Kulkarni, Parts – 1 to 5.

Environmental Studies – Author: Benny Joseph. Publication: The Mcgrow Hill Company.

Vedic Vidnyaan Va Veda Kaala Nirnaya – Dr.P.V.Vartak. Publication: Vartak Prakaashan, Shaniwaarpeth, Pune.

2013, Sustainable Approaches for Environmental Conservation
Pages **323–333**
Editors: D.R. Khanna, A.K. Chopra, R. Bhutiani, Gagan Matta & Vikas Singh
Published by: BIOTECH BOOKS, NEW DELHI

Chapter 40

Origin of Environmental Science from Vedas

Shashi Tiwari

Maitreyi College,
University of Delhi, Chanakyapuri, New Delhi – 110 021

The Vedas are the first texts in the library of mankind. They are universally acknowledged to be the most precious Indian Heritage. The antiquity to the Vedic civilization is debated to a great extent but indeed there is no civilization known to humanity with such antiquity as Vedic Aryan Civilization. The so-called Aryans would have originated in the Aryavarta. N. J. Lockyer has declared: "The Vedas, in fact, is the oldest book in which we can study the first beginnings of our language and of everything which is embodied in all the languages under the sun".1

The Vedas deal with knowledge, the knowledge of all sorts. They cover knowledge both physical and spiritual. They are source of all knowledge according to Manusmriti.2 Especially the Vedic views revolve around the concept of nature and life. The visions of the beauty of life and nature in the Vedas are extremely rich in poetic value. Perhaps nowhere else in the world has the glory of dawn and sun-rise and the silence and sweetness of nature, received such rich and at the same time such pure expression. The symbolical pictures projected there remain close to life and nature. The most authoritative among the four Vedas is called the Rigveda. Each Vedic verse has one or more sages (*Rishis*) and deities (*Devatas*) associated with it. Generally, *Rishis* are supposed to be the recipient of knowledge revealed in the verses and *Devatas* are supposed to be the gods in whose praise verses are revealed.

The oldest and simplest form of Nature-worship finds expression in Vedic texts. Many scholars have come to the conclusion that the Vedas are primarily concerned with cosmology, however, they are not in a position to show that Vedic cosmology has the solutions to the most difficult problems of modern cosmology.3 Some say, like dramas are played to remember history, the process of various *shrauta yajnas* describes the science of Cosmology.4

The Vedic hymns are full of statements, ideas and unusual images which contain truths of all sciences. Here, knowledge is couched in symbolic language and unless the symbols are decoded, the real purport of the *mantras* cannot be understood. The only point is that Vedas need to be studied and interpreted, not in a pedantic manner, but in their proper perspective and in relevant context. The tripartite model of knowledge at the basis of the hymns helps in their understanding. Generally indication of most of the principles is there in their earliest form. Often expressions of ideas are enveloped with the shade of symbolism. The approach of Vedic seers is truly comprehensive. They do not visualize in parts. They do not elaborate subjects as is done in current education. But at the same time, grandeur and brevity of the Vedas are not found in the disciplines of modern science. The Vedas and disciplines of modern science are rather complementary and not contradictory. If modern science is seen or read through Vedic eyes, the students will be much benefited. Students of science may search the earliest of the ideas about any discipline in the Vedic literature.

In recent days, environmental science and ecology are disciplines of modern science under which study of environment and its constituents is done with minute details. As Science, they are established in 20th century, but their origin can be seen long back in the Vedic and ancient Sanskrit literature. The concepts of environment differ from age to age, since it depends upon the condition, prevalent at that particular time. In this paper, an effort is made to find out the awareness of ancient Indian people about the environment. As Sanskrit literature is so wide we refer here mainly to Vedic texts, particularly the Vedic Samhitas.

The Environment (Protection) Act, 1986 defines the environment as follows: 'Environment includes water, air and land and the inter-relationship which exists among and between water, air and land and human beings, other living creatures, plants, micro-organisms and property'.5 From the above definition, it can be briefly said that environment consists of two components namely biotic (living organisms) and abiotic (non-living materials) factors. The living organisms can be grouped into three types – those living mainly on land, in water and in air. The non-living materials of the environment are land, air, water, property *etc.*

In modern Sanskrit, the word *Paryavarana* is used for environment, meaning which encircles us, which is all around in our surroundings. But in the Atharvaveda words equivalent to this sense are used; such as *Vritavrita6, Abhivarah7, Avritah8, Parivrita9 etc.* Vedic view on environment is well-defined in one verse of the Atharvaveda where three coverings of our surroundings are referred as *Chandansi:* 'Wise utilize three elements variously which are varied, visible and full of qualities. These are water, air and plants or herbs. They exist in the world from the very beginning. They are called as *Chandansi* meaning 'coverings available everywhere'.10 It proves the knowledge of Vedic seers about the basic elements of environment.

According to one indigenous theory established in the Upanishads, the universe consists of five basic elements– *viz.*,1. earth or land, 2. water, 3. light or lustre, 4. air,and 5. ether11. The nature has maintained a status of balance between and among these constituents or elements and living creatures. A disturbance in percentage of any constituent of the environment beyond certain limits disturbs the natural balance and any change in the *natural balance* causes lots of problems to the living creatures in the universe. Different constituents of the environment exist with set relationships with one another. The relation of human being with environment is very natural as he cannot live without it. From the very beginning of creation he wants to know about it for self protection and benefit.

Vedic Approach to Environment

The Vedic Aryans were children of nature. They studied nature's drama very minutely. Sand-storm and cyclone, intense lightening, terrific thunder-claps, the heavy rush of rain in monsoon, the swift flood in the stream that comes down from the hills, the scorching heat of the sun, the cracking red fiames of the fire, all witness to power beyond man's power. The Vedic sages felt the greatness of these forces. They adored these activities. They appreciated these forces. They worshiped and prayed them due to regard, surprise and fear. They realized instinctively that action, movement, creation, change and destruction in nature are the results of forces beyond men's control. And thus they attributed divinity to nature.

Divinity to Nature

Rigvedic hymns could be divided into many parts, but their main part belongs to Natural hymns, the hymns related with natural forces. Yet Vedic gods are explained in different ways by the scholars of India and West, but speaking generally, the hymns addressed to deities (*Devata*) are under the influence of the most impressive phenomenon of nature and its aspects. The word *Devata* means divine, dignity which is bright, strong, donor, and powerful. In these hymns we find prayers for certain natural elements such as air, water, earth, sun, rain, dawn *etc.* The glorious brightness of the sun, the blaze of the sacrificial fire, the sweep of the rain-storm across the skies, the recurrence of the dawn, the steady currents of the winds, the violence of the tropical storm and other such natural energies, fundamental activities or aspects are glorified and personified as divinities (*Devata*). The interaction with nature resulted in appreciation and prayer but, indeed, after a good deal of observation. Attributes assigned to deities fit in their natural forms and activities, as Soma is green, fire is bright, air is fast moving and sun is dispenser of darkness. The characteristics of these forces described in the verses prove that Vedic seers were masters of natural science.

In Vedic view this world consists of Agni *i.e.*, fire or heat and Soma *i.e.* water12. Sun (Surya) is the soul of all which is moving and also of which is not moving.13 Indra is most powerful god who kills *Vritra*, the symbol of cloud to free waters. *Vritra* means one who covers and is Maruts are Indra's associates. Vedic seers pray boldly to these natural forces and aspects for bestowing plenty and prosperity on them. Aditi is praised as *Devamata*, the mother of all natural energies and she symbolizes the Nature. derived from the root *vri*, to cover. R.R.M. Roy opines that the main force of expansion in the Vedic cosmology is Indra, and his chief adversary, the main force of contraction, is *Vritra*.[14]

A famous geologist S.R.N. Murthy has written on the earth sciences in the Vedas. He has somehow a different opinion about Vedic gods and hence states, 'the natural geological aspects have been described as Indra, Agni, Vayu, Varun, Usas *etc.*15

Cosmic Order 'Rita' and Varuna

In the Vedas, the order of the Universe is called *'Rita'*. Rita reduces chaos to cosmos, and gives order and integration to matter. It also gives symmetry and harmony in the environment. Hence the conception of *Rita* has an aesthetic content too; it implies splendour and beauty. It is for this reason that the Vedic gods, upholding *Rita*, are all lawful, and beautiful and good. Their beauty is a significance attribute.

Rita is defined variously by scholars in different Vedic contexts, but in general sense it has been elaborated as great 'cosmic order' which is the cause of all motion and existence and keeps world in order. No one can ignore it16, even gods are abided by the *'Rita'*,they are born of *Rita*. It is controlling and sustaining power. It sustains sun in the sky.17 *Rita* as Universal Law governs everything in the cosmos. The whole of the manifested universe is working under *Rita*. S.R.N. Murthy assumes it as a law of gravitation in simple form. According to H.W.Wallis 'The principle of the order of the world, of the regularity of cosmic phenomena, was conceived by the *Rishis* to have existed as a principle before the manifestation of any phenomena. The phenomena of the world are shifting and changeable, but the principle regulating the periodical recurrence of phenomena is constant; fresh phenomena are continually reproduced, but the principle of order remains the same; the principle, therefore, existed already when the earliest phenomena appeared.'[18]

In the Vedas, Varuna is depicted as the Lord of *Rita,* the universal natural order. He is sovereign god, great king, law-maker and ruler of cosmos and even of the gods. Basically, he is regarded as the Lord of water and ocean but chiefly he controls and keeps the world in order. From his throne on high he looks down upon all that happens in the world, and into the heart of man.19 'By the law of Varuna heaven and earth are held apart. He made the golden swing, the sun to shine in heaven. He has made wide path for the sun. By his ordinances the moon shining brightly moves at night, and the stars placed up on high are seen at night but disappear by day. He causes the rivers to flow. As a moral governor Varuna stands far above any other deity.[20] Thus, the concept of Varuna represents the consciousness of Vedic seers in respect to controlling and balancing the natural forces in environment.

Division of Universe

Vedic seers have a great vision about universe. The universe is made on scientific principles, and that's why it is well measured. The universe consists of three intertwined webs, *Prithivi, Antariksha* and *Dyau.* Vedic scientists divided even the length in three calling them upper, medium and lower. The tripartite division of the universe into three regions – *Prithivi,* the earth, *Antariksha,* the aerial or intermediate region which is between heaven and earth, and *Dyau,* the heaven or sky is very well established in the Vedic literature. *Prithavi* can be given a scientific name 'observer space'. It is our space, the space in which we live and die, whatever we can see and observe. From one end of the universe to the other end is the expanse of *Prithivi,* and that is what the name *Prithivi* means: the broad and extended one. *Dyau* can be termed 'Light space' because light propagates in this space. *Antariksha* can be termed as 'Intermediate space' as this space exists in between observer space and light space. A verse from the Yajurveda states that the division of universe was done on a subtle level, and not on gross level. [21] The Vedic sages had the capability of looking at such a subtle level, which is beyond the reaches of modern science. Here, in reference to environmental study, we regard the division of the universe as the most important concept of the Vedas.

Though a large number of gods are described in the hymns, and it is very difficult to arrange them in different classes, but Yaska in his Nirukta talks about three Gods: Agni in earth, Vayu or Indra in atmosphere and Sun in heaven. Each one of them is known by various names depending on the different actions performed.[22] These three gods are three major forms of energy, fire on earth, air in intermediate space and light in upper region. Other energies of those regions are related to or under them. So generally gods are classified in three groups called upper, middle and lower, and, therefore, provide a system to study atmosphere and its all aspects. Regarding global harmony, Vedic seers always pray for the welfare of all creatures and all regions.

Concept of the Earth '*Prithvi*'

The concept of the form of the earth in the Rig-veda is most fascinating. It is mostly addressed along with the heaven into a dual conception (*Rodasi, Dyavaprithivi*). There is one small hymn addressed to *Prithivi,* while there are six hymns addressed to *Dyavaprithivi. Prithivi* is considered the mother and *Dyau* is considered the father in the Vedas, and they form a pair together. One of the most beautiful verse of the Rig-veda says,' Heaven is my father, brother atmosphere is my navel, and the great earth is my mother.' [23] Heaven and earth are parents: *Matara, Pitara, Janitara*24 in union while separately called as father and mother. They sustain all creatures. They are parents of all gods. They are great (*Mahi*) and widespread. Earth is described as a goddess in Rig-veda.

In the Atharvaveda the earth is described in one hymn of 63 verses. This famous hymn called as *Bhumisukta* or *Prithivisukta* indicates the environmental consciousness of Vedic seers. The seers appear to have advanced understanding of the earth through this hymn. She is called *Vasudha* for containing all wealth, *Hiranyavaksha* for having gold bosom and *Jagato Niveshani* for being abode of whole world. She is not for the different races of men alone but for other creatures also.[25] She is called *Visvambhara* because she is representative of the universe. She is the only planet directly available for the study of the universe and to realize the underlying truth. This is wide earth which supports varieties of herbs, oceans, rivers, mountains, hills *etc.* She has at places different colours as dark, tawny, white. She is raised at some place and lowered at some places. The earth is fully responsible for our food and prosperity. She is praised for her strength. She is served day and night by rivers and protected by sky. The immortal heart of earth is in the highest firmament (*Vyoma*). Her heart is sun. 'She is one enveloped by the sky or space and causing the force of gravitation. She is described as holding Agni. It means she is described as the geothermal field. She is also described as holding Indra *i.e.*, the geomagnetic field. The earth is described then as being present in the middle of the oceans (sedimentary rocks) and as one having magical movements'.26 The hymn talks about different energies which are generated from the form of the earth.–'O *Prithivi*! thy centre, thy navel, all forces that have issued from thy body- Set us amid those forces; breathe upon us.'27 Thus, the earth holds almost all the secrets of nature, which will help us in understanding the universe. She is invested with divinity and respected as mother - 'The earth is my mother and I am Her son'.28 The geographical demarcations on this earth have been made by men and not by nature.

Concept of Water '*Apah*'

Water is essential to all forms of life. According to Rig-veda the water as a part of human environment occurs in five forms:

1. Rain water (*Divyah*)
2. Natural spring (*Sravanti*)
3. Wells and canals (*Khanitrimah*)
4. Lakes (*Svayamjah*)
5. Rivers (*Samudrarthah*) 29

There are some other classifications also in the Taittiriya Araṇyaka30, Yajurveda31 and Atharvaveda32 as drinking water, medicinal water, stable water *etc.* Chandogya Upanishad describes about qualities of water-'The water is the source of joy and for living a healthy life. It is the immediate cause of all organic beings such as vegetations, insects, worms, birds, animals, men *etc.* Even the mountains, the earth, the atmosphere and heavenly bodies are water concretized.'[33] The cycle of water is described. From ocean waters reach to sky and from sky come back to earth.[34] Rain-waters are glorified. The rain-cloud is depicted as *Parjanya* god.

The fight between Indra and *Vritra* is celebrated story from the Rig-veda. It is explained in many ways. According to one view it is a fight for waters. Indra is called *Apsu-jit* or conquering the waters, while *Vritra* is encompassing them. *Vritra* holds the rain and covers waters and thus being faulty is killed by *Indra* through his weapon called *Vajra i.e.*, thunderbolt. The Indra-Vritra fight represents natural phenomenon going on in the aerial space. By the efforts of Indra all the seven rivers flow. The flow of water should not be stopped and that is desired by humanity. The significance of water for life was well-known to Vedic seers. They mention - Waters are nectars. [35] Waters are source of all plants and giver of good health.[36] Waters destroy diseases of all sorts.[37] Waters are for purification.[38] It seems that later developed cultural tradition of pilgrimage on the river-banks is based on the theory of purification from water. The ancient Indians knowing water as a vital element for life, were very particular to maintain it pure and free from any kind of pollution. The Manusmriti stresses on many instances to keep water clean.[39] The Padma Purana condemns water pollution forcefully saying, 'the person who pollutes waters of ponds, wells or lakes goes to hell'.[40]

Concept of Air '*Vayu*'

The observer space is the abode of matter particles, light space is the abode of energy and the intermediate space '*Antariksha*' is the abode of field. The principal deity of *Antariksha* is *Vayu*. Jaiminiya Brahmana quotes,' *Vayu* brightens in *Antariksha*.' Field is another form of energy and, therefore, Yajurveda says,' *Vayu* has penetrating brightness'. The meaning of *Vayu* is made clear in Shatapatha Brahmana in the following Mantra, 'Sun and rest of universe is woven in string. What is that string, that is *Vayu*.' This verse clearly shows that here *Vayu* cannot mean air alone.[41]

Apparent meaning of *Vayu* is air. The Vedic seers knew the importance of air for life. They understood all about air in the atmosphere and also about the air inside the body. The Taittiriya Upanishad throws light on five types of wind inside the body: *Prana, Vyana, Apana, Udana* and *Saman.* Air resides in the body as life.42 Concept and significance of air is highlighted in Vedic verses. Rig-veda mentions –' O Air! You are our father, the protector'.[43] Air has medicinal values.[44] 'Let wind blow in the form of medicine and bring me welfare and happiness'.[45] Medicated air is the international physician that annihilates pollution and imparts health and hilarity, life and liveliness to people of the world. Hilly areas are full of medicated air consisted of herbal elements. Another verse describes characteristics of air - 'The air is the soul of all deities. It exists in all as life-breath. It can move everywhere. We cannot see it. Only one can hear its sound. We pray to air God'.[46] Ancient Indians, therefore, emphasized that the unpolluted, pure air is source of good health, happiness and long life. Vayu god is prayed to blow with its medicinal qualities.

Concept of Ether '*Akasha*'

Modern environmentalists discuss sound or noise pollution. There is a relation between ether and sound. The sound waves move in sky at various frequencies. Scientist could see the sky which exists only in the vicinity of earth, but Taittirya Upanishad throws light on two types of ether *i.e.* one inside the body and the other outside the body.[47] The ether inside the body is regarded as the seat of mind. An interesting advice to the mankind is found in the Yajurveda -'Do not destroy anything of the sky and do not pollute the sky.Do not destroy anything of *Antariksha*.'[48] Sun shines in *Dyuloka* and we get light from sky. The sunrays strengthen our inner power and are essential for our life. Thus importance and care for ether is openly mentioned in the Vedic verses.

Concept of Mind '*Manas*'

Many prayers are found in Vedas requesting the God to keep the mind free from bad thoughts, and bad thinking. In this regard the *Shivasankalpa Sukta* of Yajurveda is worth mentioning.[49] considering the havoc that the polluted minds may create, our ancient sages prayed for a noble mind free from bad ideas. The logicians recognize *Manas* as one of the nine basic substances in the universe.[50] The mind is most powerful and unsteady. Although the study of mind does not appear directly under the contents of modern environmental science but in reference to cultural environmental consciousness of Vedic seers, we find many ideas discussed in Vedic literature on the pollution of mind and its precautions.[51]

Animals and Birds

Animals and birds are part of nature and environment. It is natural, therefore, that Vedic seers have mentioned about their characteristics and activities and have desired their welfare. Rig-veda classifies them in three groups -sky animals like birds, forest animals and animals in human habitation.[52] All the three types of living creatures found in the universe have distance environment and every living creature has an environment of its own. But when we look from man's perspective all of them constitute his environment. There is a general feeling in the Vedic texts that animals should be safe, protected and healthy.[53] Domestic animals, as well as wild animals along with human beings should live in peace under the control of certain deities like Rudra, Pushan *etc.* Vedic people have shown anxious solicitude for welfare of their cattle, cows, horses *etc.* The cow as the symbol of wealth and prosperity, occupied a very prominent place in the life of the people in Vedic times.[54]

Plants and Herbs '*Oshadhi*'

The knowledge about the origin and significance of plants can be traced out from Vedic Literature in detail. In Rigveda one *Aranyani sukta* is addressed to the deity of forest.[55] Aranyani, queen of the forest, received high praise from the sage, not only for her gifts to men but also for her charm. Forests should be green with trees and plants. *Oshadhi Sukta* of Rig-veda addresses to plants and vegetables as mother, 'O Mother! Hundreds are your birth places and thousands are your shoots'.[56] The plants came to existence on their earth before the creation of animals.[57] Chandogya Upanishad elaborates "water have generated plants which in turn generated food.[58] The Atharvaveda mentions certain names of *Oshadhis* with their values. Later this information became important source for the Ayurveda. The Rig-veda instructs that forests should not be destroyed.[59] The Atharvaveda talks about the relation of plants with earth, 'The earth is keeper of creation, container of forests, trees and herbs'.[60] Plants are live.[61]

There is an important quotation in a Purana which says, 'One tree is equal to ten sons'.[62] The Atharvaveda prays for continuous growth of herbs,–'O Earth! What on you, I dig out, let that quickly grow over'.[63] And another prayer says, 'O Earth! Let me not hit your vitals'.[64]

The '*Avi*' element referred in the Atharvaveda, as the cause of greenness in trees,[65] is considered generally by Vedic scholars as 'Chlorophyll'. The term '*Avi*' is derived from the root '*Av*' and thus gives the direct meaning of 'protector'. Hence, plants were studied as a part of environment and their protection was prescribed by the Vedic seers.

Concept of Sacrifice '*Yajna*'

The sacrifice '*Yajna*' is regarded as an important concept of Vedic philosophy and religion but when we study it in its broader sense, it seems to be a part of Vedic environmental science. Yajurveda and Rig-veda describe it as the 'navel (nucleus) of the whole world'.[66] It hints that *Yajna* is regarded as a source of nourishment and life for the world, just as navel is for the child.

Vedas speak highly of '*Yajna*'. Through it, seers were able to understand the true meaning of the *Mantras*.[67] All sorts of knowledge was created by *Yajna*.[68] It is considered as the noblest action.[69] In simple words, *Yajna* signify the theory of give and take. The sacrifice simply has three aspects: *Dravya* (material), *Devata* (deity) and *Dana* (giving). When some material is offered to a deity with adoration, then it becomes *Yajna*. Pleasing deity returns desired material in some different forms to the devotee. This *Yajna* is going on in the universe since beginning of the creation and almost everywhere for production and also for keeping maintenance in the world. Even the creation of universe is explained as *Yajna* in the *Purusha Sukta*. Thus, the concept of *Yajya* seems to be a major principle of ancient environmental science.

In environment all elements are inter-related, and affect each other. Sun is drawing water from ocean through rays. Earth gets rain from sky and grows plants. Plants produce food for living beings. The whole process of nature is nothing but a sort of *Yajna*. This is essential for maintenance of environmental constituents. The view that *Yajna* cleans atmosphere through its medicinal smoke, and provides longevity, breath, vision *etc*., is established in Yajurveda.[70] Few scholars have attempted to study the scientific nature of the Vedic *Yajnas*.[71] Undoubtedly, they have never been simple religious rituals, but have a very minute scientific foundation based on fundamental principles. According to Vedic thought, *Yajna* is beneficial to both individual and the community. *Yajna* helps in minimizing air pollution, in increasing crop yield, in protecting plants from diseases, as well as in providing a disease-free, pure and energized environment for all, offering peace and happiness of mind. Moreover, *Yajna* serves as a bridge between desire and fulfillment.

Coordination Between all Natural Powers

Modern Indian Scientists should be astonished and also feel proud of our ancestors for their knowledge and views about environment. Ancient seers knew about various aspects of environment, about cosmic order, and also about the importance of co-ordination between all natural powers for universal peace and harmony. When they pray for peace at all levels in the '*Shanti Mantra*' they side by side express their believe about the importance of coordination and interrelationship among all natural powers and regions. The prayer says that not only regions, waters, plants trees, natural energies but all creatures should live in harmony and peace. Peace should remain everywhere. The *mantra* takes about the concord with the universe–"peace of sky, peace of mid-region, peace of earth, peace of waters, peace of plants, peace of trees, peace of all-gods, peace of Brahman, peace of universe, peace of peace; May that peace come to me!".[72]

Conclusion

From the above detailed discussion some light is thrown on the awareness of our ancient seers about the environment, and its constituents. It is clear that the Vedic vision to live in harmony with environment was not merely physical but was far wider and much comprehensive. The Vedic people desired to live a life of hundred years[73] and this wish can be fulfilled only when environment will be unpolluted, clean and peaceful.

The knowledge of Vedic sciences is meant to save the human beings from falling into an utter darkness of ignorance. The unity in diversity is the message of Vedic physical and metaphysical sciences. Essence of the environmental studies in the Vedas can be put here by quoting a partial *Mantra* of the Ishavasvopanishad - 'One should enjoy with renouncing or giving up others' part'.[74] Vedic message is clear that environment belongs to all living beings, so it needs protection by all, for the welfare of all. Thus the study proves the origin of environmental studies from the Vedas.

References

1. N.J. Lockyer, *The dawn of Astronomy*, Massachusetts, Institute of Technology, p.432.
2. सर्वं वेदात्–प्रसिध्यति। Manusmriti 2.7.

3. Raja Ram Mohan Roy, *Vedic Physics, Scientific Origin of Hinduism*, Golden Egg Publishing, Toronto, 1999, p.6.
4. Yudhishthira Mimansaka, *Vaidika Siddhanta Mimansa*, Sonipata, 1976, p.40.
5. A.R. Panchamukhi, *Socio-economic Ideas in Ancient Indian Literature*, Rashtriya Sanskrit Sansthan, Delhi, 1998, p.467.
6. वतावता, Atharvaveda 12.1.52
7. अभीवार, Ibid, 1.32.4.
8. आवता, Ibid, 10.1.30
9. परीवता, Ibid. 10.8.31
10. त्रीणि च्छन्दांसि कवयो वि येतिरे पुरुरूपं दर्शतं विश्वचक्षणम्।
    आपो वाता औषधयस्तान्येकस्मिन् भुवन आर्पितानि।। Ibid 18.1.17
11. इमानि पञ्चमहाभूतानि पृथिवी, वायुः आकाशः आपज्योतीषि। Aitareya Upanishad 3.3
12. अग्निसोमात्मकं जगत्।
13. सूर्य आत्मा जगतस्तस्थुषश्च। Rigveda 1.115.1
14. Raja Ram Mohan Roy, *Vedic Physics, Scientific Origin of Hinduism*, Golden Egg Publishing, Toronto, 1999, p.58
15. S.R.N. Murthy, *Vedic View of the Earth*, D.K. Printworld, New Delhi, 1997, p.12.
16. ऋतं नात्येति किञ्चन। Taittiriya Brahmana 1.5.5.1
17. ऋतेनादित्यास्तिष्ठन्ति। Rigveda 10.85.1.
18. H.W. Wallis, the *Cosmology of the Rigveda*, Cosmo Publications, 1999, pp.94-95
19. Ibid. pp.91-101; Rigveda 1.25.7,9,11,
20. A.A. Macdonell, *A History of Sanskrit Literature*, MLBD, 1965, pp.61-62.
21. Yajurveda 7.5
22. Nirukta 7.5
23. द्यौर्मे पिता जनिता नाभिरत्र बन्धुर्मे माता पृथिवी महीयम्। Rigveda 1.164.33
24. मातरा, पितरा, जनितारा। Rigveda 1.159,160
25. त्वं बिभर्षि द्विपदः त्वं चतुष्पदः। Atharvaveda 12.1.15; 12.1.45
26. S.R.N. Murthy, *Vedic View of the Earth*, D.K. Printworld,Delhi, 1997, p.87
27. त्ते मध्यं पृथिवि यच्च नभ्यं यास्त ऊर्जस्तन्वः संबभूवुः।
    तासु नो धेहि अभि नः पवस्व।। Atharvaveda 12.1.12.; RTH Griffith, *The Hymns of the Atharvaveda*, D.K.Publishers, Delhi, 1995, Vol.II, P 95
28. माता भूमिः पुत्रो अहं पृथिव्याः। Atharvaveda, 12.1.12.
29. या आपो दिव्या उत वा स्रवन्ति खनित्रिमा उत वा याः स्वयंजाः।
    समुद्रार्थ याः शुचायः पावकास्ता आपो देवीरिह मामनन्तु।। Rigveda 7.49.2.
30. Taittiriya Aranyaka 1.24.1-2.

31. Yajurveda 22.25.
32. Atharvaveda 1.6.4.
33. Chandogya Upanishad 7.10.1
34. अपः समुद्राद् दिवमुद्वहन्ति दिवस्पृथिवीमधि ये सृजन्ति ।Atharvaveda, 4.27.4.
35. अमृतं वा आपः। Shatapatha Bra. 1.9.3.7.
36. आपः विश्वभेषजीः। Rigveda 1.23.20
37. भेषजीरापो अमीवचातनीः। Rigveda 10.87.6
38. पवित्रं वाऽआपः। Shatapatha Bra. 1.1.1.1
    आपः पवित्रमुच्यन्ते। Nirukta 5.6.2
39. Manusmriti 4.56
40. सुकूपानां तड़ागानां प्रपानां च परंतप।
    सरसां चैव भेत्तारो नरा निरयगामिनः।। Padmapurana, Bhimi 96.7.8
41. Raja Ram Mohan Roy, *Vedic Physics, Scientific Origin of Hinduism*, Golden Egg Publishing, Toronto, 1999, p. 84; Jaiminiya Bra. 1.192; Yajurveda 1.24; Shatapatha Bra. 8.7.3.10
42. वायुर्ह वा प्राणो भूत्वा शरीरमाविशत्। Taittiriya Upanishad, 2.4
43. उत वात पितासि नः। Rigveda 10.186.2
44. आ वात वाहि भेषजम्। Ibid, 1.37.2
45. वात आ वातु भेषजं शंभु मयोभु नो हृदे। Ibid 10.186.1
46. आत्मा देवानां भुवनस्य गर्भो यथावशं चरति देव एषः।
    घोषा इदस्य श्रृण्विरे न रूपं तस्मै वाताय हविषा विधेम।। Ibid 10.168.4
47. स य एषो अन्तर्हदय आकाशः। सुवरित्यसौ।। Taittiriya Upanshad 1.6.1; 1.5.1
48. द्यां मा लेखीरन्तरिक्षं मा हिंसीः। Yajurveda 5.43
49. Ibid, 34.1-6
50. Tarkasamgrahah 2
51. Nandita Singhavi, *Vedo Me Paryavarana*, Sonali Publications, Jaipur, 2004, pp. 313- 356
52. Rigveda 10.90.8
53. Yajurveda 19.20, 3.37; Atharvaveda 11.2.24
54. N.M. Kansara, *Agriculture and Animal Husbandry in the Vedas*, Nag Publishers Delhi, 1995, pp. 126-138
55. Rigveda 10.146
56. शतं वो अम्ब धामानि सहस्रमुत वो रुहः। Rigveda, 10.97.2
57. या ओषधीः पूर्वा जाता देवेभ्यस्त्रियुगं पुरा। Ibid, 10.97.1
58. ता अन्नमसृजन्त। तस्याद्यत्र क्वचन वर्षति तदेव भूयिष्ठमन्नं भवति। Chandogya Up. 6.2.4
59. वनानि न प्रजहितानि। Rigveda, 8.1.13

60. मन्द्राग्रेत्वरी भुवनस्य गोपा वनस्पतीनां गृभिरोषधीनाम्। Atharvaveda 12.1.57

61. येन प्राणन्ति वीरुधः। ण्इपक १०३२०१ यथा वृक्षो वनस्पतिस्तथैव पुरुषो मृषा। Brihadaranyaka Up. 3.9.28

62. दशपुत्रसमो द्रुमः। Padmapurana 1.44.455

63 यत् ते भूमे विखनामि क्षिप्रं तदपि रोहतु। Atharvaveda, 12.1.35

64. मा ते मर्म विमृग्वरी। Ibid 12.1.35
पृथिवी मातर्मा हिंसी मा अहं त्वाम्। Yajurveda 10.23

65. अविर्वै नाम देवता...
तस्य रूपेणेमे वृक्षाः हरिताः हरितस्रजः। Atharvaveda 10.8.31;
Kapil Dev Dwivedi, *Vedic Sahitya Avam Sanskriti*, Varanasi, 2000, p.337

66. अयं यज्ञो विश्वस्य भुवनस्य नाभिः। Yajurveda13.62
अयं यज्ञो भुवनस्य नाभिः। Rgveda 1.164.35

67. यज्ञेन वाचः पदवीयमायन्। Rigveda 10.71.3

68. तस्माद् यज्ञात् सर्वहुत ऋचः सामानि जज्ञिरे। Ibid, 10.90.9

69. यज्ञो वै श्रेष्ठतमं कर्म। Shatapatha Bra. 1.7.1.5

70. आयुर्यज्ञेन कल्पताम्। प्राणो यज्ञेन कल्पताम। चक्षुर्यज्ञेन कल्पताम्। Yajurveda 9.21

71. M.L.Gupta, *The Cosmic Yajna*, Samhita Books Jaipur, 1999, pp.46-47

72. द्यौः शान्तिरन्तरिक्षं शान्तिः पृथिवी शान्तिरापः शान्तिरोषधयः शान्तिः। वनस्पतयः शान्तिर्विश्वेदेवाः शान्तिर्ब्रह्म शान्तिः सर्व शान्तिः शान्तिरेव शान्तिः सा मा शान्तिरेधि।। Yajurveda 36.1; Atharvaveda 19.9.94; A.C.Bose, *The Call of the Vedas*, Bhartiya Vidya Bhavan, Mumbai, 1999, p.281

73. जीवेम शरदः शतम्। Atharvaveda 19.67.1

74. तेन त्यक्तेन भुञ्जीथाः। Ishavasyopanishad, 1

Index

C

R

S